MW01400653

CORONAVIRUS – COVID-19 PROPHYLAXIS AND TREATMENT
PATENT PENDING METHODS

The patent submitted is published in its entirety
AS INFORMATION FOR PHYSICIANS, SCIENTISTS, NIH, WHITE HOUSE TASK FORCE, CARE GIVERS, PATIENTS, AND TO THE PUBLIC

BY

Dr. T. R. Shantha, MD, PhD, FACA

in collaboration with

Usha S. Martin, BSc, PA, CPT, ECMO expert
Dr. Jessica G. Makwana, MD
Dr. Erica M. Tarabadkar, MD
Lauren A. Woodling, BSc, Ct RA
Drs. Dev & Aanad T. Shantha, MD

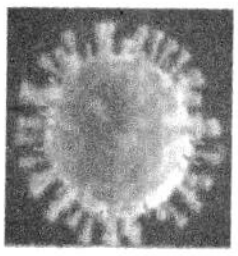

"You obviously have a brilliant mind. You have been blessed with a tremendous mind that can do great things; you will continue to use the great mind that you have hopefully to do things for the good of people."
- Honorable Richard Story, Federal Judge, 1/24/2008 about the author

Shan Publishing House, LLC
1946 Carrington Court,
Stone Mountain, GA 30087

To Monica
With all the best: you are wonderful
10.20.20

Dedication

This book is dedicated to Prof. Geoffrey H. Bourne, Chairman, Department of Anatomy, Director of Yerkes Regional Primate Research Center of Emory University; Dr. John E. Steinhaus, Chairman, Department of Anesthesiology; Dr. J. A. Evans, Columbus Medical Center; Dr. J. N. Hill of Georgia Baptist Medical Center, Dr. Morton B. Waitzman and Dr. P. Calhoun of Ophthalmology Department and Dr. R. T. Jackson of ENT department of Emory University's School of Medicine, Dr. Shadakshrappa of Victoria Hospital (Bangalore), Dr. George M. Baer of Rabies Division of the CDC; Hon. Federal Judge Richard Story; to presidents, deans, and administrators of Emory University School of Medicine, who gave me the opportunity to pursue my education, research, teaching and medical practice.

Disclaimer

Many of the methods and statements in this book have not been evaluated by the Food and Drug Administration. The products and methods described here are not intended to diagnose, treat, cure, or prevent any disease that is managed by dedicated physicians. The use of this book is not a substitute for physician, health care provider, legal, accounting or other professional services and the authors or publishers are not legally responsible for any adverse outcome.

The contents of this book reflect the scientific ideas and research thoughts of the authors. Anyone who wishes to embark on any of the protocol and method of treatment to prevent, delay, curtail and/or cure or treat a specific disease or condition should consult with and seek the to-ahead/clearance from a physician or health care professional.

Can physicians prescribe and adopt the methods of treatment enumerated? Yes, they can, because "Once the FDA approves a drug, health care providers generally may prescribe it for an unapproved use when they judge that it is medically appropriate for their patient," explains Jeremy Kahn, a spokesman for the FDA.

The author or the publisher has not performed independent authentication of each and every one of the treatment methods contained herein. They expressly disclaim legal and other responsibility for any error in the methodology, undesired outcome and in the transcript. *If I can inspire one patient to use the method, one physician to follow the protocol, and one researcher to test my methods with healing effects, my 65 years of medical practice, research, and the efforts of this patent and its publication far in advance are worth it.*

PUBLISHED BY
Shan Publishing House, LLC
1946 Carrington Court,
Stone Mountain, GA 30087

Cell: 678-640-7705 LL 678-580-5446
Email: shantha35@aol.com

www.indian-gold.com
www.wedgetherapeutics.com

Patent Office Application Submission No.: 16/928,310
Patwrite Law, Reference: 0901108-058-July 14, 2020

ISBN 978-1-7357021-0-0
Library of Congress Control Number: 2020916730

Printed in the United States of America

Acknowledgments

I give very special thanks to Ms. Nancy Addison for her excellent editing and formatting the book, without which this publication would have been difficult to publish in a timely fashion.

I sincerely thank Bob Wieden of Wedge Therapeutics for his input and project support; Totada Ramaiah and Gowramma, T. Parvathappa and Basamma, M. P. Vishalakshi (M. P. Chandramma & Murigeruddrappa), K. N. Rudrappa, Nellie Golarz de Bourne, Victoria K. S. Mcleod, the McLeods (Annette and Thomas, Tommieann, Sadie, Naomi), and Julie M. Beck, for their groundwork, and to Dev and Andy Shantha, Gail and Norman Raab (Joey, Hallie), Liane Levetan, Nina and Karim Jabr, the Martins (Usha, Bill, Samuel, Gavin), the Manis (Chelvi, Murugaiah), the Vijays (Girija, Kamala Kant, Sadhana, Sundeep, Jim), the Manochas (Swaran, Sohan, Anuj, Atul), Vishi and Palani Periasamy, Mary and Ashok Kumar (Anita and Tara), Karen and Dan Bartoli (Moni), Cass, Charles, and Bruce W. Kirbo (Lillian and the Hartfords), Paul D'Agnese, B. N. Rudresh, Frank Boyd, Krishna V. Srinivasa, Rajendar Dishpally, the Tarabadkars (Sanjivan, Archana, Neil, Angela, Anita, Bodhi), the Woodlings (Edith, Jim, Frank, Brian, Michael, Catherine), the Makwanas (Girish, Tina, Shreya, Neera), Shaam P. Sundhar, Elizabeth and Tek Ayele (Tia), Don Roe, Dr. J. Hughes, Adam Cabibi, Sue Ann Nichols, Deborah J. Hammock, Linda Mele Ramshaw, William Faloon (Editor: Life Extension), Dan Marino (Editorial director: Drug Development and Delivery), Melanie Cooper-Evans, Ariel and Santana Nottage (Santana Jr.), Prapha and P. V. Mohan, Melissa and Mark D. Torche (Patwrite Law), Kevin King, Michael Cohen, Ezra Shimshi, Mr. Conway and Don Samuel, David E. Nahmias, Dr. Rhett Bergeron, and many others who influenced, inspired, motived, and re-shaped my journey.

The image of the front page and figures in the rest of the book are courtesy of the internet images, NIH, National Institute on Aging and health and allergy and infectious diseases, internet publications, Google, PubMed, WebMD, Wikipedia, Dreams Time, and such that are modified to explain our view and methodology. We thank them to allow us to use the diagrams as modified reproductions.

Table of Contents

Part I

Submitted patent application to the US patent office is broken into separate chapters as follows for ease of reading

Part II

Innovative ideas patented, patent pending, and needing experimentation

Part III

Methods of treatment and prophylaxis for other diseases using traditional and complementary adjuvant therapies

Part IV

Abridged biography of the author

Introduction

This book is the outcome of decades of medical practice, knowledge, experimentation, and research work, and dedication to advance the science of healing the sick and needy. This and many of the projects have been the vision of the author with his medical and research knowledge for more than 6 decades.

The coronavirus is unlike other viral infections; it spreads easily, became a pandemic, and is highly virulent compared to other respiratory flu viruses. It has infected millions all over the world and killed thousands in a short span. It is not just a respiratory virus, it affects heart, blood (thrombo-embolic phenomenon), blood vessels, the gastro-intestinal track, pancreas, liver, kidneys, brain, heart and other organs. Recovery from COVID-19 in the elderly can be slow, incomplete, and costly, with a huge impact on quality of life and the economy.

Its ill effects will be felt all over the world, not only due to its effects on the COVID-19 afflicted, and the after effects, but also due to the effect on the family, finances, and its impact on commerce, politics, and the society and country as a whole due to its effect due to the quarantine—the shut down on people's movement and commerce.

But what is the difference between self-isolating and quarantine?

1. Isolation is a method used to separate an ill person from those who are healthy. Isolation restricts the movement of someone who is ill to prevent the spread elsewhere.
2. Quarantine is used to restrict the movements of a well person who may have become exposed to a communicable disease, to see whether they will become unwell and prevent further spread of the virus.

We know that there is no vaccine and no direct coronavirus antiviral drug yet, but they are coming. Here we discuss the coronavirus pandemic and how to prevent and treat coronavirus infection described in the patent filed:

1. **Wear the mask to protect yourself and protect others.** Prevent coming in contact with virus by wearing a mask and goggles, washing hands often and keeping the contact areas clean with, social isolation and distancing, use of antiviral sprays and mouth wash.

2. **Bioactive Mask**: The simple mask one wears can be modified with triple viral filters, material layers and impregnated to incorporate viral filters, viricidal such as xylitol, antimalarial, auranofin, remdesivir, Ebselen, coronavirus antibodies, titanium nanoparticles, anthelmintic such as ivermectin, antiseptics and such as described.
3. **Second hand smoke** from the infected smoking any form of tobacco including cigarettes, cigars, vaping can carry and spread the virus as smokers blow droplets. Wear mask and observe social distancing.
4. Tests for coronavirus: The RNA tests identified 80% of COVID-19 patients, but when combined with the rapid antibody tests, 100% were correctly identified.
5. Prevent the coronavirus spread to the nose, eyes, oropharynx, olfactory mucosa, olfactory bulb and brain through the olfactory mucosal spray by wearing mask, goggles, washing hands, social distancing, and wearing our antivirus ionic device (under development).
6. Prevent coronavirus binding to angiotensin receptors on healthy cells by use of angiotensin receptor blockers telmisartan and losartan, and use of antibody rich convalescent plasma from COVID-19 recovered.
7. Once coronavirus enters the host cell: prevent its multiplication by use of protease inhibitor, antiviral therapeutic agents, antimalarial, and monoclonal antibodies.
8. Once coronavirus is inside the host cells: prevent its multiplication and virion assembly in the Endoplasmic reticulum-Golgi-Lysosome complex, prevent its exits (budding) from the host cell and infect new host cells by use of protease inhibitor and antimalarials and such.
9. Prevent hyper-stimulation and multiplication of the immune system cells to prevent cytokines production and cytokine storm using corticosteroids—dexamethosone, hydrocortisone administration.
10. Prevent the leaking of blood vessels and release of cytokines to fill the alveoli that contributes to acute respiratory syndrome (ARDS) using vitamin C, thiamine hydrochloride and corticosteroids, antioxidants.
11. Treat secondary infection associated with COVID-19 using azithromycin or other appropriate antibiotics
12. Provide supplemental oxygen if patient become hypoxic through oxygen mask or endotracheal tube, positive pressure ventilation and ECMO, may need to use blood purification device (CytoSorb).
13. Manage blood pressure, body fluid volume and vital organ function.

There are more than a dozen vaccines and viricidal on the horizon, and they will be here before you can blink your eye.

Utility patient pending—filing No. 16/928,310 filed in the US patent office on July 14th 2020 and is described in its entirety. The patent application is broken down into multiple chapters for ease of reading with addition of latest developments on COVID-19 pandemic.

ABSTRACT:

The field of the present invention discloses methods of prophylactic nasal and oral spray, and mouth wash and nasal lavage as anti-coronavirus prophylactic using angiotensin receptor blockers telmisartan, chloroquine, EDTA, xylitol, protease inhibitor, superoxide dismutase, N-acetyl cysteine, and chlorohexidine intended to be used by nursing home residents, in elderly care residences, and by care givers, as well as by hospital personnel and use by the public during epidemics and pandemics. This invention includes methods of treating the COVID-19 sufferers from mild symptoms to full blown affliction of the respiratory system in the medical clinics, outpatient settings, hospitals, and intensive care with COVID-19 (mild, moderate to severe cases). Treatment includes using high dose intravenous vitamin C, thiamine hydrochloride, hydrocortisone, dexamethasone, zinc, antibiotic azithromycin, protease inhibitors remdesivir and Ebselen, antimalarial Dihydrochloroquin and chloroquine phosphate, angiotensin receptor blockers telmisartan and losartan, inti-helminthics ivermectin and Niclosamide, rapamycin, bioactive curcumin, ethylenediaminetetraacetic acid, superoxide dismutase, N-acetyl cysteine, and auranofin as antioxidant and antiviral therapeutic agents. These are combined with vital sign supportive therapies besides using supplemental oxygen, including extracorporeal membrane oxygenation (ECMO), extracorporeal blood purification (EBP) method using CytoSorb therapy, and hyperbaric oxygen therapy (HBO).

Chapter 1

Field of the invention

The field of the invention relates to the use of prophylactic nasal-oropharyngeal spray, mouth wash and administration of therapeutic agents against coronavirus infection and treatment of signs, symptoms and severe acute respiratory syndrome coronavirus 2 (SARS-C-19-COVID-19) infection by using angiotensin receptor blockers telmisartan, losartan, anti-malarial Hydroxychloroquine, chloroquine, protease inhibitors remdesivir and Ebselen, antioxidant, anti-cytokine and viricidal vitamin C, superoxide dismutase (SOD), auranofin, rapamycin, anti-helminthics such as ivermectin, Niclosamide, Ebselen, curcumin, vitamin D3, ethylenediaminetetraacetic acid (EDTA), antibody—nano particles, zinc sulfate, antibacterial and antiviral antibiotic azithromycin and other antibiotics to treat the secondary infection associated with COVID-19.

The inventive method describes therapeutic agents and methods to block the entry of the coronavirus into the mucus membrane healthy cells of the respiratory system from nose to lung alveoli when exposed to the virus from the infected person/object. The method will prevent its multiplications within the healthy cell once infected, and budding out of the infected person's cells to infect the healthy cells with various degrees of infection of cells of the respiratory and other systems, that may lead to mild to severe acute respiratory syndrome (SARS) that is resulting in increased morbidity and mortality. The intent of this invention is to use nasal and oral sprays and mouth wash, with therapeutic agents as prophylaxis to prevent the binding of the virus to nasal and oral cavity, nasopharynx, oropharynx, laryngopharynx, air passages (larynx, trachea, bronchial tree), alveoli, and to prevent coronavirus spread with multiplication in the host cells, as well as to treat the infection associated symptoms by parenteral and oral administration of above antiviral and antibacterial therapeutic agents in the nursing homes, extended care facilities, the aged living independently, and in hospital and ICU settings.

Chapter 2

Background of the invention

Background of viral afflictions

Viruses are unlike other infectious agents (bacteria, fungus, parasites). they cannot grow and reproduce on their own. They must hijack the controls of a host cell to survive and replicate like parasites. They are tiny, far smaller than microscopic bacteria and fungi. Viruses are made up of miniscule packages of protein that surround genetic material made up of mostly RNA, and some with genetic material DNA. When a virus find a suitable host cell, they invade, then use that cell's machinery to produce new copies of the virus, which can go on to infect other cells. The first step in this process occurs after the virus enters the body, through the lining of respiratory system airways or the digestive system. Now it is found out that coronavirus is not just a respiratory virus affecting the lungs, it can also affect blood vessel and its content, heart, gastro-intestinal track, pancreas, liver, kidneys, brain, spleen, and other organs besides the respiratory system. Further, recovery from COVID-19 can be slow, incomplete, needing a prolonged stay in the hospital, and costly with a huge impact on quality of life, especially in the aged.

Viruses contain surface proteins that are able to recognize specific surface features of the host cells (Chapter 21, Figures 1, 2). This allows them to attach to cells and gain entry (Chapter 21, Figures 5, 6). Once in the cell, the virus takes control of the cellular machinery to make copies of its genetic material and proteins that are then reassembled into new virus particles and exit the host cell to infect the healthy cell. Coronavirus is an RNA virus and has S-spikes which attach to the angiotensin receptors II on the host cell surface to get inside the host cell. Influenza (flu) virus is different than coronavirus with less morbidity and mortality. The flu virus mutates; hence the vaccination is only 55% effective in providing protection; the same thing may happen with coronavirus in time.

Viral infections remain a major medical challenge worldwide because of the lack of efficient therapy, prevention or vaccination strategy, their ease of propagation, and because of the rapid development of viral mutation and drug resistance that can cause major health problems worldwide in mammals including humans. Epidemic means the outbreak is confined to one geographical area (for example, malaria, sleeping sickness, Chagas disease and such), unlike pandemic which affects many countries. Endemic means confined to the certain geographic area and certain people. Now we have the coronavirus infectious disease, a 2020s pandemic—an epidemic upgraded to a pandemic once it has spread to a large number of people in other countries or continents through person-to-person contact. It is affecting 184 countries all over the world except Greenland with millions afflicted beside an untold number of morbidity and mortality of the afflicted people, besides economic disruption and its effect on the general population.

The vast majority of SARS-coV-2 transmission occurs indoors, most of it from the inhalation of airborne particles that contain the coronavirus. The best way to prevent the virus from spreading in a home or business would be to simply keep infected people away (mask use and social distance). But this is hard to do when an estimated 40% of cases are asymptomatic and asymptomatic people can still spread the coronavirus to others.

Coronavirus is a respiratory virus, meaning it infects from the nose, oral cavity, pharynx, larynx, trachea, bronchioles, all the way down to the lung's alveoli and its surrounding microvasculature (blood vessels). Recent reports show that the coronavirus is shown to afflict and inflict injury to other organs and tissues in the body and brain, not just the lungs. Is it possible that it also spreads by airborne transmission route based on the recent increase in reported cases instead of just droplets, hand contact and such in the US. It is possible that some coronavirus escapes, becomes airborne, and infects in a closed room and in crowds of people.

World Health Organization (WHO) naming the coronavirus disease

"On February 11, 2020 the World Health Organization (WHO) announced an official name for the disease that is causing the 2019 novel coronavirus outbreak, first identified in Wuhan province of China. The new name of this disease is coronavirus disease 2019, **abbreviated as COVID-19**. In COVID-19, **'CO' stands for 'corona,' 'VI' for 'virus,' and 'D' for disease**. Formerly, this disease was referred to as '2019 novel coronavirus' or '2019-nCoV' and such. We discuss the role of WHO controlling the coronavirus pandemic in a separate chapter.

Viruses and nasopharynx infections

Remember that "Most infections of the nasopharynx are caused by viruses (many by bacteria, fungus and such) and give rise to the signs and symptoms that are collectively known as the common cold. Approximately 40-50% of colds are caused by the rhinovirus (RV) group." (Schaechter M, et al., Mechanisms of Microbial Disease (3rd Edition). Lippincott, Williams, and Williams. Philadelphia, 1999, p550). Other important viral pathogens include human coronavirus, adenoviruses and influenza viruses in addition to HIV, rabies, and countless others.

Coronaviruses are zoonotic, meaning they are transmitted between animals and humans. CoV are a large family of RNA viruses that cause illness ranging from the common cold to more severe diseases such as Middle East Respiratory Syndrome (MERS-CoV) and Severe Acute Respiratory Syndrome (SARS-CoV). A novel coronavirus (nCoV) is a new strain, recently named COVID-19 by the World Health Organization, that had not been previously identified in humans.

Viral infections account for nearly 85% of acute asthmatic episodes in children and in approximately half of exacerbations in adult asthmatic subjects. RV is the most commonly detected virus (Grunberg, K et al., "Experimental rhinovirus 16 infections causes variable airway obstruction in subjects with atopic asthma", AmJ Respir Crit Care Med 1999;160:1375-80; Gem, J E et al., "The role of viral infections in the natural history of asthma", J Allerg Clin Immunol, 2000; 106:201-12). The fluctuations of hospitalization rates of patients with chronic obstructive pulmonary disease correlate strongly with seasonal variations in viral infection rates (Johnston SL et al. Am J Respir Crit Care Med. 1996; 154:654). RV infections in elderly persons (ages range from 60-90 years) cause illness severity comparable to influenza with a mean duration of viral respiratory tract symptoms of 16 days and daily activity restrictions in more than a quarter of affected individuals (Nicholson K G et al. BMJ. 1996; 313:1119). Rhinoviruses are the smallest RNA containing animal viruses which infect humans.

Epithelial cells which line the nasopharynx and bronchial airways (respiratory system)

Epithelial cells which line the nasal cavity, oral cavity, the nasopharynx and bronchial airways (respiratory system) are the primary site of rhinovirus infections, so also coronavirus and other respiratory viral and bacterial infections. While the initial site of infection is most commonly the nasopharynx, recent research demonstrates that in experiment in in vivo RV infection, the lower respiratory bronchial epithelium is commonly involved as well. (Mosser, A G et al., "Similar frequency of rhinovirus- infectible cells in upper and lower airway epithelium", J Infect Dis 2002; 185: 734-43). Entry into the epithelial cells occurs via the intracellular adhesion molecule-I (ICAM-1) which acts as the receptor for 90% of rhinoviruses. Infection of the cells induces up-regulation of the ICAM-1 receptor. (Grunberg K et al., "Experimental rhinovirus 16 infection increases ICAM-1 expression in bronchial epithelium of asthmatics regardless of inhaled steroid treatment", Clin Exp Allergy 2000; 30:1015-23). RV requires an acid environment to enter the cytoplasm via endosomal vesicles so also *coronavirus in lysosomes* to multiply, and organize into viral particles. The coronavirus multiplies and spreads rapidly throughout the respiratory system with cytokine storm and lung affliction in advanced state of infection specially in the elderly with co-morbidities.

Studies of subjects with experimentally induced or naturally acquired colds have demonstrated increased levels of several cytokines, including kinins, IL-1, IL-6, TNF-a, G-CSF and INF-y in nasal secretions during symptomatic rhinovirus infections which are released by the epithelial cells. The severity of respiratory symptoms correlates closely with increases in these cytokines. Infection of purified human respiratory epithelial cell populations with rhinovirus has been shown to induce production of a group of potent pro-inflammatory CXC chemokines, such as IP-10 and RANTES, that play a central role in initiating leukocyte recruitment. Once recruited during the acute stages of the viral infection, neutrophils, eosinophils and mononuclear cells all participate in the inflammatory cascade by releasing super- oxide, proteases and eosinophil granular proteins. Same reaction repeats during coronavirus infection, and the process is heightened many times. That is why that the signs and symptoms of coronavirus (COVOD-19) are severe with increased morbidity and mortality compared to cold or flu attacks.

Any time we have a new virus outbreak, attempts are being made to find out their source of their origin, animal to human, human to human, human to animal spread, and development of vaccines, and therapeutic agents to kill the virus and treat the symptoms they cause. Coronavirus is no exception. Coronaviridae approximately comprises 15 species, causing in humans' respiratory infections (including Severe Acute Respiratory Syndrome (SARS), enteric infections and rarely neurological syndromes. Even today, there are no cures for many viruses such as rabies of the brain, 40 million human live with HIV infection (I do believe that there will be a cure within a decade), 3% of humans are affected with HCV (now there is a cure), million affected with papilloma and herpes viruses and the list is endless. The search to fight these viral infections continues, as the new ones emerge. Now we have coronavirus infection that has become a pandemic affecting millions and killing thousands. The hunt is on by researchers all over the world to find a vaccine to prevent infection and therapeutic agents to kill the virus and treat the affliction it causes.

Types of coronavirus

Over the past 50-60 years the emergence of many different coronaviruses that cause a wide variety of human and veterinary diseases has occurred. It is likely that these viruses will continue to emerge and to evolve and cause both human and veterinary outbreaks owing to their ability to recombine, mutate, and infect multiple species and cell types. Many of the discussions herein are from Bosch, Fehr and other publications quoted herein (Bosch BJ, et al. The coronavirus spike protein is a class I virus fusion protein: structural and functional characterization of the fusion core complex. J Virol. 2003;77(16):8801–8811. Fehr AR and Stanley Perlman. Coronaviruses: An Overview of Their Replication and Pathogenesis, Methods Mol Biol. 2015; 1282: 1–23.).

Chapter 3

Coronavirus entry inside host healthy cells: signs and symptoms

How does the coronavirus enter the cell of the respiratory system?

We know that the coronavirus spreads from infected droplets transmitted by the infected person's expired air from clinically ill or from asymptomatic infected person. Could it also spread by airborne in some cases? We do not have the answer to such a spread yet, but one needs to keep it in mind. To infect a human host, coronaviruses must be able to gain entry into individual human cells. Once the human is exposed to coronavirus droplets, they enter the host cell and use the host cell's machinery to produce copies of themselves, which then spill out and spread to new cells, infecting them. On Feb. 19th 2020, in the journal Science, a research team led by scientists at the University of Texas at Austin described the tiny molecular key on SARS-CoV-2 that gives the virus entry into the cell. This key is called a spike protein, or S-protein (Chapter 21, Figures 2-6). On March 4, 2020, Zhou and his team described the rest of the puzzle: the structure of the angiotensin converting enzyme -2 (ACE2) receptor proteins (which are located on the surfaces of respiratory cells, blood vessels, heart and such. The coronavirus S spike protein binds to these angiotensin receptors II on the cell membrane and S spike protein biologically interacts at the receptor to let the viral RNA nucleoproteins inside the healthy cells. Histologically, the *type II alveolar cells are very rich in angiotensin receptors II,* that is why the coronavirus affects the alveoli more than the rest of the respiratory tract. This results in coronavirus inside the cells to multiply and spread using host cell machinery to cause COVID-19.

Research has shown that children have fewer of these angiotensin receptors in the nasal mucosa, and hence get less infection and suffer less if they come in contact with the coronavirus (Bunyavanich S, et al., Patel and Verma 2020).

"If we think of the human body as a house and 2019-nCoV [another name for SARS-CoV-2] as a robber, then ACE2 receptors would be the doorknob of the house's door. Once the S-protein grabs it, the virus can enter the house," Liang Tao, a researcher at Westlake University who was not involved in the new study, said in a statement. Coronavirus disease 2019 is an infectious disease causing severe acute respiratory syndrome coronavirus -2 (SARS-CoV-2). The disease was first identified in 2019 in Wuhan, the capital of China's Hubei province, and now it has spread globally, resulting in the 2019–2020 coronavirus pandemic causing untold number of deaths besides creating a financial disaster all over the world.

Coronavirus infection: signs and symptoms (Chapter 21, Figure 1a)

People with COVID-19 have had a wide range of signs and symptoms reported, ranging from mild symptoms to severe illness. According to the latest CDC reports, watch for these following symptoms that may appear between 2-14 days after exposure to the coronavirus:

- Fever
- Running nose
- Cough
- Shortness of breath or difficulty breathing
- Chills
- Repeated shaking with chills
- Muscle aches
- Headache, aches all over the body
- Sore throat
- New loss of taste or smell may be one of the most important self-diagnostic symptoms and signs that indicate that you may have acquired the coronavirus infection immaterial of other symptoms.

The following is a list of emergency warning signs for COVID-19 which requires medical attention immediately:

- Trouble breathing
- Persistent pain or pressure in the chest due to pleurisy (pleural lining of the lungs and chest cavity) and lung parenchymal inflammation
- New confusion or inability to arouse, mental fog
- Bluish lips or face or toe, may observe hair loss in some patients
- In severe cases, the patient develops high fever, cough with blood, decreased white blood cells, kidney failure.

This list is not all-inclusive.

Silent spreaders of coronavirus infection:

Dr. Robert Redfield of the CDC states that for every case that is reported, there are 10 other cases of coronavirus infection, indicating it is 10 times higher than what has been reported. That tells us that the US coronavirus infected are close to 23 million instead of 2.3 million as of June 2020. Dr. Marybeth Sexton, faculty at Emory University, stated that many asymptomatic people had no idea they were even positive. She also stated that long incubation period (14 days) has led to some confusion as to how "asymptomatic" is defined, and they become silent spreaders of COVID-19. It is important to note that the coronavirus can spread during incubation and from those who are asymptomatic, contradicting the earlier WHO report. Studies begin to show that those who have no symptoms or who are mildly symptomatic are possibly due to how their immune system responds by production of antibodies against the virus. These findings tell us that everyone wearing the mask will reduce the asymptomatic silent spreader disseminating the disease innocently to others.

Common symptoms of COVID-19 vary from patient to patient and appear as if it is the ***disease of the immune system***. Due to coronavirus infection, our body's immune system vigorously reacts by pouring cytokines to kill the virus, which in turn start attacking our body, besides attacking the invading virus if it is not controlled, which results in cytokine storm. Other symptoms may include abdominal pain, diarrhea, development of ***blue toe,*** dermatological hives and rashes. COVID-19 may progress to pneumonia with secondary infection and may lead to multi-organ failure. The COVID-19 in the elderly with co-morbidities can progress to become severe and deadly SARS, needing hospitalization with respiratory support. Pulse oximeter is a must in the house of the elderly and after returning from ICU to measure the oxygen levels, besides a BP monitor.

The coronavirus also affects gastrointestinal organs as angiotensin receptors II (ACE2) receptors which are richly expressed in the glandular cells of gastric, duodenal and rectal epithelium, endothelial cells and enterocytes of the small intestine. So, the COVID-19 patients can experience gastro-intestinal track pain, and diarrhea.

Many factors determine the coronavirus infection signs and symptoms and outcome. It depends on age of the afflicted, health condition, amount of virus exposed and maybe even the mutation form of the coronavirus. Some of these mutated forms may be less virulent or even hypervirulent than the original virus, though the mutation coronavirus is *not noted yet* but will show up. Further, it is possible that old people have more angiotensin receptors II on their respiratory system lining cells with depressed immune system than the younger age group, hence more coronaviruses gets into the respiratory system cells from nose to the alveoli, causing severe symptoms—acute toxic shock like syndrome. It is possible that the some of these cases may have fewer of these coronavirus binding angiotensin II receptors and the immune system is less developed to respond severely as may be the case in children, young adults, and maybe even in some aged.

Our own observation shows that angiotensin receptor II blockers (ARBs) such as telmisartan and losartan, prescribed to treat blood pressure, lowers the incidence of flu and cold. I had hardly any flu since I started taking ARBs for decades. This is also true of people who take antacid medications such as cimetidine, Famotidine and such. I take ARBs, and have recommended them at lower doses to many of my family, friends, and some patients during the flu season to ward off the flu, not to treat blood pressure; of course, if you have hypertension/sauna, it has added benefits. *There are reports from China that the patients on angiotensin receptor blockers anti-hypertensive medications did better than others on blood pressure medications when infected with coronavirus.*

While the COVID-19 symptoms can go from bad to worse quickly, the more "prolonged course" of COVID-19 makes it unlike influenza or other similar respiratory illnesses, said William Schaffner, a professor of medicine in the Division of Infectious Diseases at the Vanderbilt University School of Medicine. Patients on ventilators (for longer than comparable patients who had the flu) are seriously ill and the outcome is life-threatening.

Spread of coronavirus to the baby by breast feeding, and feeding donor breast milk—is it safe?

The latest studies published in JAMA (8-19-20) by Christina Chambers, co-principal investigator of the study, from the University of California (UC) San Diego School of Medicine, found that breast milk is not likely to transmit COVID-19. This study found that the coronavirus was unable to replicate and cause infection in the breastfed infants although one sample tested positive for viral RNA. Hand hygiene and sterilizing pumping equipment after each use are important when one breast feeds the baby.

The researchers also mimicked conditions of the Holder pasteurization process commonly used in human donor milk banks by adding SARS-CoV-2 to breast milk samples from two different donors who were not infected. The samples were heated to 62.5 degrees Celsius for 30 minutes and then cooled to 4 degrees Celsius. Following pasteurization, infectious virus was not detected in either sample.

Research studies show that breastfeeding is associated with a reduced risk of sudden infant death syndrome (SIDS and obesity in children, as well as improved immune health and performance on intelligence tests. In mothers, breastfeeding has been associated with lower risks for breast and ovarian cancer, cardiovascular disease and type 2 diabetes.

"We hope our results and future studies will give women the reassurance needed for them to breastfeed. Human milk provides invaluable benefits to mom and baby," Aldrovandi (one of the co-investigators) said. This study gives the reassurance and encouragement needed for mothers to breastfeed without the fear of transmitting the coronavirus and other infections from the breast feeding (also noted in our book published in 1972: “Breast and Its Care” from Fredrick Fell, Emory University).

Children and coronavirus infection showing up with signs and symptoms akin to Kawasaki disease: “Pediatric multisystem inflammation syndrome”

Hundreds of cases of COVID-19 in young children have been reported, some with critical conditions and outcome. The patients have fever, rashes on the back, lymph node enlargement in the neck, abdominal pain, redness of the eyes, irritation and inflammation of the mouth, lips and throat. hand and feet swelling, and respiratory distress with bradycardia and even cardiac arrest. multiorgan (e.g., cardiac, gastrointestinal, renal, hematologic, dermatologic and neurologic) involvement, and elevated inflammatory markers. Respiratory symptoms were not present in all cases.

With passage of time, many more will be reported. They may present as patient with shock and Kawasaki disease. Kawasaki disease (KD), also known as Kawasaki syndrome, is an acute febrile illness of unknown cause that primarily affects children younger than 5 years of age. I believe this syndrome is not restricted to those less than 5 years old; it affects children older and even teens; it won’t be a surprise. It has been labelled as a “multisystem inflammatory syndrome in children” abbreviation as MIS-C. The disease was first described in Japan by Tomisaku Kawasaki in 1967. Clinical signs include fever, rash on the back and all over, swelling of the hands and feet, irritation and redness of the whites of the eyes, swollen lymph glands in the neck, and irritation and inflammation of the mouth, lips, and throat. KD is a leading cause of acquired heart disease in the United States. Serious complications include coronary artery dilatations and aneurysms. The standard treatment, intravenous immunoglobulin and aspirin, substantially decreases the development of these coronary artery abnormalities.

In the year 2000, approximately 4248 hospitalizations with KD were reported among children under 18 in the US. Testing for a substance called B-type natriuretic peptide (BNP) that's released when the heart is under stress may be helpful in diagnosing Kawasaki disease. However, more research is needed to confirm this finding. Electrocardiogram is used to measure the electrical impulses of the child's heartbeat. Kawasaki disease can cause heart rhythm problems. Ultrasound images are used to show how well the heart is working and can help identify problems with the coronary arteries.

Treated with Gamma globulin immune protein through a vein (intravenously) can lower the risk of coronary artery problems. High doses of aspirin may help treat inflammation. Aspirin can also decrease pain and joint inflammation, as well as reduce the fever, so also other anti-inflammatories. Aspirin has been linked to Reye's syndrome, a rare but potentially life-threatening condition. With coronavirus positive, these children should be treated with anti-coronavirus therapy and may need ventilatory and cardiovascular support. They will need observation after recovery for cardiovascular disease.

New syndrome due to coronavirus infection in children: "Multisystem Inflammatory Syndrome in Children (MIS-C)" pediatric inflammatory multisystem syndrome (PIMS-TS)

- An individual aged <21 years presenting with fevers, laboratory evidence of inflammation, and evidence of clinically severe illness requiring hospitalization, with multisystem (>2) organ involvement (cardiac, renal, respiratory, hematologic, gastrointestinal, dermatologic or neurological)
- No alternative plausible diagnoses
- Positive for current or recent SARS-CoV-2 infection by RT-PCR, serology, or antigen test; or COVID-19 exposure within the 4 weeks prior to the onset of symptoms
- Fever >38.0°C for ≥24 hours, or report of subjective fever lasting ≥24 hours
- Including, but not limited to, one or more of the following: an elevated C-reactive protein (CRP), erythrocyte sedimentation rate (ESR), fibrinogen, procalcitonin, elevated d-dimer, ferritin, lactic acid dehydrogenase (LDH); or interleukin 6 (IL-6), elevated neutrophils, reduced lymphocytes and low albumin (Royal College

of Pediatrics and Child Health Guidance: Pediatric multisystem inflammatory syndrome temporally associated with COVID-19, https://www.rcpch.ac.uk1;
https://www.cdc.gov/kawasaki/index.html).

The children who have developed this syndrome need immediate attention and need to be hospitalized immediately and may need to be cared for in the ICU. There are many reported cases of encephalopathy with seizures, fever, hallucinations, and patients who suffer from flashbacks and significant memory problems which may be due to coronavirus entering the brain through the olfactory mucosa as described below.

CNS effects are a consequence of either "the post-infectious multisystem inflammatory syndrome," or if these cases are "directly attributable to the neurological involvement of the virus" transmitted as we describe. COVID-19 may have lasting effects on some children's brains. According to the researchers, including those from King's College London in the UK by Dr. Shankar Hari, the rare disease, referred to as pediatric inflammatory multisystem syndrome (PIMS-TS), has emerged in a small number of children during the COVID-19 pandemic and could be due to cell-cell signaling molecules called cytokines, and reduced levels of the immune system's white blood cells called lymphocytes.

Children morbidity and mortality rate with COVID-19 is very low. Why?

The death rate among COVID-19 positive children has been way below other age groups, including those in the prime of their life, though some cases result in "Multisystem Inflammatory Syndrome in Children (MIS-C)." Such a low incidence is attributed to:

1. Children have a paucity or low level of angiotensin receptors protein ACE-2 found on the cell surface compared to the adults. Hence a smaller number of coronavirus enter the healthy cells to multiply and spread. The number of these receptors are high on the old people respiratory system cells; hence more virus enters resulting in high morbidity and mortality.
2. These angiotensin receptors in the young are under developed, hence coronaviruses are unable to firmly attach to healthy cells to enter and multiply.
3. The lung as well as the rest of the respiratory system may produce micro and macromolecules which are anti-coronavirus, prevent their growth and expansion on the surface and inside the cells.

4. Further, the immune system is very active in the young, thus eliminating the virus as the immune system attacks them as they enter our systems due to active T and other immune system cells.
5. The thymus gland is the largest by the time you reach puberty. As you age, by age 75, the active thymus gland is replaced by fat and is no longer active. The thymus gland is the primary lymphoid organ responsible for production and maturation of T cells up to age of 15. It is functional, producing protective T cells needed to attack the virus compared to adults, in whom the thymus gland is nonfunctional, and hence does not produce T cells any more.
6. The thymus gland is instrumental in the production and maturation of T-lymphocytes or T cells, a specific type of white blood cell that protects the body from certain threats, including coronaviruses and infections.
7. The thymus produces / secretes thymosin, a hormone needed for maturation of T-cells in the center of the thymus gland, where the T-cells complete their process of maturing and are then released into the bloodstream. After puberty, the thymus gland involutes and no longer produces thymosin to keep T cells active and functional and is replaced by fat by the time you reach 75. This once again is responsible for maintaining good health in the young compared to the aged, reducing the morbidity and mortality even with coronavirus infection.
8. In the young, the trachea-bronchial tree lining cells have highly active cilia which traps the virus in the mucosa and disposed in the cough without allowing the virus getting to healthy cells to replicate and spread, which is not the case in the elderly with comorbidities.
9. I believe that the BCG (tuberculosis) vaccine, and MMR (measles-mumps-rubella) vaccine provide nonspecific protective effects in dealing with unrelated viruses such as coronavirus as I have seen for the flu.

Now you can see how and why COVID-19 results in reduced morbidity and mortality compared to the elderly.

Diagnosis of coronavirus infection

The standard method of testing is real-time reverse transcription polymerase chain reaction (rRT-PCR). The test is done on respiratory samples obtained by a nasopharyngeal swab; a nasal swab or sputum sample may also be used. Results are generally available within a few hours to two days. The FDA in the United States approved the first point-of-care test on 21 March 2020 for diagnosis on the spot within minutes. Such tests will continue to evolve and will be made available for use.

Chapter 4

Classification: Different stages of COVID-19; men versus women

Our classification of "Different Stages of COVID-19" signs and symptoms

The coronavirus infected patients can show from mild to severe symptoms depending upon what stage of the coronavirus infection the person is suffering, location of the coronavirus, level of infection, age, sex and health condition of the patients. The stages of COVID-19 are as follows:

First stage of coronavirus infection:

Once coronavirus in the nose and/or mouth, it latches on to cells that express a protein called ACE-2 to angiotensin receptors II; located deep in the back of the throat and high up in the nasal passage. The S- spikes on the surface of the coronavirus have found a way to unlock by latching to angiotensin receptors to get inside the cell and use host machinery to replicate, make its many copies. Develop mild to no symptoms

Second stage of localized coronavirus spread:

The coronavirus begins to multiply in the affected host cells with various symptoms of sore throat, mild cough, headache, loss of smell sensation, maybe mild fever, and such. In the meantime, the body's immune system inflammatory response is becoming more active, trying to combat the virus and the patient develops sore throat, cough, and fever as a manifestation of early infection.

Third stage spreading of the coronavirus down to the air passages:

The new viruses then exit the infected cells and the infections spreads further down on to infect more and more cells spreading to larynx, trachea bronchial mucosal lining, with fever, production cough, burring in the chest, fatigue, headache, due to inflammation by the immune system to rid of the virus.

Fourth stage of coronavirus affecting the lungs and virus battle with immune system:

The coronavirus has descended to alveolar lining cells with severe symptom that is because the alveoli and the surrounding blood vessels are richly endowed with angiotensin receptors II, for coronavirus virus to attach and cause severe signs and symptoms of the COVID-19 (Chapter 21, Figure 3). The immune system is unable to wipe out the coronavirus and the immune system response accelerates. Inflammatory cells and fluid start pouring into air sacs from the microvasculature (which are also attacked by coronavirus) surrounding the alveoli. This often develops into Acute Respiratory Distress Syndrome (SARS-ARDS), which is when “inflammation and fluid begin to accumulate in the spaces around and in the air sacs,” with hypoxia and fever. Due to production of more pro-inflammatory cytokine and immune system cell, the fluid can’t drain normally, and the air sacs then fill with inflammatory fluid giving a feeling of a “drowning feeling” in patients, and the blood’s oxygen levels drop. The need for a ventilator assisted support and supplementary oxygen arises quickly before or after admission to the hospital. Some of these patients may need positive pressure ventilatory support and may need Extracorporeal membrane oxygenation (ECMO) in the intensive care unit (ICU) to save the life of the patient. The immune system has difficulty controlling the viruses that continue to multiply and spread far and wide from the respiratory system to other parts of the body through pulmonary circulation with increased morbidity and mortality.

Why women face less severe COVID-19 symptoms than men

It is observed that men may have a greater risk than women for more severe symptoms and worse outcomes from COVID-19 regardless of age. According to Leanne Groban, a professor at Wake Forest School of Medicine in the US, the difference is due to female sex hormone estrogen and its role in lowering the level of angiotensin-converting enzyme2 (ACE2) receptors in the heart and lungs, which may modulate the severity of COVID-19 in women (study published in Current Hypertension reports Aug. 28, 2020)."We know that coronavirus affects the heart and we know that estrogen is protective against cardiovascular disease in women, so the most likely explanation seemed to be hormonal differences between the sexes." The researchers said the published literature indicated that ACE2, which is attached to cell membranes in the heart, arteries, kidneys and intestines, is the cellular receptor of the coronavirus responsible for COVID-19 infections, and helps bring the virus into the cells of those organ systems. Men have more of them, conversely, higher levels of ACE2 and viral load in tissues could account for why symptoms are worse in men than women, Groban said. Estrogenic hormones in ACE2 expression and regulation may explain the gender differences in COVID-19 infection and outcomes, which means less morbidity and mortality in women.

Chapter 5

"Cytokine storm syndrome" in COVID-19

Cytokine release syndrome (CRS) or cytokine storm syndrome (CSS) or simply cytokine storm in COVID-19 leading to ARDS/SARS

Cytokine release syndrome or cytokine storm syndrome or cytokine storm is a form of systemic inflammatory response syndrome that can be triggered by a variety of factors such as infections, now coronavirus, and certain drugs, which may lead to excessive inflammation, organ failure and death [Shimabukuro-Vornhagen A, Gödel P, Subklewe M, et al. 2018. "Cytokine release syndrome". J Immunotherapy Cancer. 6: 56]. CRS occurs when large numbers of white blood cells are activated by massive coronavirus infection, which results in the release of massive inflammatory cytokines, which in turn activates yet more white blood cells into action.

CRS in COVID-19 occurs when large numbers of white blood cells, including *B cells, T cells, natural killer cells, macrophages, dendritic cells, and monocytes* are activated and release inflammatory cytokines, which activate more white blood cells and microvasculature around the alveoli in a positive feedback loop of pathogenic inflammation in a vicious circle.

This can occur when the immune system is fighting coronavirus pathogens. As cytokines produced by immune cells recruit more effector immune cells such as T-cells and inflammatory monocytes (which differentiate into macrophages) to the site of inflammation or infection. In addition, pro-inflammatory cytokines binding their cognate receptor on immune cells results in activation and stimulation of further cytokine production. This process, when dysregulated, can be life-threatening due to systemic hyper-inflammation, hypotensive shock, and multi-organ failure (Chapter 21, Figure 14).

Prophylactic treatment for COVID-19 and cytokine morbidity and mortality is to take vitamin D3, 5000 IU 2-3 times a day, which is low in these patients. Treatment of symptoms like fever, muscle pain, fatigue, hypoxia and such are needed in critical care ICU setting. Moderate CRS requires oxygen therapy, maybe intubation and giving fluids and anti-hypotensive agents to raise blood pressure and maintain vital signs. Doctors may need to use immunosuppressive agents, like rapamycin and its analogues, and corticosteroids with antibiotics to control the lung infection. Tocilizumab (brand name Actemra and RoAcemtra), commonly used to treat rheumatoid arthritis, an anti-IL-6 monoclonal antibody, was FDA approved for steroid-refractory CRS based on retrospective case study data. It should be combined with remdesivir with or without auranofin and could be very effective in coronavirus infected patients to prevent extreme inflammation in those gravely ill. Lenzilumab, an anti-GM-CSF monoclonal antibody, may also be effective at managing cytokine release by reducing activation of myeloid cells and decreasing the production of IL-1, IL-6, MCP-1, MIP-1, and IP-10. Our therapy is high dose vitamin C, thiamine hydrochloride and hydrocortisone therapy with use of antiviral therapies described in this invention in addition to above measures. Consider using anti-helminthic such ivermectin, Niclosamide and protease inhibitor cum anti-oxidant Ebselen.

The blood purification device CytoSorb of Biocon Biologics may need to be used for extracorporeal blood purification (EBP) to reduce viral and pro-inflammatory cytokines load in COVID-19 patients who have been admitted to the intensive care unit with imminent respiratory and systemic failure.

Skin rashes in the coronavirus infected

There are reports of coronavirus infected showing the a "dengue fever-like rash" that forms large red patches, hives-like rash that either spread or remain the same in size, measles-like rash that creates "little dots all over," discoloration of the skin that forms in a tree pattern, small, red to purple bumps or nodules on the hands and feet (J of AA of Dermatology - March 2020). The children may show rash on the back. When the doctor is examining the patient, they should look for these dermatological signs. Blue toe may indicate thrombosis of blood vessels to the toe, indicating thrombo-embolic episode due to COVID-19.

Chapter 6

How COVID-19 spreads to the brain: Route of coronavirus spread

Route of coronavirus spread from olfactory mucosa to olfactory bulb and the brain. It can directly invade the brain. (Chapter 21, Figures 11, 12, 13)

New, immediate or sudden loss of smell and taste during COVID-19 epidemic or pandemic may be one of the most important self-diagnostic symptoms and signs that indicates that you may have been infected with coronavirus infection though you are not tested or tested negative. There are many reports of CNS (brain) signs and symptoms due to COVID-19. We were not surprised based on our 6 decades of work on the membranes of the nervous system and olfactory nerves as being the route of viral (now coronaviruses), bacterial and amoebic infection of the brain. We believe that coronavirus can infect the brain entering through the olfactory mucosa from the nose based on basic research from Emory University (Shantha TR, Nakajima Y. Histological and histochemical studies on the rhesus monkey (Macaca Mulatta) olfactory mucosa. Yerkes Regional Primate Research Center, Emory University, Atlanta, Georgia. Z. Zellforsch.1970; 103:291-319. Shantha TR. CNS Delivery bypassing BBB: Drug Delivery from the Olfactory Mucosa to CNS. Drug Development and Delivery, 2017, V17, 1, 32-37. Shantha TR, Bourne GH. Perineural epithelium; structure and function of nervous tissues. Academic Press. 1969;1: 379-458. Baer G, Shantha TR, Bourne GH. The pathogenesis of street rabies virus in rats. Bulletin World Health Org. 1965; 33:783-794).

We know that the infected person breathes in and out the coronavirus laden droplets. Study shows loud talking can generate over 1,000 respiratory droplets which may contain coronavirus — but experts say 6 feet is still sufficient to protect against the coronavirus. We also know that about ±10-15% of breathing air comes in contact with olfactory mucosa as the air passes through the upper part of the nose. The coronavirus can easily attach to the olfactory mucosa and olfactory mucosa receptor cells on their cell projection with coronavirus S-spikes to attach, as we find in the rest of respiratory system cells. There is another part of the CNS such as blood vessels have

ACE2 receptors. Hence, the virus gets attached to thick olfactory mucus secretion on the sticky mucosal lining then held by olfactory cell cilium. From there, it can pass between the olfactory mucosal receptor and supporting cells, then to the sub perineural epithelial space that cover 20 olfactory nerves fasciculi (Chapter 21, Figures 11, 12, 13), then enter the olfactory bulb through glomeruli and sub arachnoid space cerebro-spinal fluid. Coronavirus enters the cerebro-spinal fluid around the olfactory bulb as well as to the interpeduncular cistern and then affect any and all parts of the brain (Chapter 21, Figure 13).

As the olfactory mucosal receptor cells come in direct contact with coronavirus infected breath (through droplets or airborne), it is the sense of smell which is affected first, which may also act on taste if the patients have nose breathing problems. Looking at 214 cases of severe coronavirus illness treated in Wuhan city during the early phase of the global pandemic, doctors reported that 36.4% of patients displayed neurological symptoms. There are reports of encephalopathy, maybe even necrotizing encephalopathy which may result in permanent brain damage if the virus spreads over the brain. COVID-19 may cause patients to develop subclinical seizures, cognition impairment, acute stroke with other thrombo-embolic phenomenon such as DVT, pulmonary emboli, cardio-vascular episodes (due to increased blood clotting), peripheral neuropathy, confusions, tingling, numbness and such.

The coronavirus encephalopathy may lead to delirium, altered mental state resulting in impairment of the cognition, attention, orientation, sleep–wake cycle and consciousness, headache, epileptic seizures, myoclonus (involuntary twitching of a muscle or group of muscles) or asterixis ("flapping tremor" of the hand when wrist is extended), subtle personality changes, and an inability to concentrate. Other neurological signs may include dysarthria, hypomimia, problems with movements (they can be clumsy or slow), ataxia, tremor, involuntary grasping and sucking motions, nystagmus (rapid, involuntary eye movement), jactitation (restless picking) at things characteristic of severe infection, and respiratory abnormalities such as Cheyne-Stokes respiration (cyclic waxing and waning of tidal volume), apneustic respirations and post-hypercapnic apnea. That is why it is important to examine all the COVID-19 patients for neurological signs and symptoms during and after the affliction. In such patients' blood tests (for clotting factors included), cerebrospinal fluid examination by lumbar puncture (also known as spinal tap), brain imaging studies,
electroencephalography (EEG), and similar diagnostic studies may be used.

Chapter 7

Effects of diabetes on coronavirus infection; angiotensin receptor blockers, hypertension and what is "herd immunity" Immune passport, blood test for the disease

Effect of diabetes and COVID-19: thromboembolic phenomenon

Many physicians are discovering diabetics, those with elevated blood pressure, old age, obese and such develop abnormal clotting, known as thrombosis, which may also play a major role in lethal COVID-19 if the clots are detached, and travel as emboli to other organs, especially to lungs and brain. Doctors are seeing blood clots in large and small blood vessels all over the body, including deep vein thrombosis_(DVT) in the legs, lungs, (PE) brain blood vessels causing stoke, and small clots in tiny blood vessels in organs throughout the body. Early autopsy results are also showing widely scattered clots in multiple organs. Such COVID-19 patients may need to take anticoagulants, and bioactive curcumin with vitamin D3, 5000 IU.

We believe that the coronavirus invades the systemic circulation blood from the lungs microcirculation and pulmonary veins, that creates an inflammatory response in the blood. This results in sticking of platelets, white blood cells, thrombin, proteins and red blood cells forming micro – macro embolic mass in the beginning. They grow bigger due to the inexhaustible supply of clotting factors from the blood flow, which grows larger resulting in thromboembolic phenomenon noticed in COVID-19.

High blood pressure and its relation to COVID-19

Early research shows that people with hypertension may be more likely to get COVID-19, have worse symptoms, be slower to recover, recover with complications, and many succumb to the infection. In Italy, they found that the hypertensive patient may have cancer, diabetes, lung diseases, obesity comorbidities and such increasing the morbidity and mortality. More than 99% of people who have died from the virus had one or more of these conditions; 76% of them had high blood pressure. People with high blood pressure are also more likely to die from coronavirus. Their risk is about 6% higher than the overall population. They should be treated with angiotensin receptor blockers such as telmisartan and losartan which lower the blood pressure and prevent coronavirus attacking the mucus membrane of the respiratory system to enter the healthy cells to cause infection.

Deep breathing and its effect on COVID-19

There are some health care providers claiming that lung exercises help to ease their discomfort and kept their symptoms from progressing in COVID-19. Is there any evidence behind these claims? The short answer is no. One should have the healthy habit of practicing yoga and breathing exercise, before and after COVID-19 if one develops it, to maintain proper lung function as the oxygenator of the trillions of cells of our body to keep them humming.

"Immunity passport" and "risk-free certificates" after COVID-19

The World Health Organization stated that there is currently no evidence that people who have recovered from COVID-19 have antibodies that protect them against a second infection. They are cautioning against the idea of "immunity passports" or "risk-free certificates" that has been floated as a way of allowing people protected against reinfection to return to work after getting COVID-19. Probably it can, but the Geneva-based UN health agency said in a scientific brief released, that more research is needed. It said "at this point in the pandemic, there is not enough evidence about the effectiveness of antibody-mediated immunity to guarantee the accuracy of an "immunity passport" or "risk-free certificate."

A small US study published July 20, found the antibodies that fight the coronavirus may only last a few months in people with mild illness, suggesting people could become susceptible again, i.e. re-infection. If it can occur, that could undermine the idea of *"immunity passports"* for returning back to workplaces.

Second time coronavirus infection: COVID-19 development after recovering from first

It takes about three weeks to build up a sufficient quantity and quality of antibodies, and even then, they may provide protection for only a few months. There are reports of new reinfection months after full recovery from COVID-19. Though occasional and sporadic, it could be a fresh new infection or coronavirus not totally left from the body and came back as the immune system and other conditions became favorable to reappear. It maybe also due to a new mutated virus, and the immunity developed after the first infection may not be effective. Indeed, several cases from South Korea—one of the first countries hit by the new coronavirus—found that patients who recovered from COVID-19 later tested positive for the virus. According to Genetics Institute at University College London Francois Ballou, the virus never completely disappeared in the first place and remains—dormant and asymptomatic—as a "chronic infection," like herpes. I think it can also re-emerge as another COVID-19.

Cases reported from Reno, Nevada showed that the virus from the second infection represented a genetically different strain, indicating it is a new infection. Could it be the original virus in the body got genetically modified? The reinfection with the new virus is probably rare, but said the findings imply that initial exposure to the virus may not result in full immunity for everyone.

What is "herd immunity" for COVID-19

For the moment, it is also unclear whose antibodies are more potent in beating back the disease: someone who nearly died, or someone with only light symptoms or even no symptoms at all. And does age make a difference in production of antibodies against coronavirus? Faced with all these uncertainties, some experts have doubts about the wisdom of pursuing a "herd immunity" strategy such that the virus — unable to find new victims — is out by itself when a majority of the population is immune. At the same time, laboratories are developing a slew of antibody tests to see what proportion of the population in different counties and regions have been contaminated and developed herd immunity.

Blood test for coronavirus infection: white blood cell count, ESR, D-dimer test

Those infected with coronavirus show that the white blood cell count in the blood test will be high with increased erythrocyte sedimentation (ESR) rate. D-dimer (or D dimer) is a fibrin degradation product (or FDP), a small protein fragment present in the blood after a blood clot is degraded by fibrinolysis. It is so named because it contains two D- fragments of the fibrin protein joined by a cross-link. During the COVID-19, there may be elevated D-dimer in blood test during coronavirus infection, indicating possible thromboembolic phenomenon taking place, needing anticoagulant therapy. A negative test indicates that there is no presence of any unusual clotting activity occurring anywhere in the body. On the other hand, a positive D-dimer test indicates the presence of an underlying medical condition which is affecting the normal clotting process of the body. The blood test of these COVID-19 may reveal soaring levels of inflammatory markers including C reactive proteins. Additionally, people with COVID-19 and acute respiratory distress syndrome (ARDS) have classical serum biomarkers of cytokine release syndrome (CRS), including elevated C-reactive protein (CRP), lactate dehydrogenase (LDH), D-dimer, and ferritin. Now the instant coronavirus antibody testing is available to diagnose whether they had coronavirus infection.

Breath analyzer for coronavirus infection diagnosis

It is only a matter of time until breath testing to diagnose coronavirus infection will be developed for rapid diagnosis and ease of testing. The breath analyzer is already being tested that was developed in Sweden. It is cheaper and faster. This is akin to breath test for tuberculosis.

A high-level team of Israeli researchers arrived in Delhi on July 25, 2020, to determine the effectiveness of four cutting-edge coronavirus testing technologies. The potentially ground-breaking rapid testing kits are based on Israeli technology and Israeli-Indian scientific research.

If the tests prove effective, people will be able to find out if they are COVID-19 positive just by spitting onto a specified surface, breathing into a device, or speaking into a mobile phone.

Saliva Direct rapid test developed by Yale University researchers

Federal officials have given emergency approval to a coronavirus saliva test that Yale University researchers used on NBA players and staff. Stephen Hahn, commissioner of the Food and Drug Administration, called the test "groundbreaking," partly because it doesn't need additional components, which have been prone to shortages, necessary with the standard nasal swab COVID-19 test. The test, known as Saliva Direct, "is simpler, less expensive, and less invasive than the traditional method for such testing," Yale said in a news release Saturday 8-15-20. Saliva Direct doesn't rely on proprietary technology, the makers don't intend to commercialize, and it will be made available to others. Yale expect the labs that use the test to charge $10 dollars per test.

Latest rapid test for coronavirus with a swab

Scientists have developed a new low-cost nasal swab test that can accurately diagnose the presence of SARS-CoV-2 virus, which causes COVID-19, in just 20 minutes. The test called N1-STOP-LAMP, is a 100 percent accurate molecular test in diagnosing samples containing SARS-

CoV-2 at high loads. The results were fast, with an average time-to-positive of 14 minutes for 93 of those clinical samples. "The test requires a small shoebox-sized machine, as well as reagents, but everything is portable," said Tim Stinear, a professor at the University of Melbourne in Australia. Findings are published in the Journal of Medical Microbiology (August 2020).

Loss of smell and taste—first telltale symptoms of COVID-19

New, immediate or sudden loss of smell and taste during COVID-19 epidemic or pandemic may be one of the most important first self-diagnostic symptoms and sign that indicate that you may have been infected with coronavirus infection whether you have other symptoms or not, according to lead researcher Carl Philpott, a professor at UEA's Norwich Medical School, findings published in Rhinology (Aug 2020).

I add to this list ear ache, sinus symptoms, and frontal headache.

Chapter 8

Coronavirus immunopathology of COVID-19

Although SARS-COV-2 has a tropism for ACE2-expressing epithelial cells of the respiratory tract, patients with severe COVID-19 have symptoms of systemic hyperinflammation, as well. Clinical laboratory findings of elevated IL-2, IL-7, IL-6, granulocyte-macrophage colony-stimulating factor (GM-CSF), interferon-γ inducible protein 10 (IP-10), monocyte chemoattractant protein 1 (MCP-1), macrophage inflammatory protein 1-α (MIP-1α), and tumor necrosis factor-α (TNF-α) indicative of cytokine release syndrome (CRS) suggest an underlying immunopathology. Systemic inflammation results in vasodilation, allowing inflammatory lymphocytic and monocytic infiltration of the lung and the heart. In particular, pathogenic GM-CSF-secreting T-cells were shown to correlate with the recruitment of inflammatory IL-6-secreting monocytes and severe lung pathology in COVID-19 patients. Lymphocytic infiltrates have also been reported at autopsy.

Increased incidence of thrombo-embolic phenomenon of the COVID-19 in the aged

There are reports that the COVID-19 patients suffer from increased incidence of blood clotting associated with thrombo-embolic phenomenon seen in many autopsies. Hence the stroke, pulmonary emboli, deep vein thrombosis, blue toe, cardio-vascular diseases and such. Any of these signs and symptoms of thrombosis and embolic phenomenon show elevated blood D-dimers in the blood test. Such patients should be put on appropriate *anticoagulants* therapy. Bioactive curcumin and vitamin D3, act as anticoagulants, hence we recommend them in the coronavirus treatment protocol with reduced doses of anticoagulants. This is specially seen in the elderly, could be due to accumulation many *senescent dysfunctional cells, depressed immune system, arteriosclerotic vascular disease and blood biological / biochemical clotting factor changes enhanced by stasis of venous blood. Being bedridden and having lack of bodily movement also amplifies this process.*

Chapter 9

"Post intensive care unit syndrome" after recovering

"Post intensive care unit syndrome" after recovering from severe COVID-19 with intubation and ventilatory support

The following are some of the signs and symptoms of *"post intensive care unit syndrome,"* which is also known as post ventilation or post intubation syndrome. The longer one stays in the ICU with ventilator support, the longer the time to recover, which is associated with increased morbidity and mortality. After a patient recovers from SARS, they may need nasal oxygen in the home, experience lack of breath with exertion, change in voice due to intubation tube in the larynx, weakness, loss of muscle mass, fatigue, confusion, loss of memory and thinking. They may suffer from PTSD and organ damage to the kidney, heart, liver, and brain. The lungs may be permanently scarred, and may develop chronic bronchitis and emphysema. The older patients with underlying health conditions may take months to recover, or some of them may never come back to a normal healthy state after ICU intubated ventilation treatment, and experience "post COVID-19 syndrome" symptoms after recovery for the rest of their life. Look also for intra and post COVID-19 associated thrombosis and embolism due to increased clotting effects of the COVID-19, as well as lungs, heart, kidney and brain dysfunction.

Incubation period of COVID-19 and coronavirus infectivity after infection

The time from infection to onset of symptoms, known as incubation period, is 5-8 days on average among all coronavirus infected. The scientists noted, adding that the average duration of symptoms was eight days. They said the length of time patients remained contagious after the end of their symptoms ranged from one to eight days and maybe many weeks or more in some patients.

"If you had mild respiratory symptoms from COVID-19 and were staying at home so as not to infect people, extend your quarantine for another two weeks after recovery to ensure that you don't infect other people," suggested study co-author Lixin Xie from the Chinese PLA General Hospital in Beijing. "COVID-19 patients can be infectious even after their symptomatic recovery, so treat the asymptomatic/recently recovered patients as carefully as symptomatic patients," the scientists said. The researchers said all of these patients had milder infections and recovered from the disease, and that the study looked at a small number of patients. They noted that it is unclear whether similar results would hold true for more vulnerable patients such as the elderly, those with suppressed immune systems, and patients on immunosuppressive therapies and afflictions associated with comorbidities associated with severe COVID-19.

Chapter 10

Coronavirus: COVID-19 morbidity and mortality
Effect of poverty, survival of coronavirus outside the body
Stealth transmission, shedding of virus after infection

As of March 25, 2020, the overall rate of deaths per number of diagnosed cases is 4.5 percent, ranging from 0.2 percent to 15 percent according to age group and other health problems, such as underlying health conditions, treatment modalities available, besides the effects of poverty. The mortality rate is very high in patients residing at nursing homes and extended care facilities, compared to the general population. Some healthy patients may not experience any symptoms at all except maybe a mild cold or flu; they may not get tested and become silent spreader of the disease to others.

Underprivileged and lack of medical care of the COVID-19 afflicted population

COVID-19 has demonstrated a darker side of threats to health; inequality in healthcare and its relation to impoverishment, or, plainly said, poor people. While the battle against COVID-19 rages on all over the world, it is likely that many of the poor people will perish silently, unattended and forgotten, due to lack of care, not knowing how to prevent this preventable and treatable disease compounded by lack of financial resources, knowledge, education and facilities to seek treatment.

The governments should play an active role in providing proper nutrition, preventive cost-free diagnosis and free or subsidized treatment with educating the masses of the poverty-stricken population all over the world who are afflicted with COVID-19 besides other afflictions and needs. Even in a rich country like the United States, there are underserved areas where the poor live and suffer increased incidence of morbidity and mortality after COVID-19. Afro-American hospital admissions and death rate are disproportionately higher in COVID-19 stricken compared to other racial and socio-economic groups. It needs further study, attention and help from the local, federal and central governments as well as by charitable institutions, religious institutions and social welfare groups.

Coronavirus spread and how long does it live outside our body? — "stealth transmission"

The coronavirus is mainly spread during close contact and via respiratory droplets produced when people cough, sneeze, talk loudly, breathe rapidly or otherwise transmit by droplets (Chapter 21, Figure 3 #12). Respiratory droplets may be produced during breathing but the virus is not considered airborne to spread to others 6 feet away. A forceful sneeze may expel the virus as far away as 27 feet. People may also catch COVID-19 by touching a contaminated surface and then touching their face (Chapter 21, Figure 3). *Now there is evidence to support that it can also spread by aerosolization of infected droplets.* It is most contagious when people are symptomatic, although spread may be possible before symptoms appear as the virus is multiplying in the nasal-oral-pharyngeal-laryngeal and respiratory mucosa. It is during this phase of the virus in which we describe the method to prevent or inhibit spread of free-floating virus the breathed air droplets (Chapter 21, Figure 4) to healthy cells in this patent using prophylactic sprays and treatments. Coronavirus infection is not an airborne infection; it spreads by a coronavirus containing droplet spewed by the infected person's expired air.

The virus in the droplets can live on surfaces up to 72 may hours maybe more on some surfaces. The study published in The New England Journal of Medicine examined how long the coronavirus lived in various contexts. As an aerosol gas, coronavirus suspended in air, the virus could hang around for up to three hours. Some medical procedures such as intubation and cardiopulmonary resuscitation (CPR) may cause *respiratory secretions to be aerosolized* and thus result in an airborne spread to care givers and those next to the patients. Respiratory droplets, as mentioned, tend to land quickly on surfaces. "On plastic and stainless steel, the virus was viable for up to three days," said Kirsten Hokeness, professor and chair of the Department of Science and Technology at Bryant University at Toledo and an expert in immunology, virology, microbiology, and human health and disease. The virus seemed to disintegrate a little faster on stainless steel than on plastic (think of doorknobs, handles and kitchen surfaces). Coronavirus does not survive for long when it lands or comes in contact with a copper surface. Copper is viricidal according to the latest study, so also is silver nitrate. The researchers found that the virus sticks around for much less time on copper — about four hours. "On cardboard, some virus was found for 24 hours," Hokeness said. "Some people get concerned about parcel and post-delivery, but the likelihood of transmission is low." She noted that porous surfaces “don’t tend to lend themselves well to letting viruses live very long,” and suggested paper is also probably less likely to host the virus. ***Sunlight, hot air with humidity of summer and ultraviolet rays*** are viricidal, hence the coronavirus may be killed in a flash of movement as it is expelled in the infected breath. If the virus was only transmitted outdoors, the sunlight and hot air would make the number of cases end or be hardly noticeable as summer approaches. Unfortunately, indoors, out of the sun’s rays, the outdoor heat will have no effect.

According to a model from researchers out of China, “stealth transmission” contributed to the rapid rise in cases. “Stealth transmission refers to the fact that people who have mild, non-discrete symptoms or are asymptomatic are contributing greatly to the spread of the virus,” and also “aerosolizing medical procedures,” like during intubation, suctioning the oral and nasal secretions, attending to the patient needs in close proximity can cause these virus droplet particles to spread to the care givers.

Shedding of COVID-19 after the infection

In the study, researchers, including scientist Lokesh Sharma from Yale University, analyzed samples recently collected of throat swabs taken from all the patients on alternate days (2020). They said the patients were discharged after their recovery and confirmation of negative viral status by at least two consecutive polymerase chain reaction (PCR) tests. "The most significant finding from the study is that half of the patients kept shedding the virus even after resolution of their symptoms," said co-lead author Sharma. "More severe infections may have even longer shedding times," he said.

Chapter 11

COVID-19 incubation period and preventive measures
Air borne spread, cluster cases, pool testing and lockdown

Time from exposure to onset of symptoms is generally between two and fourteen days, with an average of five days. The standard method of diagnosis is by reverse transcription polymerase chain reaction (rRT-PCR) from a nasopharyngeal swab. The infection can also be diagnosed from a combination of symptoms, risk factors and a chest CT scan showing features of pneumonia and serological test for anti-coronavirus immune bodies such as immunoglobulins (Ig).

Recommended measures to prevent infection include frequent hand washing, social distancing (maintaining physical distance of 6 feet from others, especially from those with symptoms), covering coughs and sneezes with a tissue or inner elbow, and keeping unwashed hands away from the face (Chapter 21, Figures 3, 4). The use of masks is recommended by national health authorities. There is no vaccine or specific antiviral treatment for COVID-19 yet. Management involves treatment of symptoms, supportive care, isolation, and experimental measures. Self-isolation for period of 2-4 weeks is the most effective way to control infection and its spread. We want to add to this, prophylaxis methods in addition described in this invention.

The World Health Organization (WHO) declared the 2019–20 coronavirus outbreaks a Public Health Emergency of International Concern (PHEIC) on January 30, 2020 and a pandemic on March 11, 2020. Local transmission of the disease has been recorded across all six continents and in all countries except isolated island chains in the South Pacific and such as of now. Our inventive methods nasal and oral spray, mouth wash may prevent, curtail, stop local progression and other stages as described above and / or cure the infection.

Mitigation efforts to prevent spread of coronavirus

Dr. Anthony Fauci and Dr. Deborah Brix of the National Institute of Allergy and Infectious Diseases and members of the White House coronavirus task force, state that to mitigate the spread of virus is the key to control

COVID-19 to prevent its spread all over the world. Their key methods to this mitigation protocols are:

1. Detection of coronavirus infection
2. Wear a mask all the time, isolate the infected to protect self, isolate from others to prevent infection from undiagnosed carriers, maintain proper social distancing of 6 feet or more, keep away from gatherings and large crowds. Hand washing is important to prevent contact spread as shown in Chapter 21, Figures 3. 4.
3. Contact tracing of those who are exposed innocently who may be spreading the coronavirus.
4. Rapid coronavirus testing development is very important for contact tracing and prevention of spread. Testing by breath analyzer or naso-pharyngeal secretions, fingerstick on the spot test and such are coming. Development of testing by analyzing the breath may be one of the best methods to test rapidly to prevent contact spreading. Such test will be available with lapse of time,

Hand washing and maintaining social distance are the main measures recommended by the CDC and the World Health Organization (WHO) to avoid contracting COVID-19. Unfortunately, these measures do not prevent infection by inhalation of small droplets exhaled by an infected person that can travel a distance of meters or tens of meters in the air and carry their viral content. Science explains the mechanisms of such transport and there is evidence that this is a significant route of infection in indoor environments. Despite this, no countries or authorities consider airborne spread of COVID-19 in their regulations to prevent infections transmission indoors. It is therefore extremely important that the national authorities acknowledge the reality that the virus spreads through air, and recommend that adequate control measures be implemented to prevent further spread of the SARS-CoV-2 virus, in particularly removal of the virus-laden droplets from indoor air by ventilation and coronavirus Ion zapper devices (under development).

Airborne spread: There is a possibility that the coronavirus can spread by airborne. If you take it for granted that airborne spread can happen in some cases, still the above mitigation methods can prevent the spread of coronavirus and reduce the chances of getting infected.

Pool testing: Test 10 people by swabs as one test. If it is negative, you do not have to test the rest of the 10 people. If the pool test become positive, then you have to test all of them individually. This can help to facilitate mass testing inexpensively. This can only work if the test results can be immediately obtained on the spot.

A cluster of cases with symptoms in a constellation of locations can highlight an emerging hot spot for the virus within the population which becomes difficult to approximate the true extent of the pandemic in such cases. With introduction of rapid self-testing in 5 minutes, the isolation ban and shut down can be eased as the coronavirus wanes off.

Coronavirus "lockdown" and the consequences of early or premature lifting

Lockdown has affected the world, to limit the coronavirus from spreading through contact. There are attempts to lift the coronavirus lockdown, due to demand to get back to work. Japan's northern island of Hokkaido offers a grim lesson in the next phase of the battle against COVID-19. It acted quickly and contained an early outbreak of the coronavirus with a 3-week lockdown. But, when the governor lifted restrictions, a second wave of infections may even hit harder. Twenty-six days later, the island was forced back into lockdown. Experts say restrictions were lifted too quickly and too soon because of pressure from local businesses, coupled with a false sense of security in its declining infection rate. So, I state that the *lifting of the lockdown should be based and founded on science not on business or politics.* ***Don't forget that when science wins everyone wins***. If everyone wears a mask, goggles, and stops mixing in large groups, the coronavirus spread will diminish drastically.

How many people can COVID-19 spread to from one infected person

The basic reproduction number of people coronavirus infects from single infected persons has been estimated to be between 1.4 and 3.9. (Li Q, et al. 2020. "Early Transmission Dynamics in Wuhan, China, of Novel Coronavirus-Infected Pneumonia". The New England Journal of Medicine. 382 (13): 1199–1207. Riou J, Althaus CL 2020. "Pattern of early human-to-human transmission of Wuhan 2019 novel coronavirus (2019-nCoV), December 2019 to January 2020". Euro-surveillance. 25 (4). doi:10.2807/1560-7917.ES.2020.25.4.2000058). This means that each infection from the virus is expected to result in 1.4 to 3.9 new infections when no members of the community are immune and no preventive measures are taken. By airborne spread, the numbers by which it spreads are much higher.

Incubation period and contact spreading of coronavirus infection (COVID-19)

The virus spreads easily from person to person and it is a virulent infection. Now it is estimated the that the incubation period of COVID-19 ranges from one to four to fourteen days, with most cases occurring approximately five to six days after exposure. Therefore, exposed individuals are quarantined for two weeks from exposure (the maximum incubation time). The long incubation period explains why people exposed to infection on a plane don't fall sick right away. This might be one of the methods by which the coronavirus has spread all over the world. Infected people may wind up in countries all over the world while they are still in the incubation period phase of infection. They may infect many other people in their flight and in destination countries.

The incubation period leads us to another important concept in blocking coronavirus spread to others. Isolating those who test positive for coronavirus and following up with meticulous contact tracing can end the pandemic. A single missed coronavirus positive contact can spread the disease to many unknowingly. Contact tracing and disease control is difficult, especially in poorer countries, akin to difficulties faced in infection spreading in plane travel, because passengers may have reached different countries and the infection can rapidly spread across the world. Some coronavirus sick people act as super-spreaders, infecting a large number of other people, but recent evidence proves that the coronavirus can also be transmitted by apparently healthy people who do not know that they are infected with coronavirus. These infected people appear healthy; they are asymptomatic, but they can still infect others. Children can get infected but are asymptomatic or have a very mild disease, but they can spread the infection to their vulnerable grandparents who come in physical contact with them. Social isolation of 6 feet is a must to contain its spread. Asymptomatic coronavirus infected person many be infective for about 14 days, and maybe more.

Indoor ventilation and coronavirus spread: The best ways to reduce the risk of COVID-19 indoors is to maintain low levels of CO2

The vast majority of SARS-CoV-2 transmission occurs indoors, most of it from the inhalation of airborne particles that contain the coronaviruses. The World Health Organization and the U.S. Centers for Disease Control and Prevention say that poor ventilation increases the risk of transmitting the coronavirus many times. Every time we breathe, we release carbon-dioxide to the air around us, with coronavirus if the occupant is positive for the test or is a silent carrier.

A carbon dioxide monitor (CO_2) gives you an indication if there is enough ventilation. If CO_2 levels start going up, open some windows and take a break outside. As the coronavirus is most often spread by breathing, coughing, sneezing or talking, you can use CO_2 levels to see if the room is filling up with potentially infectious exhalations. This is specially so in class rooms, meeting places, bars, restaurants, social gatherings and such. The CO_2 level lets you estimate if enough fresh outside air is getting in or not. Outdoor CO_2 levels are 400 prats million (ppm). A healthy thoroughly ventilated room will have around 800 ppm of CO_2. Any higher than that and it is a sign the room might need more ventilation. Experts consider approximately six air changes an hour to be good for a 10-foot-by-10-foot room with three to four occupants in it.

If you can't get enough fresh air into a room, an air cleaner might be a good alternative. If you do get an air cleaner, *be aware that they don't remove CO_2*, so even though the air might be safer, CO_2 levels could still be high in the room. The best option is an air cleaner that uses a high-efficiency particulate air (HEPA) filter, as these remove more than 99.97% of all particle sizes, so also the coronavirus-laden floating droplets circulating in the enclosed air space.

Last year, researchers in Taiwan reported on the effect of ventilation on a tuberculosis outbreak at Taipei University. Many of the rooms in the school were under-ventilated and had CO_2 levels above 3,000 ppm. When engineers improved air circulation and got CO_2 levels under 600 ppm, the outbreak completely stopped. According to the research, the increase in ventilation was responsible for 97% of the decrease in transmission. Now you know that all classrooms, meeting places, bars, restaurants, churches and such should have CO_2 monitors and the level should be maintained below 600 ppm to mitigate the spread of coronavirus, to make indoor spaces safe during this pandemic.

In commercial buildings, outside air is pumped in through heating, ventilating, and air-conditioning (HVAC) systems. In homes, outside air gets in through open windows and doors, in addition to seeping in through various nooks and crannies—cracks, gaps, splits, crevices.

Since the coronavirus is spread through the air, higher CO_2 levels in a room likely mean there is a higher chance of transmission if an infected person is inside. Based on the study above, I recommend trying to keep the CO_2 levels below 600 ppm. You can buy good CO_2 meters for around <$100. All of the air in a room should be replaced with fresh, outside air at least six times per hour if there are many people inside, and also use HEPA filter air purifiers.

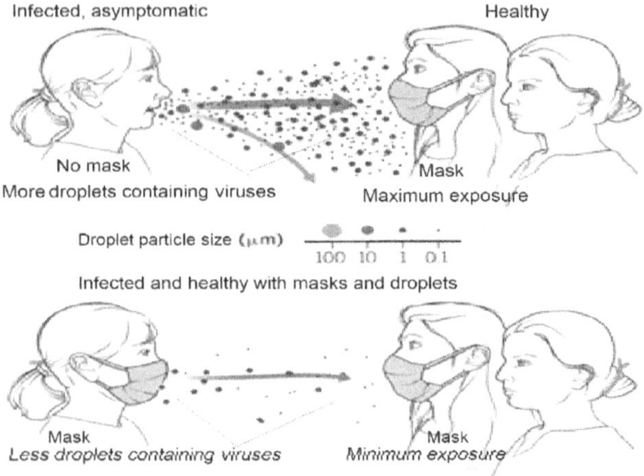

The above illustration shows that wearing the mask by infected and noninfected protective against coronavirus infection.

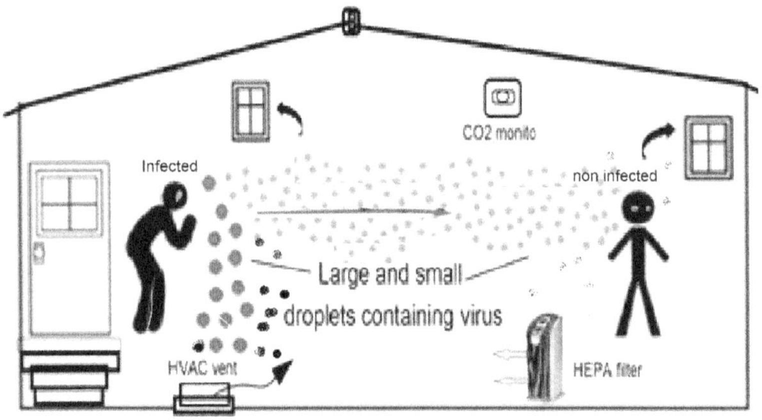

The above drawing shows viral spread inside the closed space. The infected have a good chance of infecting the healthy due to lack of circulation of air. Heavy droplets from the exhaled infected drop on the floor and small droplets get airborne and infect the other persons farther away. It shows CO2 monitor, HAVC vent and HEPA filter to monitor and clean the air in closed spaces or rooms to reduce the airborne infection from droplets from the infected persons. Exhaled droplets can carry the coronavirus 20 feet.

Chapter 12

Are bats a reservoir of coronavirus?
S-spike on COVID-19, pathophysiology

Coronaviruses are zoonotic, meaning they are transmitted between animals and humans. As it stands now, the bats appear to be an important major reservoir for these COVID-19 and related viruses that have resulted in the pandemic.

1. First, it will be interesting to determine how they seem to avoid clinically evident disease and become persistently infected and act as hosts.
2. Second, many of the non-structural and accessory proteins encoded by these viruses remain uncharacterized with no known function, and it will be important to identify mechanisms of action for these proteins as well as defining their role in viral replication and pathogenesis to develop prophylaxis and treatments. These studies should lead to a large increase in the number of proper therapeutic targets to combat infections. Furthermore, many of the unique enzymes encoded by coronaviruses, such as ADP-ribose-1"-phosphatase, are also present in higher eukaryotes, making their study relevant to understanding general aspects of molecular biology and biochemistry.
3. Third, gaining a complete picture of the intricacies of the reverse transcriptional complex (RTC) will provide a framework for understanding the unique RNA replication process used by these viruses (Chapter 21, Figures 3-6) as observed in other RNA viruses.
4. Finally, defining the mechanism of how coronaviruses cause disease and understanding the host immune-patho-physiological response will significantly improve our ability to design vaccines, therapeutic agents and reduce or treat disease burden.

S-spike on COVID-19 TMPRSS2 priming for entry of virus to healthy cells

Coronaviruses contain a receptor-binding site—a part of the S-spike protein necessary to bind to the cell membrane—that is important for human infection (Chapter 21, Figures 1-6 #6). This binding site makes it possible to affix to a host cell surface protein that is abundant on human respiratory and intestinal epithelial cells, endothelial cells, and kidney cells as well as the other blood vessels of the body and brain.

Studies have demonstrated that initial spike protein priming on the coronavirus (Chapter 21, Figures 2, 3) by transmembrane protease serine 2 (TMPRSS2) is essential for entry of SARS-CoV-2 via interaction with the ACE2 receptor on the healthy cells they come in contact with healthy cells to enter the host cell. These findings suggest that the TMPRSS2 inhibitor *camostat* approved for use in Japan for inhibiting fibrosis in liver and kidney disease might constitute an effective off-label treatment [Hoffmann M, Kleine-Weber H, et al. 2020. "The novel coronavirus 2019 (2019-nCoV) uses the SARS-coronavirus receptor ACE2 and the cellular protease TMPRSS2 for entry into target cells". bioRxiv (preprint). doi:10.1101/2020.01.31.929042. Iwata-Yoshikawa N, Okamura T, Shimizu Y, Hasegawa H, Takeda M, Nagata N (March 2019). "TMPRSS2 Contributes to Virus Spread and Immunopathology in the Airways of Murine Models after Coronavirus Infection". Journal of Virology. 93 (6). doi:10.1128/JVI.01815-18]. In February 2020, favipiravir was being studied in China for experimental treatment of the emergent COVID-19, so also other protease inhibitors.

Coronavirus: its infection of respiratory system and lung damage: pathophysiology

Once again, we state that coronavirus is a respiratory virus and it primarily infects epithelial cells of the nasal-oral-pharyngeal-laryngeal mucosa, air passages, and then the lung alveolar cells and blood vessels that surround alveoli (Chapter 21, Figure 1). The virus is capable of entering macrophages and dendritic cells but only leads to an abortive infection [Peiris JS., et al. Clinical progression and viral load in a community outbreak of coronavirus-associated SARS pneumonia: a prospective study. Lancet. 2003; 361(9371):1767–1772. Spiegel M, et al. Interaction of severe acute respiratory syndrome-associated coronavirus with dendritic cells. J Gen Virol. 2006; 87(Pt 7):1953–1960.]. Despite this, infection of nasal-oral-pharyngeal-laryngeal mucosa, the immune system cells may be important in *inducing pro-inflammatory cytokines, chemokines and prostaglandins* that may contribute to the severe disease of lungs besides other signs and symptoms [Law HK, et al. Chemokine upregulation in SARS coronavirus infected human monocyte derived dendritic cells. Blood. 2005; 106:2366–2376]. In fact, many cytokines and chemokines are produced by these immune system cell types and are elevated in the serum of COVID-19 stricken patients [Lau YL, Peiris JSM. Pathogenesis of severe acute respiratory syndrome. Curr Op Immunol. 2005;17, :404–410].

The exact mechanism of lung injury and cause of severity of the disease in humans remains undetermined and is also under intense research investigation. Viral titers seem to diminish when severe disease develops in both humans and experimental animals (Fehr and Perlman Page 9 Methods Mol Biol. PMC 2016). The rodent adapted COVID-19 strains show similar clinical features to the human disease, including an age-dependent increase in disease severity [Roberts A, et al. Aged BALB/c mice as a model for increased severity of severe acute respiratory syndrome in elderly humans. Journal of virology. 2005; 79(9):5833–5838]. These animals also show increased levels of proinflammatory cytokines and reduced T-cell responses, suggesting a possible immunopathological mechanism of disease.

It is worth noting that young people have a healthy immune system, besides having fewer angiotensin receptors on the cell surface, and in most instances can overcome the coronavirus infection. Whereas, elderly people have compromised immune systems, have more angiotensin receptors compared to the young, have more senescent cells accumulated in the body, and may succumb to the infection. That is why we advocate the use of angiotensin receptor blockers such as telmisartan, losartan and/or ivermectin to mitigate the lung damage in all cases of coronavirus infections. I recommend this to all those with elevated blood pressure and the elderly. One of the other effects of angiotensin receptor blockers is that they may have anti-aging effects, also.

Coronavirus enters the nose and respiratory system through target cells, ciliated lining, alveolar lining type II pneumocytes cells, absorptive enterocytes of the gastro-intestinal track which have angiotensin II receptor proteins and have another protein transmembrane protease serine 2 (TMPRSS2) that facilitates the entry of coronavirus once it attaches to the receptors on the cell surface. Hence the gastro-intestinal track gets coronavirus by swallowing the mucous secretions of the respiratory system and the virus from the blood through the capillaries surrounding the alveoli.

Spike of coronavirus attaches to the angiotensin receptors 2, then the TMPRSS2 activates the spike protein, allowing the coronavirus to gain entry to a target cell and deliver its genetic protein to the healthy cell which will allow them to divide, produce many viral RNA particles which regroup Golgi lysosomal organelle to form a new virus with its covering and exit the cell to infect the next cell. Due to virus using the cell organelle and energy to make new copies of the virus, the infected cell succumbs to the infection due to loss of its life support system. As the cell succumbs, to the viral onslaught, the immune system cells pour in to attack the virus, setting up the inflammatory reaction of the affected tissue. The bodily inflammatory reaction continues, with poring of cytokines, setting up for cytokine storm.

Chapter 13

Coronavirus physical and microscopic structure

Coronavirus viruses are spherical with diameters of approximately 125 nm (1nm=one billionth of a meter) as depicted in recent studies by cryo-electron tomography and cryo-electron microscopy. The most prominent feature of coronaviruses is the club-shape spike projections emanating all-around the surface of the virion. These spikes are a defining feature of the virion and give them the appearance of a solar corona, prompting the name, coronaviruses. Within the envelope of the virion is the nucleocapsid (Chapter 21, Figures 3-6). Coronaviruses have helically symmetrical nucleocapsids, which is uncommon among positive-sense RNA viruses, but far more common for negative-sense RNA viruses.

Coronaviruses are enveloped RNA viruses like HIV and rabies. They are endowed with a *large RNA genome, and an exceptional replication ability*. Coronaviruses cause a variety of diseases in mammals and birds ranging from enteritis in cows and pigs and upper respiratory disease in chickens to potentially lethal human respiratory infections such as adult or adolescent respiratory distress syndrome type lung condition. Here we provide a brief introduction to coronaviruses discussing their method of attachment to healthy cells, mode of replication, pathogenicity, and current prevention and treatment strategies. We will also discuss the outbreaks of the highly pathogenic Severe Acute Respiratory Syndrome Coronavirus (SARS-CoV) now named by WHO as COVID-19 and the recently identified Middle Eastern Respiratory Syndrome Coronavirus (MERS-CoV).

COVID-19 virion structure

To formulate any prophylaxis and treatment for coronavirus, understanding the structure is imperative. Coronavirus particles contain four main structural proteins. These are the spike (S), membrane (M), envelope (E), and nucleocapsid (N) proteins (Chapter 21, Figures 2-6), all of which are encoded within the 3′ end of the viral genome. The S protein (~150 kDa), utilizes an N-terminal signal sequence to gain access to the endoplasmic reticulum (ER), and is heavily N-linked glycosylated (Chapter 21, Figures 2-

6). Homotrimers (homotrimer is a protein which is composed of three identical units of polypeptide) of the virus encoded S protein make up the distinctive spike structure on the surface of the virus and mediates attachment to the host receptors [Beniac DR, et al. Architecture of the SARS coronavirus prefusion spike. Nature structural & molecular biology. 2006;13(8):751–752. Delmas B, Laude H. Assembly of coronavirus spike protein into trimers and its role in epitope expression. Journal of virology. 1990;64(11):5367–5375.]. The trimeric S glycoprotein (Chapter 21, Figures 2, 3, 6) is a class I fusion protein [Bosch BJ, et al. The coronavirus spike protein is a class I virus fusion protein: structural and functional characterization of the fusion core complex. J Virol. 2003;77(16):8801–8811]. In most, but not all, coronaviruses S is cleaved by a host cell furin-like protease into two separate polypeptides noted S1 and S2. S1 makes up the large receptor-binding domain of the S protein while S2 forms the stalk of the spike molecule (Chapter 21, Figures 2, 3).

At present there is no vaccine, and no therapeutic agents to kill the virus. Treatments are being developed to treat the patients who are infected with coronavirus and develop signs and symptoms. The only measure available to prevent its infection and spread is self-imposed social isolation. This virus preys on those who are immunocompromised with co-morbidities. We all know that COVID-19 infection in the nursing and retirement homes is lethal because of the compromised immune system, age, lack of proper nutrition, sedentary nature of life with lack of exercise, and other diseases, comorbidities, and underlying health conditions of these residents.

Chapter 14

COVID-19 antiviral drugs and research projects under study

There is a rush to find prophylaxis and treatment for COVID-19.

The below description shows the therapeutic agents undergoing clinical trials or have trials planned regarding COVID-19, due to the single-stranded RNA virus 2109-nCoV (SARS-CoV-2-COVID-19) and many more will be added to the list with passage of time. They were not listed in the actual patent application due to lack of space, but I have included them in the book.

Type of medicine and mechanism of their action are as follows:

COVID-19 Antiviral drugs, non-specific

a) Immunoglobulin: Contains nonspecific antibodies; proposed to block viral Fc receptor activation by boosting endogenous neutralizing antibodies and preventing antibody-dependent enhancement of infection
b) Interferons: Activate cytoplasmic enzymes affecting viral messenger RNA translation and protein synthesis; evidence of minor efficacy in MERS-CoV in combination with ribavirin and remdesivir
c) Interleukin-2: Multiple anti-inflammatory actions

COVID-19 Antiviral drugs, broad spectrum

a) Favipiravir: Inhibitor of viral RNA-dependent RNA polymerase; typically used in treating influenza, but inhibits the polymerase of other viruses in vitro (Ebola, yellow fever, chikungunya, norovirus, and 2019-nCoV)
b) Triazavirin: Non-nucleoside antiviral drug; effective against tick-borne encephalitis virus and forest-spring encephalitis virus
c) Umifenovir: Membrane haemagglutinin fusion inhibitor in influenza viruses; active against influenza virus infections, in which it may reduce the risk of bacterial pneumonia

Antiviral drugs, antiretrovirals

a) ASC09: HIV protease inhibitor; to be used in combination with ritonavir
b) Azvudine: Azidocytidine nucleoside analogue; HIV reverse transcriptase inhibitor
c) Danoprevir: Hepatitis C virus NS3 protease inhibitor; to be used in combination with ritonavir
d) Darunavir: HIV protease inhibitor; used in combination with cobicistat, a CYP3A inhibitor
e) Lopinavir + ritonavir: Both HIV reverse transcriptase inhibitors; ritonavir is mainly used to enhance the action of other drugs by inhibition of CYP3A4; in vitro and possible clinical efficacy in SARS-CoV
f) Remdesivir: Nucleotide analogue; inhibitor of RNA-dependent RNA polymerase; used to treat Ebola and Marburg viruses; effective in vitro against SARS-CoV-1 and MERS and blocks infection with 2019-nCoV in vitro

Additional COVID-19 antiviral drugs

a) Chloroquine and hydroxychloroquine anti-malarial: In addition to its antimalarial actions, chloroquine has some efficacy in HIV-AIDS and blocks infection by 2019-nCoV in vitro, perhaps by inhibiting glycosylation of viral ACE-2 or inhibition of quinone reductase 2, reducing synthesis of viral sialic acid; in one study to be combined with lopinavir/ritonavir
b) Dihydroartemisinic: Interaction between its peroxide bridge and haem iron may underlie its antimalarial action; mechanisms of action against viruses, e.g. zika virus, are not well understood; to be combined with the antimalarial drug piperaquine

COVID-19 antibiotics, anti-parasitic and immune suppressants

a) Corticosteroids: Many anti-inflammatory actions; experience with corticosteroids in other infectious diseases has not been uniformly beneficial; in SARS corticosteroids may worsen the disease. But when combined with vitamin C and thiamine hydrochloride, it becomes an effective therapeutic agent given only for a short period.
b) Fingolimod: Immunosuppressant; sphingosine 1-phosphate receptor modulator; used in multiple sclerosis

c) Leflunomide: Immunosuppressant; thalidomide analogue; inhibitor of dihydro-orotate dehydrogenase and tyrosine kinases: causes degradation of intracellular transcription factors; used to treat rheumatoid arthritis and psoriatic arthritis
d) Thalidomide: Immunosuppressant; mode of action poorly understood; inhibits production of excess tumor necrosis factor-alfa (TNF-α), down-regulates cell surface adhesion molecules involved in leukocyte migration, has anti-angiogenic activity, and modulates the thalidomide-binding protein, cereblon, mediating protein degradation; used to treat multiple myeloma and many cancers - I have used it

COVID-19 Kinase inhibitors

a) Jakotinib hydrochloride: Inhibitor of Janus-associated kinases
b) Ruxolitinib: Inhibitor of Janus-associated kinases (JAK1 and JAK2); used to treat myelofibrosis, polycythemia vera, and graft-versus-host disease; to be combined with mesenchymal stem cell infusion

Monoclonal antibodies, to be used in combination with protease inhibitors in COVID-19

a) Adalimumab: Target: TNF-alfa; used to treat rheumatoid arthritis, juvenile idiopathic arthritis, ankylosing spondylitis, psoriatic arthritis, psoriasis, hidradenitis suppurativa, Crohn's disease, ulcerative colitis, and uveitis
b) Camrelizumab: Target: receptors for programmed cell death protein 1 (PD-1); used in China to treat Hodgkin's lymphoma and nasopharyngeal carcinoma, alone or in combination with gemcitabine + cisplatin
c) Eculizumab: Target: C5 complement; used to treat paroxysmal nocturnal haemoglobinuria, atypical haemolytic uremic syndrome, refractory generalized myasthenia gravis, and neuromyelitis optica spectrum disorder
d) Mepolizumab: Target: interleukin-5 (IL-5); used to treat eosinophilic asthma
e) PD-1 mAb: Target: receptors for programmed cell death protein 1 (PD-1)
f) Tocilizumab: Target: soluble and membrane-bound IL-6 receptors (sIL-6R and mIL-6R); used to treat rheumatoid arthritis, systemic juvenile idiopathic arthritis, juvenile idiopathic polyarthritis, giant -

cell arteritis, and chimeric antigen receptor (CAR) T cell-induced severe or life-threatening cytokine release syndrome

Miscellaneous others for COVID-19 infection

a) Acetylcysteine: Mucolytic; symptomatic relief; oral N-acetylcysteine ameliorated the intestinal effects of porcine epidemic diarrhea coronavirus in piglets
b) Angiotensin receptor blockers: Antagonists at angiotensin receptors, telmisartan and losartan
c) Angiotensin converting enzyme (ACE) inhibitors: To be used in at least one study in individuals with hypertension in addition to COVID-19; ACE-2, which facilitates the entry of 2019-nCoV into cells is not inhibited by ACE inhibitors
d) Bismuth potassium citrate: Inhibits growth of Helicobacter pylori; used in peptic ulceration; may inhibit SARS-CoV1 helicase
e) Bromhexine hydrochloride: Mucolytic; symptomatic relief; small trials in other respiratory disorders show only modest benefit.
f) Diammonium glycyrrhizinate: Derivative of glycyrrhetinic acid used in Chinese medicine; a competitive inhibitor of high mobility group box-1; causes moderate inhibition of the porcine epidemic diarrhea coronavirus in vitro, with inhibition of entry of the virus into Vero cells and replication
g) Dipyridamole: Antiplatelet action by inhibition of adenosine uptake; also inhibits cGMP-phosphodiesterase, augmenting the increase in cGMP produced by EDRF (endothelium-derived relaxing factor, nitric oxide); there is a suggestion that antiplatelet agents, including dipyridamole, slow the progress of idiopathic pulmonary fibrosis.
h) Ebastine: Histamine H1 receptor antagonist (antihistamine); presumably intended for symptomatic relief
i) Hydrogen peroxide: Non-specific supposed antiviral action in the throat
j) Inhaled gases: Oxygen, hydrogen, or the combination; symptomatic relief of hypoxia by oxygen; molecular hydrogen is proposed to have an anti-inflammatory effect. ± Ozone.
k) Pirfenidone: Various mechanisms proposed; reduces fibroblast proliferation, production of fibrosis-associated proteins and cytokines, and increased biosynthesis and accumulation of extracellular matrix in response to cytokine growth factors such as transforming growth

factor-beta (TGF-β) and platelet-derived growth factor (PDGF); used to treat idiopathic pulmonary fibrosis

l) Polyinosinic-polycytidylic acid: Analogue of double-stranded RNA; reportedly has agonist actions at: toll-like receptors TLR-3; retinoic acid inducible gene I (RIG-I)-like receptors; and melanoma differentiation-associated gene 5 (MDA5); reduces the amount of mRNA in vitro in PK-15 cells infected with transmissible gastroenteritis coronavirus; used in China to treat tumors; to be used in combination with vitamin C

m) rhG-CSF: Recombinant human granulocyte colony-stimulating factor; its counterpart, granulocyte macrophage-colony stimulating factor, directs the activation, proliferation, and differentiation of myeloid-derived cells and causes maturation of antigen-presenting cells, thus affecting adaptive immune responses; the 3C-like protease of SARS-CoV-1 reduces expression of granulocyte-macrophage colony-stimulating factor (GM-CSF) in transfected cells in vitro

n) Thymosin: A group of polypeptides, initially isolated from the thymus. They are involved in T-cell differentiation. There have been trials of their Inhibitor of viral RNA-dependent RNA polymerase; typically used in treating influenza, but inhibits the polymerase of other viruses in vitro (Ebola, yellow fever, chikungunya, norovirus, and 2019-nCoV)

o) Non-nucleoside antiviral drug: effective against tick-borne encephalitis virus and forest-spring encephalitis virus Membrane haemagglutinin fusion inhibitor in influenza viruses; active against influenza virus infections, in which it may reduce the risk of bacterial pneumonia. Used in cancer treatment, with little evidence of benefit.

p) Tranilast: Tryptophan analogue: anti-allergic, inhibits NLRP3 in the inflammasome pathway; used to treat bronchial asthma, atypical dermatitis, allergic conjunctivitis, keloid, and hypertrophic scars

q) Vitamin C: Antioxidant; no evidence to support purported efficacy in any condition other than vitamin C deficiency. On the other hand, high doses of vitamin C given intravenously can be viricidal, used in many viral infections for more than 7 decades. Combine with thiamine hydrochloride and Hydrocortisone as we describe here.

Combination studies not included above for COVID-19

a) Adamumab + tozumab: Names not recognized; these may be adalimumab and tocilizumab (qqv)

b) Interferon α: Combined with antiviral drugs (lopinavir/ritonavir, ribavirin, umifenovir)
c) Lopinavir/ritonavir: Combined with other antiviral drugs (emtricitabine and tenofovir)

(Modified from a chart from The Centre for Evidence-Based Medicine develops, promotes and disseminates better evidence for healthcare).

To quote the conclusion of the authors of a recent (2019-2020) Chinese systematic review of 75 trials listed on the Chinese Clinical Registration Center's website and clinicaltrials.gov, "based on the poor quality and small sample size[s] and long completion period[s] [of the registered trials], we will not be able to obtain reliable, high-quality clinical evidence about COVID-19 treatment for quite a long time in the future."

Advanced therapy medicinal products under investigation (n=24) (wording taken from the titles of the registered studies)

a) Aerosol inhalation of vMIP: viral macrophage inflammatory protein, Ankylosaurus; M1 macrophages target; Anti-2019-nCoV inactivated convalescent plasma; Anti-SARS-CoV-2 inactivated convalescent plasma. We add to this list, disease specific antibody tagged nanoparticles, and plasma from COVID-19 recovered patients, and monoclonal antibodies.
b) Biological preparation of human placenta; Plasma treatment; Infusion of convalescent plasma; Convalescent plasma treatment, Therapeutic antibody from recovered coronavirus pneumonia patients; Cord blood mesenchymal stem cells, Mesenchymal stem cells; Umbilical cord mesenchymal stem cells (hucMSCs); Human menstrual blood-derived stem cells; Immunoglobulin (Ig) from cured patients; Umbilical cord blood mononuclear cells; Umbilical cord Wharton's Jelly derived mesenchymal stem cells.
c) Inactivated Mycobacterium vaccine such a BCG vaccination; Mesenchymal stem cells exosomes atomization; mRNA-1273; Natural killer cells (NK cells); Recombinant cytokine gene-derived protein injection.
d) Regulating intestinal flora; Washed microbiota transplantation.

I like to add to this list, spraying or aerosol delivery of ***Super Oxide Dismutase***, a type of natural body antioxidant through inspired air and ethylenediaminetetraacetic acid (EDTA) on to the lung surface to prevent the attachment of the coronavirus to the lungs alveolar surface. It may enhance the antioxidant output from alveoli and their blood supply.

Chapter 15

Post coronavirus infection syndrome, future research

"Post coronavirus infection syndrome" after coronavirus infection—long term effects of the COVID-19

The CDC defines ***recovery from COVID-19*** as an absence of fever, with no use of fever-reducing medication, for three full days; improvement in other symptoms, such as coughing and shortness of breath; a period of seven full days since symptoms first appeared. Two negative swab tests on consecutive days are considered as the all-clear, meaning self-isolation can end and a patient can theoretically begin having contact with others, including at work.

Many healthy and young, and those with good immune system and health without any health problems (no comorbidities) recover completely and may not suffer from any long-time effects as the elderly. The aged with underlying health conditions (comorbidities) suffer from ***"Post coronavirus infection syndrome***" which can affect any organ in the body. The Hong Kong Hospital Authority observed that two out of three recovering patients had lost 20-30% of lung function needing physiotherapy. Kidney, liver, heart and gastro-intestinal track can also be affected and show signs and symptoms of further problems. Those aged on ventilators may develop delirium and need to give attention to stroke, seizures, cognition changes, confusion, dementias, cardio-vascular afflictions, kidney failure, thrombo-embolic afflictions from legs to lungs, heart, gastro-intestinal track, and brain; secondary infections of the lungs and other organs. Loss of smell and maybe taste though transient may be a first diagnostic symptom that one might be infected with coronavirus, though the coronavirus tests are inconclusive or negative during this pandemic. Their life may never be the same as before COVID-19 in the aged and it may be cut short with reduced quality of life.

Future research on coronavirus

Future research on coronaviruses needs to investigate many aspects of viral attachment to the nasal-oral-pharyngeal-laryngeal-tracheal-bronchial-alveolar mucosa as well as olfactory mucosa, its replication, pathogenesis and finally how to counter it. First, understanding the propensity of these viruses to jump between species, to establish infection in a new host, and to identify significant reservoirs of coronaviruses will dramatically aid in our ability to predict when and where potential epidemics may occur and how to control them. The future is bright to contain and treat coronavirus infection. Vaccines are on the way and we can hope that COVID-19 will be like flu with decreased morbidity and mortality and becomes a seasonal disease.

Chapter 16

Alternative natural therapies for coronavirus infection

Can alternative treatments help prevent or treat coronavirus? *So far, this objective has not been achieved.* There were little evidence that natural or alternative remedies could either prevent or cure the illness caused by the virus. Chinese and Indians have used various herbs in combination to enhance the immune system to treat coronavirus and such viral infections. All I can say is: ***stick to traditional medical treatments*** and seek alternative therapies only if your heart desires, as a supplemental therapy. The following are some of the therapies attempted by alternative therapists:

a) Melatonin enhances the immune system; it has value.
b) Logo's Iodine Nasal Solution will kill bacteria and other microbes. It can be helpful.
c) Hydrogen Peroxide Inhalation Therapy has been used and needs to have a device to make the mist. Excessive use may damage respiratory system. Needs an expert to guide.
d) Colloidal Silver is an antiseptic, and it has been used to kill bacteria in treatment of chronic wounds. Can clean the nose with it but I would avoid using internally.
e) BHT, Butylated Hydroxytoluene, is a food additive approved by the FDA. It is often found in cereal ingredients, and some alternative therapies claim it as an effective natural remedy for a number of viral conditions, such as hepatitis B, hepatitis C, the Epstein Barr virus, the herpes virus, and even warts. BHT can be taken with coconut oil, vitamin C, Ted's Alkalizing Tonic, and herbs.
f) Indian government has a special govt. department called the AYUSH Ministry. AYUSH Ministry just means Ministry of Natural Medicines (more or less)—incorporating the Siddha, Unani, Ayurvedic traditional medicines and Homeopathy etc. They claim that eating a tablespoon of Chywanprash daily enhances immunity and it may help prevent the spread of the virus, according to Ayurveda experts. In an article in the Hindustan Times, here are some recommended ayurvedic remedies for coronavirus using basil,

tumeric, Tinospora Cordifolia, fennel, putting sesame oil in the nostril, eating a piece of fresh ginger or drinking ginger tea, mint tea, cinnamon tea, and fennel tea, besides yoga and breathing exercises: https://www.hindustantimes.com/lifestyle/here-s-what-ayurveda-has-to-offer-in-fight-against-coronavirus/story-s9dm6hnh1mol304fXQhkvN.html

g) Indian Prime Minister Modi recommended that everyone should use the ayurvedic natural remedies against the coronavirus besides traditional therapies.

h) Lung remedies helped their lungs heal when they got sick with COVID-19 symptoms. These include simple tips like sleeping on your stomach, sleeping on the left side of the body (SLSD) and soothing herbs like mullein that will help ease congestion in the lungs, deep breathing exercises and yoga practice.

i) Salt and calcium carbonate protocol: one (1) Gram of Iodized Table Salt, one 500 mg tablet of carbonate. At the first sign of the chills, put the salt on your tongue and then slowly chew a calcium carbonate tablet. When sodium and calcium are increased the sodium/potassium pumps are inhibited from spinning out of control. This cuts off the virus's fuel and the virus starve. Experimental data needed, otherwise off the shelf therapy.

j) Potassium in old people to maintain brain and heart function

k) Vaccines for COVID-19: It will come in a year. To enhance the immune system, lower the infections take 2 intradermal BCG vaccination, 14 days apart (Faustman DL edit. The value of BCG and TNF in Autoimmunity. 2018. Academic Press). This is a long-term prophylactic measure. Immunity is not specific to coronavirus, but it a general immune protection.

l) Lemon & aspirin tea is said to cure flu within hours, I am not sure it works for COVID-19.

m) Lysosome (Italy): In 1922, 6 years before he made the discovery of penicillin, Alexander Fleming discovered the enzyme lysozyme by studying nasal secretions from a cold patient. These secretions were found to inhibit the growth of bacteria.

n) Turpentine poultice may relieve lung congestion.

With the introduction of modern antibiotics, antiviral medicines and vaccines, many of the alternative therapies are relegated to a lesser role.

Chapter 17

Methods to prevent COVID-19 morbidity and mortality

The only measure that works is to prevent COVID-19 infection entering to the respiratory system and prevent coronavirus getting inside the healthy cells is *not coming in contact with the infected, through social isolation with mask wearing*. Though we concentrate now on the coronavirus effect on the lungs, the organs next to it such as the heart, gastro-intestinal track, brain, liver and kidneys are also get affected. We need to understand that someone dying from bad pneumonia will ultimately die from cardiac arrest. Further it is known that a patient cannot get enough oxygen in severe acute respiratory distress syndrome due to vulnerability in our systems.

The coronavirus pneumonia can cause widespread inflammation in the body that in turn can lead to a plaque in the coronary arteries becoming instable causing a heart attack. Inflammation can also cause myocarditis leading to weakening of the heart and heart failure. Further, the coronavirus itself can directly attack the heart muscle because the coronavirus attaches to a type of angiotensin II receptors in the lungs. Those receptors are also found in the heart (maybe not in the olfactory mucosa and neurons?); the virus can find its way to the heart through massive pulmonary circulation from the lungs. This can result in cardiac attack of the coronavirus leading to complications such as myocarditis, myocardial failure, irregular heart beat and cardiac arrest. From the heart it can spread all over the body by blood.

Studies show that those who had heart disease before coronavirus infection are much more likely to show heart damage and a cardiac episode afterwards. Further it was noted that some of the patients with the New York coronavirus infection had symptoms that mimicked a heart attack, but upon examination doctors found a coronavirus infection, not a heart attack. EKG, ultrasound and cardiology consultation may be in order to diagnose and treat heart afflictions from the lung problem of coronavirus infected patients.

Once the virus is inside the nasal cavity-nasopharynx-oral cavity-oropharynx-and larynx, isolation is ***needed to protect yourself and protect others, and prevent contact spreading of the virus.*** We want to use the COVID-19 viral receptor and their binding sites angiotensin receptors on the mucosal lining of these air passages, using therapeutic agents in prophylactic sprays of our invention, use of therapeutic agents such as telmisartan and losartan to prevent coronavirus attaching to the healthy cells angiotensin receptors, and destroy the virus within the cell once inside the cells along with the inhibition of its multiplication within the cells, infecting other cells and cytokine storm in the lungs using the method we describe. The present invention relates to the use of said compounds to prevent entry of the virus to healthy cells and destroy the virus once inside the cells and then prevent them to continue their journey to infect the other adjacent cells and other people.

Use our patented prophylactic and treatment methods described herein to prevent the attachment of the coronavirus to healthy cells, then multiplication within the coronavirus infected cells, its shedding to infect other cells, its effects, to inhibit or prevent cytokine storm and their effect on the lungs, and to treat other signs and symptoms COVID-19 produces. These include hypoxia, thromboembolic phenomenon, pneumonia due to secondary bacterial infection, besides maintaining vital signs and life support in ICU and hospital setting. Our invention describes the prophylaxis, killing of the multiplying virus within the cells as the infection takes place, as well as treatment of signs and symptoms it causes once inside our body and management of hypoxia it creates in the aged with comorbidities residing in the nursing homes, extended care facilities, and the aged living independently.

Coronavirus affects nasal-oral-pharyngeal-laryngeal, tracheal, bronchial, alveolar mucosa. It also affects blood vessels and endothelial cells that surround the alveoli by direct viral attack and by the cytokine it creates by immune systems and endothelial cells around the alveoli resulting in leaking cellular and plasma to alveoli with invasion of white blood cells that causes severe acute respiratory distress syndrome (SARS-ARDS). We want to describe methods of how to prevent infection, its spread to healthy cells within the body once infected, and how to prevent the development of severe acute respiratory distress syndrome (SARS) by: inhibiting cytokine, cytokine storm, using angiotensin II receptor blockers to prevent coronavirus entry into healthy cells, viricidal dihydrochloroquin and chloroquine, protease inhibitor remdesivir, antioxidant and viricidal such as vitamin C, thiamine hydrochloride and corticosteroids, auranofin, EDTA, curcumin, vitamins, cimetidine, anti-helminthic such ivermectin and Niclosamide, protease inhibitor cum antioxidant Ebselen to kill the virus within the cells, and mini dose of rapamycin to prevent cytokine storm) and suppress the hyperactive immune system and treatment of symptoms as they develop to reduce morbidity and mortality including thrombo-embolic phenomenon.

Chapter 18

Summary of the invention

When we say coronavirus, it refers to the particular virus that causes severe acute respiratory syndrome coronavirus 2 (SARS-Cv2), named by WHO as **COVID-19**. This coronavirus spreads easily and is highly virulent. The terms dihydrochloroquin and chloroquine are used interchangeably so also angiotensin receptors and angiotensin receptors II.

The present invention relates to prophylaxis and treatment of signs and symptoms of coronavirus infection that results in coronavirus infectious disease of 2019 (COVID-19).

The present invention provides novel therapeutic, pharmaceutical, biochemical, and biological agents or compounds that bind to the free-floating coronavirus in the nasal-oral-pharyngeal-laryngeal mucosa, air passages, and then the lung alveolar cells and blood vessels that surround alveoli to prevent their entry into healthy cells of these respiratory system, the primary target of coronavirus infection. This is accomplished by using nasal and oral sprays, and mouth and nasal washes compounded using angiotensin receptor blockers telmisartan and losartan, protease inhibitor remdesivir, ethylenediaminetetraacetic acid (EDTA), antimalarials dihydrochloroquin and chloroquine phosphate, to prevent coronavirus adherence to respiratory mucosa including nasal and oral cavity by using xylitol, and chlorohexidine and dimethyl sulfoxide.

The present invention provides novel compounds/therapeutic agents in the form of oral and nasal sprays, nasal and mouth washes that prevent the binding of the coronavirus attached to the healthy cell in the nasal-oral-pharyngeal-laryngeal mucosa, air passages, lung alveolar cells and blood vessels that surround alveoli. This is to prevent the coronavirus infection of healthy cells and also by using parenteral and oral use of angiotensin II receptor blockers such as telmisartan, losartan, zinc, dihydrochloroquin and chloroquine phosphate, auranofin, vitamin C, EDTA, superoxide dismutase, chlorohexidine, xylitol, anti-oxidants and cimetidine.

The present invention provides novel compounds/therapeutic agents that prevent the biding of the coronavirus attached to the healthy cell and block them entering into the nasal-oral-pharyngeal-laryngeal mucosa, air passages, lung alveolar cells by using superoxide dismutase (SOD), antibody nano particles, and antibodies against coronavirus derived from patients serum, aerosolized sprays of the respiratory system including lungs.

The present invention provides novel compounds/therapeutic agents that prevent the biding of the coronavirus attached to the healthy cell and block them entering into the nasal-oral-pharyngeal-laryngeal mucosa, air passages, lung alveolar cells to deliver their genetic material inside the healthy cells by blocking the angiotensin II receptors telmisartan and losartan to prevent the damage to the lung lining cells.

The present invention provides novel combinations of therapeutic agents that bind to the coronavirus attached to the olfactory mucosa and block them entering into the olfactory bulb and brains by inhibiting the coronavirus viral multiplication on the surface of the olfactory mucosa by dihydrochloroquin and chloroquine phosphate, protease inhibitor remdesivir, auranofin, EDTA and xylitol.

The present invention provides novel compounds/therapeutic agents that prevent the biding of the coronavirus attached to the healthy cell and block them entering into the nasal-oral-pharyngeal-laryngeal mucosa, air passages, lung alveolar cells to deliver their genetic material inside the healthy cells by blocking the formation of biofilm and growth of microorganisms in the nose and the mouth using nasal spray, oral spray, and mouth wash and nasal lage by using EDTA, angiotensin II receptor blockers, telmisartan and losartan, superoxide dismutase, zinc, xylitol, SOD, anti-oxidants and chlorohexidine

The present invention provides novel therapeutic agents that bind to the coronavirus inside the healthy cell in the nasal-oral-pharyngeal-laryngeal mucosa, air passages, lung alveolar cells and blood vessels that surround alveoli after coronaviral entry by using chloroquine, protease inhibitor Remdesivir, angiotensin II receptor blockers telmisartan and losartan, protease inhibitor remdesivir, and antioxidant vitamin C, and auranofin and use of intravenous EDTA.

The present invention provides novel therapeutic agents that bind to the coronavirus inside the healthy cell once they enter the host cells in the nasal-oral-pharyngeal-laryngeal mucosa, air passages, and then the lung alveolar cells and blood vessels that surround alveoli to prevent coronavirus multiplication by using herein described antimalarial, angiotensin receptor blockers, protease inhibitor, antioxidants.

The invention further relates to the use of a composition comprising one or more inventions that also relates to a method of treatment of viral infections inside the cell and their formation of viral particles to spread and infect other cells by using high dose intravenous vitamin C, thiamine hydrochloride, hydrocortisone, zinc, EDTA and auranofin.

The invention further relates to the use of therapeutic agents comprising one or more methods of treatment of viral infections inside the cell and their formation of viral particles with their glycoprotein envelope in the lysosomes using dihydrochloroquin and chloroquine phosphate, EDTA, superoxide dismutase, and zinc.

The invention further relates to the use of therapeutic agents comprising one or more methods of treatment of this viral infection inside the cell and their formation of viral particles with inhibition of their proteolysis of RNA protein once the viral RNA enters the cells by using dihydrochloroquin and chloroquine phosphate and remdesivir.

The invention further relates to the use of therapeutic agents comprising one or more methods of treatment of viral infections inside the cell and their formation of viral particles with inhibition of their transcription of RNA by suing herein described antimalarials and protease inhibitor.

The invention further relates to the use of therapeutic agents comprising one or more methods of treatment of viral infections inside the cell and their formation of viral particles with inhibition of their translation of RNA by using herein described antimalarials and protease inhibitor and zinc.

The invention further relates to the use of therapeutic agents comprising one or more methods of treatment of viral infections inside the cell and their formation of viral particles with inhibition of their assembly and budding of RNA in the lysosomes, and Golgi intracellular organelle by using dihydrochloroquin and chloroquine phosphate, zinc, EDTA, auranofin and remdesivir.

The invention further relates to the use of therapeutic agents comprising one or more methods of treatment of viral infections inside the cell and their formation of viral particles with inhibition of their protease activity by using protease inhibitor remdesivir with EDTA,

The invention further relates to the use of therapeutic agents comprising one or more methods of treatment of viral infections inside the cell and their formation of viral particles with inhibition of their messenger RNA formation by suing herein described antimalarials and protease inhibitor.

The invention further relates to the use of therapeutic agents comprising one or more methods of treatment of viral infections inside the cell and their formation of viral particles with inhibition of virus release through the cell membrane-budding after formation of fully formed virus inside the cell by using herein described antimalarial, angiotensin receptor blockers, protease inhibitor, antioxidants, superoxide dismutase and EDTA.

The invention further relates to the use of therapeutic agents comprising one or more methods of treatment of viral infections with inhibition of released virion attacking and attaching to new cells on the angiotensin receptor by using angiotensin receptor blockers, protease inhibitor remdesivir, EDTA, to prevent, curtail, stop the progression of and/or cure the disease.

The invention further relates to the use of therapeutic agents comprising one or more methods of treatment of viral infections inside the cell and their formation of viral particles with their RNA replication and packaging using by suing herein described antimalarials, protease inhibitor, superoxide dismutase and EDTA.

The invention further relates to the use of therapeutic agents comprising one or more methods of treatment of viral infections inside the cell and their formation of viral particles with their glycoprotein envelope using antimalarials, protease inhibitor, EDTA, high dose intravenous vitamin C, and auranofin.

The invention further relates to the use of therapeutic agents comprising one or more inventions which also relates to a method of treatment of viral infections inside the cell, lysosomal organelle multiplication of the virus due to acid media and formation of viral particles with their glycoprotein, more specifically of infections with enveloped viruses by using therapeutic agents for binding small molecules or agents that changes lysosomal milieu alkaline media for inhibiting the viral formation using dihydrochloroquin and chloroquine phosphate, protease inhibitor, vitamin C, EDTA, dimethyl sulfoxide, and zinc.

The invention further relates to the use of therapeutic agents comprising one or more inventions which also relates to a method of treatments of viral infections inside the cell by inhibiting from formation of messenger RNA and transported to the endoplasmic reticulum for assembly, packaging and budding by using protease inhibitor.

The invention further relates to the use of therapeutic agents comprising one or more inventions which also relates to a method of treatments for prevention of viral infections inside the cell by inhibiting the RNA replication using protease inhibitor remdesivir and dihydrochloroquin and chloroquine phosphate.

The invention further relates to the use of therapeutic agents comprising one or more inventions which also relates to a method of treatments to prohibit viral infections inside the cell RNA by disrupting their assembly and packaging in the Golgi complex and lysosomes.

The invention further relates to the use of therapeutic agents comprising one or more inventions which also relates to a method of treatments viral infections inside the cell by inhibiting RNA assembly, packaging and then budding off the host cell membrane using angiotensin receptor blockers, protease inhibitor and dihydrochloroquin, chloroquine phosphate and utilizing the viricidal effect of antiparasitic ivermectin.

The invention further relates to the use of therapeutic agents comprising one or more inventions which also relates to a method of treatments by inhibiting assembled RNA virus budding by exostosis through the cell membrane to infect healthy cells by using protease inhibitor and Dihydrochloroquin and chloroquine phosphate, superoxide dismutase and ethylenediaminetetraacetic acid.

The invention further relates to the use of therapeutic agents comprising one or more inventions which also relates to a method of treatments to prevent formation of cytokines, chemokines, prostaglandins to prevent cytokine storm of the lungs by using rapamycin, vitamin C, vitamin B1, auranofin, angiotensin receptor blockers telmisartan and losartan described herein.

The invention further relates to the use of therapeutic agents comprising one or more inventions which also relates to a method of treatments to prevent formation of free radicle by providing antioxidants to prevent damage to infected and protect healthy cells by vitamin C, thiamine hydrochloride and hydrocortisone, auranofin, vitamin A, EDTA, cimetidine, SLD, vitamin D3, and Colchicine.

The invention further relates to the use of therapeutic agents comprising one or more inventions which also relates to a method of treatments to prevent formation of new virus and their killing as they are formed by providing hydrogen peroxide media which is viricidal with use of high dose intravenous vitamin C, auranofin, and remdesivir.

It is the intent of the invention to administer vitamin C, vitamin D3, vitamin B1 (thiamine hydrochloride in the form of Benfotiamine) and such as a prophylaxis to reduce viral infection and to ward off a cytokine storm by augmenting the immune system and improving the alveolar function as well as reducing its damage by the coronavirus to the respiratory air passages.

The invention also provides for the use of intravenous thiamine hydrochloride with high dose vitamin C and hydrocortisone/dexamethasone to augment the effect of vitamin C and stabilize the alveolar and endothelial lining of blood vessel cells to prevent their plasma leakage into lung alveoli and suppress the activity of immune system, suppress cytokine production and their damage of the respiratory system, especially lungs..

The invention also provides for the use of intravenous hydrocortisone with high dose vitamin C to augment the effect of vitamin C, and antioxidants curcumin, vitamin D3, auranofin, telmisartan and losartan, stabilize the alveolar and endothelial lining of blood vessel cells and inhibit the cytokine production.

The invention also relates to pharmaceutical therapeutic agents of many prophylactic nasal and oral sprays as the coronavirus is floating in the expired and inspired breath by using angiotensin receptor blockers telmisartan and losartan, antioxidants vitamin C, protease inhibitor remdesivir, ethylenediaminetetraacetic acid, auranofin, dimethyl sulfoxide.

The intent of the invention relates to prevention of damage to the lungs by coronavirus, and production of cytokines by using pharmaceutical therapeutic agents of many prophylactic nasal and oral sprays and mouth wash as the coronavirus is floating in the expired and inspired breath by using angiotensin receptor blockers telmisartan and losartan, antioxidants vitamin C, protease inhibitor remdesivir, antimalarials dihydrochloroquin and chloroquine phosphate, and superoxide dismutase.

The intent of the invention is relating to prevention of damage to the lungs by coronavirus, and production of cytokines and the secondary infection that follows due to bacterial infection of the respiratory system including lungs by using azithromycin and related antibiotics appropriate to the secondary microbial infection and use of monoclonal antibodies Tocilizumab, with remdesivir, vitamin C, vitamin D3, with supportive therapies in ICU with proper hypoxia treatment due cytokine storm. It also relates to cleaning the blood using extracorporeal blood purification (EBP) method by means of a device developed by Bicon Biologics of India named as CytoSorb therapy and providing the needed oxygen using ECMO device when the lungs are failing to oxygenate the venous blood.

The invention thus relates to the use of envelop-cas antiviral compounds, more particularly compounds active against enveloped viruses such as nucleoside, non-nucleoside and nucleotide reverse transcriptase inhibitors such as for instance, dideoxy adenosine, stavudine, zalcitabine, zidovudine, lamivudine, didanosine, nevirapine, delavirdine, efavirenz, tenofovir, foscamet sodium and the like HIV protease inhibitors such as saquinavir, ritonavir, indinavir, nelfinavir, remdesivir, amprenavir and the like.

Chapter 19

Brief description of the drawings

The purpose of the present invention will become readily valued and understood from deliberation of the following comprehensive descriptions of the preferred embodiments when taken together with the accompanying drawings in Chapter 21, in which:

1. Figure 1a is the diagram 100a of the human body showing various organs affected in the COVID-19 (modified from CDC and USA today publications, so also many of the below illustrations).
2. Figure 1 shows the diagram 100 of the coronavirus based on the present studies which show spike glycoprotein.
3. Figure 2 from the internet publications shows the diagram 200 of the coronavirus structure based on the present studies which show spike glycoprotein.
4. Figure 3 is the diagrammatic presentation 300 of the coronavirus from the droplets in the expired air into the respiratory system.
5. Figure 4 is the diagrammatic presentation 400 which shows how the coronavirus from the eye area and the route of travel.
6. Figure 5 is the diagrammatic presentation 500 which shows it attaches to the healthy cells, migrates inside the healthy cell to propagate its genetic material, multiplies and propagates.
7. Figure 6 is the diagrammatic presentation 600 which shows how the coronavirus enters and multiplies, and the effect of therapeutic agents.
8. Figure 7 is the diagrammatic presentation 700 which shows how the coronavirus activates the immune system.
9. Figure 8 is the diagrammatic presentation 800 which shows the coronavirus binding to the healthy tissue angiotensin receptor.
10. Figures 9 and 10 are diagrams 900 and 1000 of the spray 35 for nose and oral cavity.
11. Figures 11, 12 and 13 are the diagrams 1100, 1200 and 1300 showing the olfactory mucosa and the spread of coronavirus to the brain.

12. Figure 14 is the diagram 1400 showing the effect of cytokine storm due to COVID-19 on various organs of the body (modified after Alexander Shimabukuro-Vornhagen et al. Journal for ImmunoTherapy of Cancer, 2018. 6:56).

Chapter 20

Detailed description of the invention

The present invention relates to methods of providing prophylaxis and treatment against coronavirus infectious agent by topically, orally, and parenterally administering compound to the respiratory system of a subject.

"Cytokines" refers to polypeptide signaling proteins of the immune system, which are numerous hormonelike, low-molecular-weight proteins, secreted by various immune system cell and endothelial cells around the alveoli, which regulate the intensity and duration of immune response and mediate cell-to-cell communication such as interferon, interleukin, lymphokine, chemokines. They are intercellular PROTEIN or GLYCOPROTEIN signaling molecule, secreted by many cell types and involved in cellular regulation and proliferation. Cytokines exert their effects by binding to specific RECEPTORS on the membrane of target cells. They include GROWTH FACTORS, INTERLEUKINS and LYMPHOKINES.

"Induce" and variations thereof refer to any measurable increase in cellular activity. Thus, as used herein, "induce" or "induction" may be referred to as a percentage of a normal level of activity, for example, a cytokine, chemokine, or co-stimulatory marker; "induce" also may refer to any measurable increase in the amount of the cell product that is produced in response to a stimulus.

"Prophylaxis" refers to any degree of limiting an infection by an infectious agent viral or bacterial including (a) preventing or limiting an initial infection, (b) preventing or limiting the spread of an existing infection, or both. "Prophylaxis" may be used interchangeably with "reduce infection" and variations thereof.

"Respiratory tract" or "respiratory system" refers, generally, to the major structures and passages that permit or provide air flow between the environment and the lungs of a subject.

"Upper respiratory tract" refers, generally, to the nasal cavity, paranasal sinuses, nasopharynx, oral cavity, pharynx, and larynx.

"Lower respiratory tract" refers, generally, to the trachea and lungs, including the bronchi, bronchioles, and alveoli.

“Receptors” located on cell surface (membrane receptors, transmembrane receptors) are receptors that are embedded in the plasma membrane of cells. They act in cell signaling by receiving (binding to) extracellular molecules. They are specialized integral membrane proteins that allow communication between the cell and the extracellular space. The extracellular molecules may be virus, bacteria, hormones, neurotransmitters, cytokines, growth factors, cell adhesion molecules, or nutrients; they react with the receptor to induce changes in the metabolism and activity of a cell. With regards to coronavirus, angiotensin receptors II are the receptors on the host cells for the virus to attach and enter the host cells to multiply, propagate and cause COVID-19.

“Receptor blocker” also named “receptor antagonist” is a type of receptor ligand or drug that blocks or dampens a biological response by binding to and blocking a receptor rather than activating it like an agonist. Receptor blockers binding will disrupt the interaction and inhibit the function of an agonist or inverse agonist at receptors such as receptor blockers on the cell surface to prevent coronavirus entering the host cells. Antagonist activity may be reversible or irreversible depending on the longevity of the antagonist–receptor complex.

“Angiotensin-receptor blockers” known as ARBs are a class of drugs (as telmisartan, losartan, valsartan and such) that block the effects of angiotensin II by inhibiting access to a subset of its receptors and that are used especially in the treatment of hypertension. The newer angiotensin II receptor blockers also relax blood vessels.

Graphic and diagrammatic embodiments of our inventive methods of treatments of COVID-19 are shown below. In the interest of clarity, not all features of actual application are described in this specification. It will of course be appreciated that in the development of any such actual embodiment, numerous application specific choices must be made to achieve the inventor’s detailed goals, such as compliance with technology or business-related restraints, which may vary from one application to another. Moreover, it will be appreciated that the effort of such a development might be complex and time-consuming but would nevertheless be a routine undertaking for those of ordinary skill in the art having the benefit of this disclosure.

In the following detailed description of the inventive methods to treat COVID-19. It will be understood by those of ordinary skill in the art that the present invention may be practiced without these specific details. In other instances, well-known methods, procedures, components and structures may not have been described in detail so as not to obscure the present invention.

Chapter 21

Explanation of figures of coronavirus prophylaxis and COVID-19 treatment based on our invention

Referring now to the drawings, the following explanation/description of our methods of treatment of COVID-19 and other similar viral infections are merely exemplary in nature and is in no way intended to limit the invention, its applications, presentation or uses. Numbering and their explanation in individual figures may not be identical in all the figures and drawings. The invention will now be illustrated and explained by methods of illustration with reference to the following drawings, of which:

Figure 1a, below, is the diagram 100a of the human body showing various organs affected in the COVID-19 and the signs and symptoms of the diseases The signs and symptoms vary as the severity of the COVID-19 afflictions vary depending upon the severity, age, sex, and other associated health conditions—comorbidities such as lung, heart, or kidney diseases; diabetes, blood pressure, autoimmune diseases, also shown in Figure 14.

Figure 1, below, shows the diagram 100 of the coronavirus based on the present published studies of the coronavirus, which show spike glycoprotein 6 projecting all around the virus. Like other coronaviruses, SARS-CoV-2 that causes COVID-19 affliction has four structural proteins, known as the S (spike), E (envelope), M (membrane), and N (nucleocapsid) proteins. It shows spike protein S 6 which attaches to the receptors on the host cell membrane where it comes in contact blocked by dihydrochloroquin and chloroquine phosphate, protease inhibitor, angiotensin receptor blockers and antioxidants such as vitamin C, and curcumin. S protein 6 attaches to angiotensin receptor on the cell surface of healthy cells as shown in Figure 5 and 6.

It shows M protein 7, envelope 8, E-protein 10 and the envelopes that encloses the cell RNA and nucleoproteins 9 that forms the virus when it enters the healthy host cell using the cell machinery. The virus attaches to cell membrane by S protein to angiotensin receptors and proteases break open the cell membrane and delivers the viral RNAs for production of more viruses using the host cell machinery. The angiotensin receptor blockers, protease inhibitor and anti-malarial described below block the viral entry. S protein is cleaved by a host cell furin-like protease into two separate polypeptides noted S1 and S2. S1 makes up the large receptor-binding domain of the S protein while S2 forms the stalk of the spike molecule (Figures 2, 3).

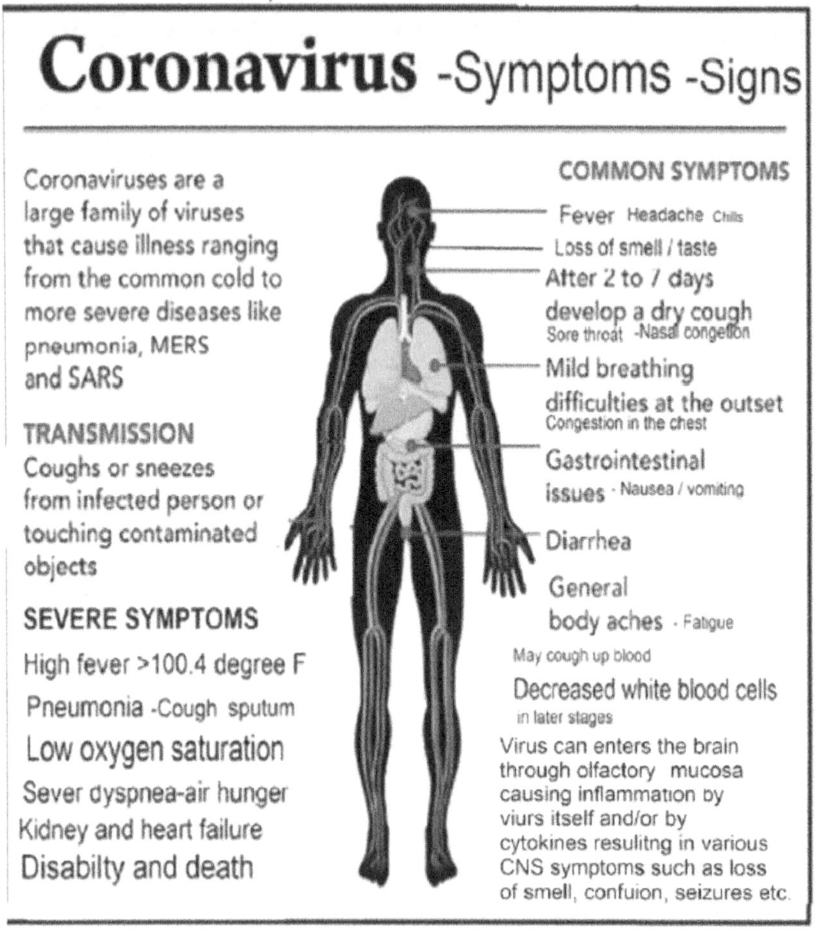

Coronavirus -Symptoms -Signs
Coronaviruses are a large family of viruses that cause illness ranging from the common cold to more severe diseases like pneumonia, MERS and SARS
TRANSMISSION
Coughs or sneezes from infected person or touching contaminated objects
SEVERE SYMPTOMS
High fever >100.4 degree F
Pneumonia -Cough sputum
Low oxygen saturation
Sever dyspnea-air hunger
Kidney and heart failure
Disabilty and death
COMMON SYMPTOMS
Fever Headache Chills
Loss of smell / taste
After 2 to 7 days develop a dry cough
Sore throat -Nasal congeflon
Mild breathing difficulties at the outset
Congestion in the chest
Gastrointestinal issues · Nausea / vomiting
Diarrhea
General body aches · Fatigue
May cough up blood
Decreased white blood cells
in later stages
Virus can enters the brain through olfactory mucosa causing inflammation by viurs itself and/or by cytokines resulitng in various CNS symptoms such as loss of smell, confuion, seizures etc.

Figure 1a

The angiotensin receptor blockers telmisartan and losartan, protease inhibitors remdesivir act by blocking this cleaving of spike protein, thus inhibits or prevents spikes biding with angiotensin receptors on the healthy cells and thus block coronavirus entry. Dihydrochloroquin and chloroquine phosphate, and angiotensin receptor blockers also help to prevent the attachment of the S- proteins to host cell receptors, so also xylitol, may be ethylenediaminetetraacetic acid (EDTA). Intravenous chelation therapy using, EDTA to augment the effect of coronavirus antiviral therapeutic agents and to act as antibacterial.

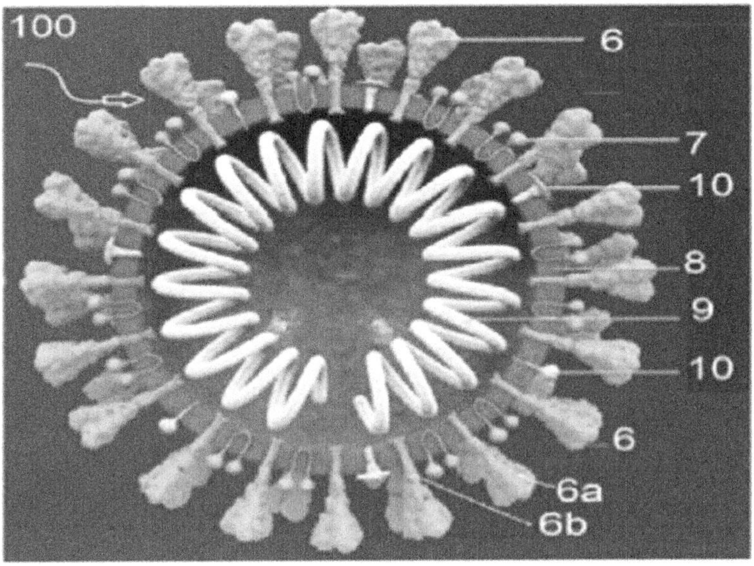

Figure 1

Figure 2, below, shows another diagrammatic presentation 200 of the coronavirus based on the published studies which show spike glycoprotein projecting all around the virus. Like other coronaviruses, SARS-CoV-2 has four structural proteins, known as the S (spike), E (envelope), M (membrane), and N (nucleocapsid) proteins. It shows S protein 6 which attaches to angiotensin receptor on the cell surface. Its membrane also shows M protein 7, small envelope protein 7a, envelope 10, hemagglutinin acetyl esterase glycoprotein 11 which is part of the cell membrane. The membrane structures enclose the viral RNA 12 and nucleoproteins 9 needed for reproduction of the virus within the host cell and they need to delivered to host cell for coronavirus reproduction and propagation.

This virus is parasitic and needs a host cell to sustain and reproduce. All the components facilitate the virus entry into healthy cells and their multiplication. The S spike protein 6 plays a major role by its attachment to healthy host angiotensin receptors on the cell surface, without which the virus will not survive. The virus attaches to the cell membrane through S protein, and proteases break open the cell membrane to deliver the viral RNA for production of more viruses using the host cell machinery. The angiotensin receptor blockers, protease inhibitor and anti-malarial described below block the viral entry. The angiotensin receptor blockers, protease inhibitor, Dihydrochloroquin and chloroquine phosphate, angiotensin receptor blockers and antioxidants such as vitamin C, vitamin D3, 5000 IU, and curcumin; all play a role in preventing the entry of virus into healthy cells and the viral reproduction withing the host cells. The vitamin C produces hydrogen peroxide around the cells, that prevents the biding of the coronavirus to angiotensin receptors and at the same time killing the virus (viricidal) as the coronavirus is attempting to attach and enter the host cell.

S protein is cleaved by a host cell furin-like protease into two separate polypeptides noted S1 and S2. S1 makes up the large receptor-binding domain of the S protein while S2 forms the stalk of the spike molecule (Figures 2,3). The protease inhibitors act by blocking this cleaving of spike protein, thus inhibits or prevents spikes biding with angiotensin receptors on the healthy cells and thus block their entry. Dihydrochloroquin and chloroquine phosphate, and angiotensin receptor blockers also help to prevent the attachment of the S proteins to host cell receptors, thus preventing their entry. The free-floating viral particles unless attaches to host cell and it cannot enter, will not survive.

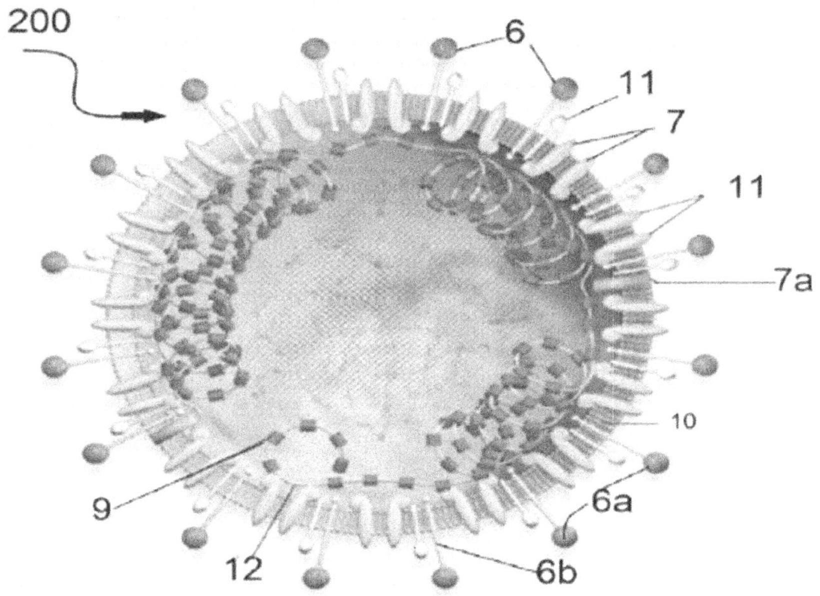

Figure 2

Figure 3, below, is the diagrammatic presentation 300 showing how the coronavirus from the droplets 21 from the infected person are passed as droplets in the expired air 21 from the respiratory system that are then transported to the healthy individuals as they are breathing this coronavirus infected droplets. It shows the route of travel of the coronavirus from the nasal and oral passages to the, pharynx, trachea, bronchial tree and to alveoli. When the infected person coughs, sneezes, or otherwise transmits the droplets, the coronaviruses are expelled into the atmosphere 21 as droplets probably surrounded by expired air humidity though the expired air, and they can travel up to 6 feet or more and infect the healthy person.

These droplets 21 enter through the nasal 14, oral 15 passages then enter into nasal-oral-laryngeal pharynx 16 with inspired air or contact with infected hand or objects. From here, coronaviruses are carried by inspirator air to larynx and trachea 17 to lungs and their alveoli 18. 20, 22. The diagram shows the virus inside the lung's alveoli 18, 20 and detail structure of the virions 20, 22 and how the droplets come out when a person coughs or sneezes 21 as droplets and its passage into lung alveoli as droplet infection. coronavirus containing droplets can also attach to the olfactory mucosa on the thick mucus coating on the receptor cells that can transmit the infection to the CNS.

Our antiviral spray we describe herein will kill, disable or inhibit the free-floating virus and its attachment to the healthy cells of the respiratory system lining cells and olfactory mucosa by using angiotensin receptor blockers telmisartan and losartan, vitamin C, dihydrochloroquin or chloroquine phosphate, protease inhibitor remdesivir, antioxidants such as curcumin, vitamin D3, auranofin and such with xylitol and EDTA.

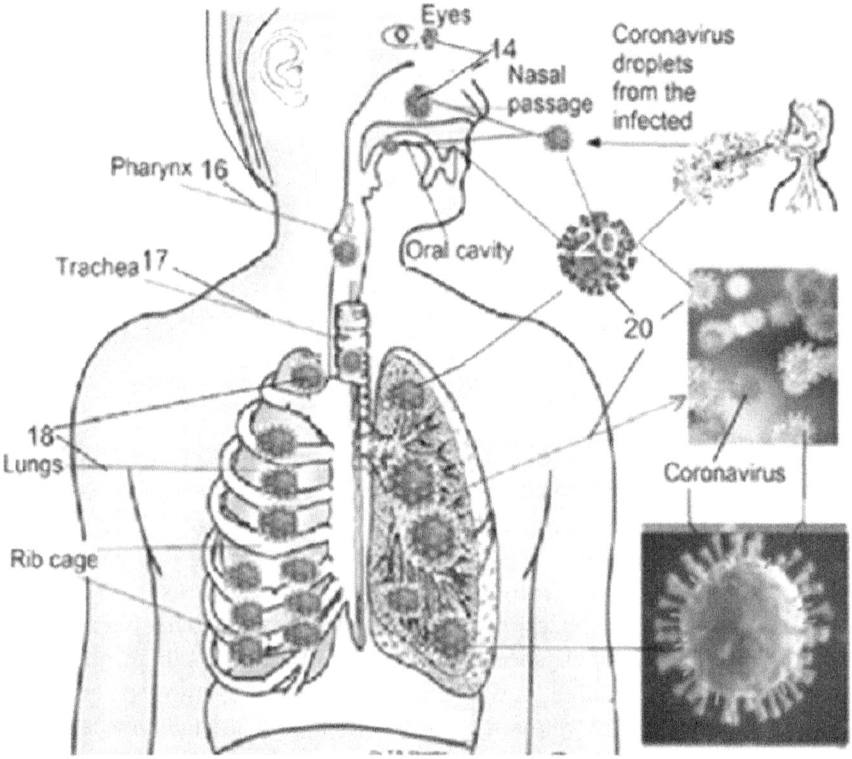

Figure 3

Figure 4, below, is the diagrammatic presentation 400 showing how the coronavirus in the eye's conjunctival sac 23, in oral and nasal cavities 55,58 due to touching of these sites or face by infected fingers and hands after coming in contact with the infected objects, saliva or breath is transported the respiratory system. From touching eyes with infected fingers, the coronavirus are transported to the eyes' tears in the conjunctival sac 23, the coronaviruses are transported to the nasal mucosa through the tear duct 24 and tears. There is a direct communication from the eyes to the nose through the tear ducts 25, and hence the coronavirus can travel from the conjunctival sac 23 infection, carried by tears to the lacrimal canaliculi 24 to the lacrimal sac 25 then to lacrimal duct (tear duct) and to the opening under the inferior nasal turbinate 26 then spread to the other mucosal surface of the nose 27.

Recently there was a case reported of eye infection of a care giver who was not wearing safety goggles. Hence the care giver in direct contact with the patients should wear eye protective goggles to prevent coronavirus infection of the eyes also. The diagram also shows that touching the nose 55 and the mouth 58 can transfer the coronavirus from the infected hands and objects. From nasal, oral, and nasal cavities, the coronavirus is carried to the nasal mucosa 27, 55 and to oral mucosa 58 to the nasal, oral and laryngeal - pharynx then to the trachea, bronchial tree and lung alveoli and to the alveolar blood and lymph supply. This diagram depicts even touching the eyes besides touching the face, nose and mouth with infected hands or objects can infect the respiratory system. From the nasal cavity, coronavirus can infect air sinuses contributing to the headache, to the eustachian tube to the middle ear and mastoid sinus contributing to the ear pain and to the olfactory mucosa then to brain to cause CNS symptoms.

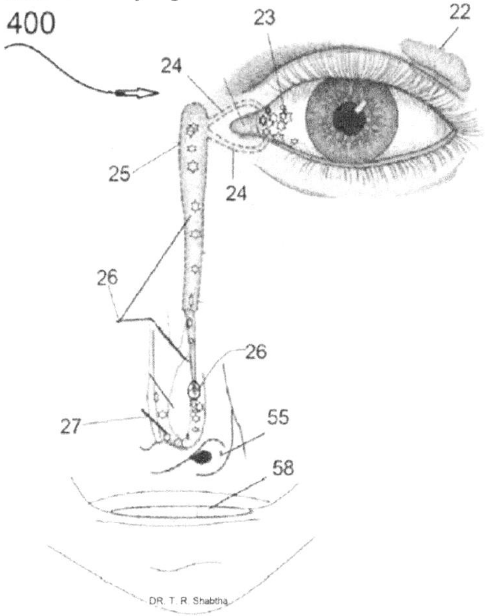

Figure 4

Figure 5, below, is the diagrammatic presentation 500 showing how the coronavirus from the patient breathing droplet-containing coronavirus air comes in contact with the air passages mucosal lining cells and attaches to the healthy cells, migrates inside the healthy host cell to propagate its genetic material, and multiplies inside the host cell to spread far and wide. It also shows the target of various sites on the coronavirus and on the host cells which can be used to block its infection by using appropriate therapeutic agents. It depicts the mechanism involved in its attachment to host cell surface membrane 45 angiotensin's receptors II 41 and enters inside the healthy cell, empties its genetic material and begins to multiply inside the host cell and spread. It shows the S-spike protein 40 of the coronavirus 42 and attaches to the 41 Angiotensin receptors II on the surface of the healthy mucosal cells. It shows SARS virus 42, anti-ACE-2 antibody 43, attaches to the angiotensin receptors -transmembrane ACE-2 receptors, of the host cell membrane 44 to enter the cell.

Once the virus "unlocks the receptor," and breaks the host cell membrane with the help of the cell protease, it can get inside the cell and use the host cell machinery to replicate, and makes many copies. The new viruses within the cell exit that cell and go on to infect more and more cells healthy cells. In some cases, the virus may stay more localized in the oral and nasal mucosa and upper airways and in other cases, the virus replicates and then makes its way down into the lungs and starts to infect the lung alveolar cells, as well as its microvasculature.

The angiotensin receptors II, 41, to which the virus latches on, is blocked by angiotensin receptor blockers such as telmisartan and losartan, protease inhibitor remdesivir, antimalarial dihydrochloroquin and chloroquine phosphate and others to prevent the entry of the virus into healthy cells. Infection of the coronavirus breaks the healthy cell membrane 45 by using proteases and transfers its nucleoprotein RNA 12, 46 of the coronaviruses into cytoplasm of the host cell for breakdown and multiply using the host cell machinery as described in Figure 6.

Antimalarial Dihydrochloroquin and chloroquine phosphate 48, 49 and protease inhibitor remdesivir 44, angiotensin receptor blocker such as telmisartan, block the entrance of coronavirus 42 and inhibit coronavirus multiplications inside the host cells. Protease inhibitors 48 also prevent the binding and entrance of the coronavirus inside the host cells. ARBs, angiotensin receptor blockers II, therapeutic agents 50 bind to angiotensin receptors I 41, and the angiotensin receptors II 41 located adjacent to each other on the cell surface membrane like conjoined twins and thus help to block the viral entry of coronavirus inside the healthy cell. Xylitol, vitamin C, vitamin D3, curcumin with DMSO 50, superoxide dismutase and EDTA bind to the receptor on the host cells and coronavirus spikes with therapeutic agents to prevent the binding and entry of the virus and transport RNA 12 inside the healthy host cell 46.

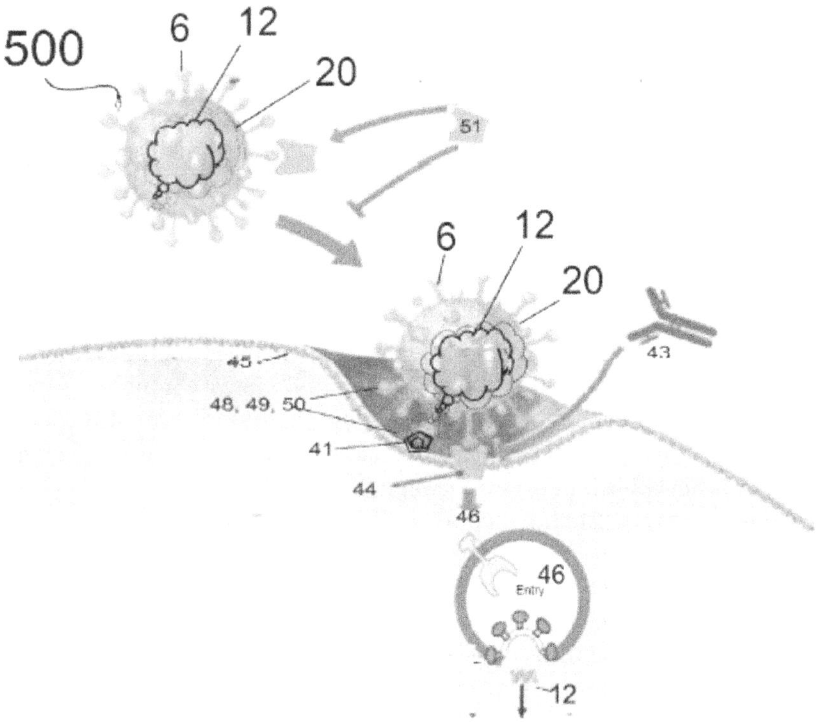

Figure 5

Figure 6, below, is the diagrammatic presentation 600 showing how the coronavirus from breathing infected air droplets and/or objects (hands and finger) comes in contact with the air passages mucosal lining cells and is attached to the healthy cells, migrates inside the healthy cell, and multiplies inside the host cell using its structural mechanism to multiply and propagate its genetic material to spread. It shows how the target of various sites on the coronavirus and on the host cells to block its infection using a combination of therapeutic agents. It shows the mechanisms involved in its attachment coronavirus to host cell membrane, breach the cell membrane and empty its genetic RNA, replicate inside the healthy cell, multiply, exit from the host cell and spread to new host cells. It displays the effect of various therapeutic agents and their site of actions.

Figure 6 shows the 41 SARS coronavirus attaches with the S spike protein 40 to host cell to the angiotensin receptors 49 on the cell wall. Coronavirus breaks the cell membrane with the help of proteases to deliver its RNA inside the host cell for further multiplication. This process is blocked by antimalarial Dihydrochloroquin / chloroquine phosphate, angiotensin receptor II blockers telmisartan and losartan, monoclonal antibodies, and proteases inhibitors remdesivir 47, 48. Note that the chloroquine 48 facilitates the entry of zinc 62, inside the cell which act to inhibit the coronavirus multiplication mechanisms inside the host cells. Note that the RNA 12 template is delivered inside the host cell cytoplasm after breaching the cell membrane 45, and then it is broken down by RNA polymerase also known as replicase 84 by a process called proteolysis 94 to make more RNA template needed for the multiplication of the coronavirus. Protease inhibitors, specific coronavirus antibodies, monoclonal antibodies 47 described in this invention can slow or stop this process of RNA breakdown and reassembly. Though the angiotensin receptor blockers drugs attach to the angiotensin receptors II, it also has an effect on the binding of the coronavirus to the angiotensin receptors I, also. I believe these angiotensin receptor blockers inhibit or slow down or partially or completely block the coronavirus S-spike getting attached to the angiotensin receptors 41on the healthy cells.

Coronaviruses replicate by copying their genetic material using an enzyme known as the RNA-dependent RNA polymerase (Figure 6 #84, 82, 93, 98). Matthias Götte, a virologist and professor at the University of Alberta, found that the enzymes can incorporate remdesivir, which resembles an RNA building block, into new RNA strands. Shortly after adding remdesivir, the enzyme stops being able to add more RNA subunits. This halts genome replication. RNA containing remdesivir takes on a strange shape that does not fit into the enzyme, thus stops viral replication as described in this invention incorporated into nasal sprays besides administering parenterally. Researchers suggest using the viral RNA polymerase of coronaviruses as a target in finding therapeutic agents to prevent the coronavirus infection.

These broken-down RNA templates which are passed from the coronavirus undergo transcription proteolysis 94, and messenger RNA 93 is formed which is inhibited by chloroquine, remdesivir, monoclonal antibody and coronavirus specific antibodies. It is a single strand RNA which has genetic information in a sequence of nucleotides to form RNA coronavirus replication. These are passed on to the endoplasmic reticulum (ER) 50 for further viral development, then they pass through the ER-Golgi intermediate compartment (ERGIC) 51 then into Golgi complex 52. The dynamic nature and functional role of the ERGIC 51 have been debated for quite some time. In the most popular current view, the ERGIC clusters 51 are mobile transport complexes that deliver secretory cargo from ER exit 50 sites to the Golgi 52 complex situated adjacent to ER. It is a compartment of protein 51 trafficking shuttle from nucleus 54 and endoplasmic reticulum 50. Protease inhibitors and antimalarial therapeutic agents described in this invention can slow down or block the trafficking at this site 51. This slows down the final assembly of the coronavirus at Golgi 52 complex and lysosomes 53. RNA particles 41 are further modified in Golgi complex 52 and in the lysosomes 53. High dose vitamin C with other therapeutic agents described in this invention can slow down or block the formation of new coronavirus.

From the Golgi apparatus, the virions are released for RNA replication and packaging 98, which takes place in lysosomes 53 acidic media. Dihydrochloroquin / chloroquine phosphate and such pH changing therapeutic agents inside the lysosomes and even in the Golgi complex changes the acidity to more alkalinity, inhibiting the viral final assembly, thus inhibiting coronavirus formation for propagation. The virion needs acid media to assemble and get membrane component to enclose them to from a complete virus. From this assembly site (lysosomes) 98, they are released 60 out through the cell membrane 45, a process called viral shedding/budding 78, and to infect new cells 86. Once the viral replication has been completed, the host cell is exhausted of all resources in making viral progeny and succumbs. This is how this coronavirus infection destroys our internal organs and their function.

The newly formed coronavirus exits the host cells by several methods. The term is used to refer as shedding from a host cell, shedding fully formed coronavirus 60 from one part of the body cells into another part of the body cells 86, and shedding from our respiratory system affected bodies into the environment through breathing air as droplets, where the viruses infect other people. This is how the virus expelled from the respiratory system lining cells gets into expired air and infects other people. Shedding of the virus through budding uses up the cell membrane and cell contents which eventually perish due to viral multiplication using vital contents of the host cell. Zinc 62 in oral spray with other anti-viral therapeutic agents and anti-oxidants helps to heal the healthy cells as the viruses are extruded from the cell, thus restoring the damaged cell. We know that zinc blocks the entry of rhino virus that causes the common cold, as well as flu virus and hence it prevents the easy binding of the coronavirus on the cell membranes as well. Once inside the host cell, the zinc inhibits the viral multiplication also whose effect is augmented by dihydrochloroquin and chloroquine phosphate and protease inhibitor remdesivir.

Figure 6 shows various stages of coronavirus entry into host cell, its reproduction within the host cell and extrusion, and how our inventive therapeutic agents prevent, curtail, stop the progression of, kill coronaviruses, and/or cure the disease as described in the invention:

a) 40= S-spike on coronavirus floating to find a host cell to attach to angiotensin receptors and get in the host cell. The vitamin C produces the hydrogen peroxide which kills the coronavirus before it attaches and enters the host cell and the angiotensin receptor blockers such as telmisartan and losartan block the docking of the virus to the angiotensin receptors. Ethylenediaminetetraacetic acid and superoxide dismutase help vitamin C.
b) 41= coronavirus floating to attach to the angiotensin receptors on the host cells. Angiotensin receptors 41. The virus cannot produce the COVID-19 if it does not get into healthy host cells.
c) 46= attachment of coronavirus to the host cell membrane angiotensin receptors, breaking the cell membrane and unloading nucleocapsid phosphoprotein RNA 12 into the host cell. This process is blocked by angiotensin receptor blockers telmisartan and losartan, and inhibited by protease inhibitor such as remdesivir as well as by anti-malarias we describe in this invention.
d) 47= dihydrochloroquin / chloroquine phosphate inhibits assembly of viruses making the media alkaline in the lysosomes 53 of the cells which is inhibitory to coronavirus assembly,
e) 48= protease inhibitor of site action on the cell membrane and inside the cell, the sites of protease activity, and the protease inhibitor will prevent the entry of coronavirus inside the host cell, their multiplication and their assembly inside the host cell machinery.
f) 49= angiotensin receptor converting enzyme ACE2 receptors where the telmisartan and losartan block the entry of coronavirus inside the host cell when sprayed or taken orally as described in the invention.
g) 62= Zinc in the spray of our invention or taken orally and/or given parenterally, binding to cell membrane and viral spikes to prevent

coronavirus enters the healthy cells and promotes healing of the afflicted cell. As the zinc enters the cell by iontophoresis, it is inhibitory for viral multiplication inside the host cell.

h) 78= coronavirus virion once formed inside the host cell, are then released to infect other healthy cells. This takes place through the cell membrane 45, it is prevented by antimalarials, protease inhibitor such as remdesivir and high dose vitamin C

i) 82= RNA replication, and packaging in the lysosomes is affected by anti-malarial Dihydrochloroquin and chloroquine phosphate which make lysosomes less acidic which inhabit the viral assembly.

j) 84= RNA replicase, [RNA-dependent RNA polymerase (RdRP, RDR)] that breaks down the viral RNA into multiple viral particles templates to make more coronavirus. It is inhibited by protease inhibitor remdesivir, and high dose vitamin C

k) 86= coronavirus attacks other cells budding off from the healthy cells 78. This reinfection and affinity to get attached is changed by angiotensin receptors using angiotensin receptor blockers telmisartan and losartan, ivermectin, ethylenediaminetetraacetic acid, and superoxide dismutase and vitamin C

l) 94= RNA particles undergo proteolysis to form new virions template within the infected cells.

m) 93= messenger RNA formed to further production processes, and its movement within the cell is affected by protease inhibitor, antimalarials, and vitamin C

n) 93= translation of messenger coronavirus RNA is inhabited by protease inhibitor remdesivir which prevents viral multiplication withing the host cells.

o) 94= transcription inhibited by protease inhibitor and antimalarials, angiotensin receptor blockers and vitamin C to prevent the reproduction of coronavirus.

p) 95= proteolysis of RNA, inhibited by protease inhibitor such as remdesivir and anti-malarial, as well as by vitamin C.

q) 98= assembly and budding of coronavirus at Golgi apparatus and lysosomes, inhibited by Dihydrochloroquin and chloroquine phosphate, and protease inhibitor to produce free budding off free floating coronavirus which are killed by hydrogen production by the high dose vitamin C.

r) 50-ER= endoplasmic reticulum helps to assemble coronavirus particles, probably by adding more proteins derived from the nucleus.
s) 51-ERGIC= endoplasmic reticulum Golgi interphase compartment affected by protease inhibitor and antimalarials. This is a transport route for emerging coronavirus from endoplasmic reticulum to Golgi complex is important for further maturation of virions into coronavirus.
t) 52-Golgi= Golgi complex organelle helps the assembly of coronavirus particles, inhibited by protease inhibitor and antimalarial. I have published 12 research papers on the subject from Emory University School of Medicine on the Golgi apparatus, which is a dynamic organelle that participates in secretions and further viral maturation.
u) 53=lysosomes: final assembly plant, where the viral particles transported from the Golgi complex are built into virus finally with its membrane. Virus needs acid media for this maturation. It is made more alkaline to prevent viral assembly by suing dihydrochloroquin and chloroquine phosphate with alkaline diet and infusion of sodium bicarbonate.
v) 54=nucleus with its chromosomes control the all the cell activities needed for coronavirus to replicate and many of the anti – coronavirus therapeutic agents we describe will affect the nucleus and it content of chromosomes to prevent further formation of viral particles.

Figure 6, below, also shows the content inside the host cell that plays a role in reproduction of coronavirus virion as the RNA content 12 of the coronavirus are released within the cytoplasm 46 of the host cell. They are broken down by replicase 84. Further, note where the antimalarial – Dihydrochloroquin and chloroquine phosphate 47 and 48, protease inhibitors such as remdesivir work inside the cell at various phases of viral RNA break down and assemble to reproduce and multiply coronavirus RNA and angiotensin receptor blockers and remdesivir prevent and/or inhibit or slow down the attachment of the virus to host cell surface of the new cells.

Researchers exploring the interaction between the coronavirus and its hosts have discovered that when the SARS-CoV-2 virus infects a human cell, it sets off a ***morbid transformation***. Obeying instructions from the virus within, the newly infected cell sprouts multi-pronged tentacles studded with viral particles (see Figure 6). These cells appear to be using those gushing filaments, or ***filopodia***, to reach healthy neighboring cells. These protuberances appear to bore into the cells bodies and inject their viral poison-particles directly into those cells' genetic command centers — thus creating another viral producing (insert Figure 6 of filopodia is courtesy of Dr. E. Fisher of NIAIDS/NIH).

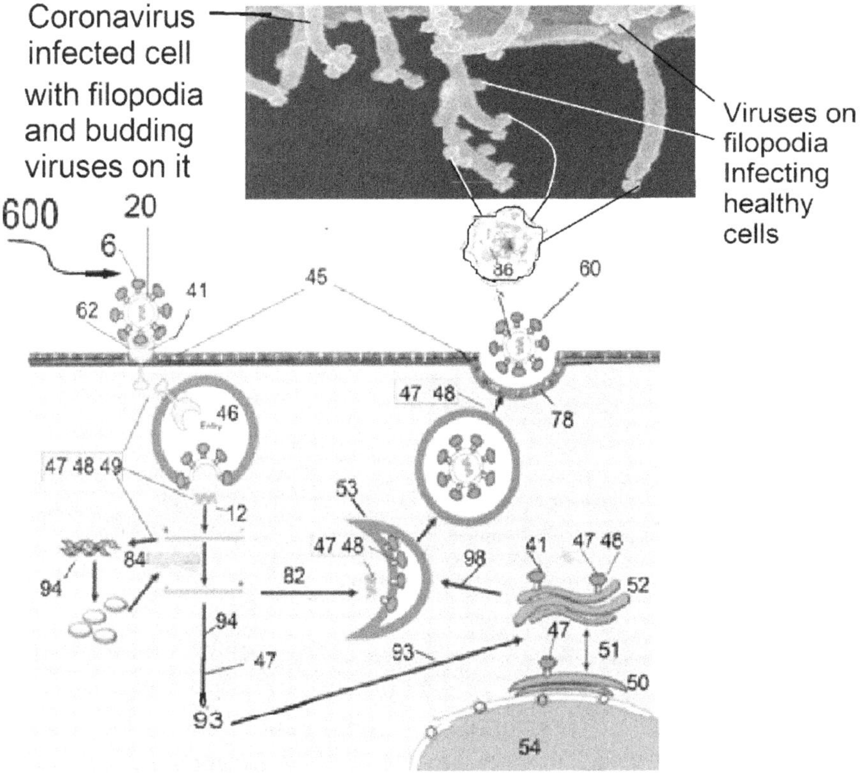

Figure 6

Figure 7, below, is the diagrammatic presentation 700 showing how the coronavirus inside the host cell multiplies and activates the immune system cytokine system that affects lungs and blood vessels by production of the inflammatory cytokines, chemokines, prostaglandins and the effect of antimalarials such as dihydrochloroquin and chloroquine phosphate, protease inhibitors, vitamin C, vitamin D3, hydrocortisone, auranofin, bioactive curcumin, superoxide dismutase, ethylenediaminetetraacetic acid (EDTA) and such inhibit the production of the these cytokines, free radicals, ROS and their adverse effects on the lungs and other organs (see Figure 14).

The flow chart shows potential inhibitory effect of Dihydrochloroquin / chloroquine phosphate, vitamin C, vitamin D3, vitamin B1, bioactive curcumin, zinc ethylenediaminetetraacetic acid (EDTA), superoxide dismutase and such, on the immunopathogenesis by cytokines resulting in severe acute respiratory distress syndrome (SARS) and COVID-19.

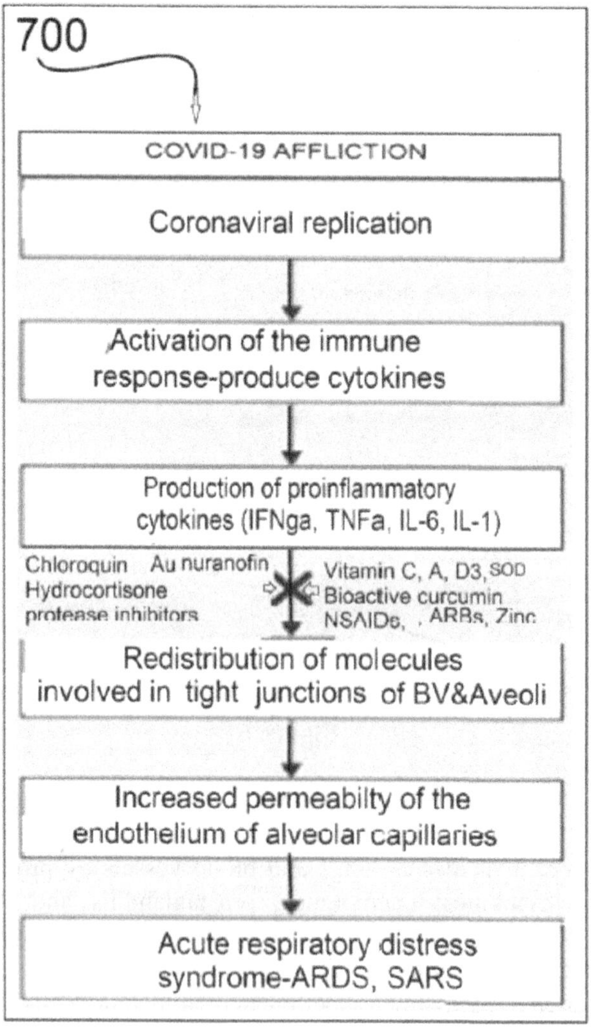

Figure 7

The cytokines, chemokines, and prostaglandins make the rich alveolar blood vessels endothelium leaky and spill their content into alveolar sacks. The chart shows that the proinflammatory cytokines are responsible for respiratory and other organ failure needing ventilation support with 100% oxygen and other vital sign managements. The angiotensin receptor blocker telmisartan and losartan, anti-malarial drugs, protease inhibitor, vitamin C, thiamine hydrochloride, hydrocortisone and auranofin, block the production of cytokines by reducing viral production and their attack on healthy cells. The Dihydrochloroquin and chloroquine phosphate, protease inhibitor, vitamin C by inhibiting the TNFα and interleukin 6 (IL6) production, block the subsequent cascade of events, which leads to inflammation of the lungs and other organs in the body leading to adult respiratory distress syndrome (ARDS). Angiotensin receptor blocker telmisartan and losartan also have such a blocking effect preventing the severe development of SARS.

Vitamin C, vitamin D3, vitamin B1 (thiamine hydrochloride in the form of Benfotiamine) to reduce viral infection and cytokine storm

In this invention we describe that all vulnerable patients in nursing homes, extended care facilities, and the aged living independently, and those ***suffering from coronavirus infection should get vitamin C (1000-3000-mgs), vitamin D3 (5000-10000 IU), thiamine hydrochloride in the form of Benfotiamine (250-500 mg)*** to improve immune system function and decrease the chances that an infected person will have a catastrophic cytokine storm, which may improve the outcome of COVID-19 cases and decrease the overall morbidity and mortality. During viral pandemics and flu seasons, these supplements are a must to beef up the immune system and can provide relief from the ravages and viral infections. One can also add zinc supplement to this regimen. All these are available over the counter, and do not need a doctor to prescribe it.

Figure 8, below, is the diagrammatic presentation 800 showing the effect of angiotensin receptor II blocker (ARBs) and angiotensin converting enzyme inhibitors (ACEi) on the respiratory system cell surface and how the coronavirus binding to the healthy tissue angiotensin receptor on the cell surface can be blocked by use of angiotensin receptor blocking drugs (ARBs) such as telmisartan and losartan, dihydrochloroquin and chloroquine phosphate, protease inhibitors such as remdesivir. Without such an attachment-docking facilitation, the coronavirus cannot infect the host healthy cells.

Though it is thought that the ARBs drugs block only type I of angiotensin receptors on the cell wall, we believe that as these type I and II angiotensin receptors are twin receptors situated next to each other like conjoined twins, hence the ARBs therapeutic agents also block partially or completely the Angiotensin receptors type II also thus inhibiting, delaying or blocking the coronavirus entering the healthy cell. That is why using ARBs during flu season, prevents the development of flu and inhabits the symptoms if one gets infected. Further, the ethylenediaminetetraacetic acid may change the receptor nature in such a way, thus the S-spike is unable to dock on to it tightly, thus block the attachment of coronavirus to angiotensin receptors to enter the host cell.

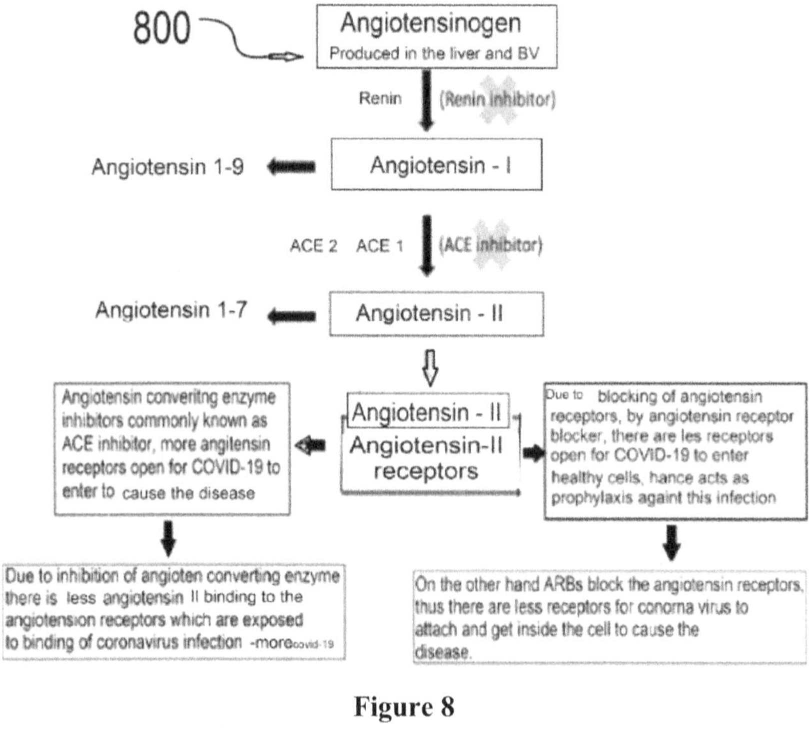

Figure 8

Figures 9 and 10, below, are diagrams 900 and 1000 of the spray 35 for nose and oral cavity showing sprayer in the nasal cavity above the superior surface of the hard palate 21, soft palate 22 and uvula 23 using the therapeutic agents in a spray 99 as described below can be delivered through the nose nozzle to deliver therapeutic agents to the nasal mucosa with anti-coronavirus viricidal and viral anti-adhering agent to prevent the virus passage to the rest of the respiratory system and olfactory mucosa during inspiration and expiration of the viral droplets. Content of the sprays are described above and can also contain antibody nano particles, monoclonal antibias, antibody containing serum obtained from COVID-19 patients. Therapeutic agents bind to the nasal as well as to olfactory mucosa and the spray contents prevent the coronavirus airborne in the inspired air. These sprays also block the coronavirus multiplication and entry into nasal, oral and olfactory mucosa and then to the CNS. Figure 10 is the diagram of another simple squirt nasal spray. The spray content and mouth wash we describe used as mouth wash and for nasal irrigation to counter coronavirus, and microbial infection and against the formation of biofilm on the oral and nasal respiratory passages.

We prepare saline and such diluent solution sprays mixed with specially prepared nanoparticles of coronavirus antibodies, or diluted plasma from patients recovered from the COVID-19 containing coronavirus antibodies, or diluting with monoclonal antibodies specific to the coronavirus, and sprayed (Figure 9) deep in to the throat and nasal cavity including olfactory mucosa when a person takes a deep breath to coat the entire trachea-bronchial tree and alveolar lining cell. The coronavirus antibodies will bind to coronavirus S-spikes to block the virus attaching to the respiratory tract lining cells and alveoli angiotensin receptors II and prevent the coronavirus binding and entering healthy lung cells, thus preventing the coronaviral infection and reduce the production of cytokines.

This prevents the pulmonary damage and cytokine storm. These biological agents to coat the respiratory system can also be delivered through the ventilator system, endotracheal tube, tracheostomy tubes, and oxygens rebreathing delivery system. Such coating of the lining of lungs can be repeated 2-3 times a day. It is an important method to prevent the coronavirus attack and attach to healthy lung lining cells, because these cells are rich in angiotensin receptors II to which virus tries to attach and create havoc. We can also use angiotensin receptor blockers telmisartan and losartan also to bloc these receptors and thus inhibits and or prevents lung infection especially elderly which is associated with high morbidity and mortality. This method can be used with ethylenediaminetetraacetic acid and superoxide dismutase to coat the respiratory system to prevent the virus attach to the respiratory mucosa.

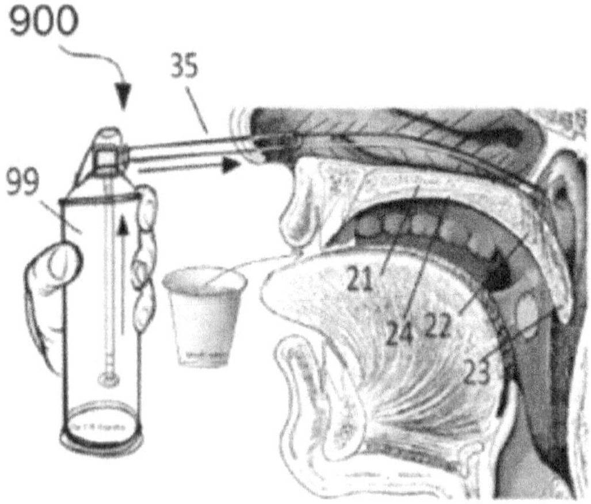

Figure 9

Figure 10

Figure 11, below, is the diagram 1100 showing the olfactory mucosa of the nose 78 on the nasal septum. It shows cranial nerves CN1 olfactory nerves form the olfactory mucosa entering the cribriform plate of the ethmoid bone and joining the olfactory bulb 82 inside the cranial cavity surrounded by cerebro-spinal fluid 79 which connects to the rest of cerebro-spinal fluid surrounding the brain. It shows CN- V (trigeminal nerve) also, which plays a minor or no role in transport of coronavirus to the olfactory bulb 82.

Note the coronavirus on the olfactory mucosa on the nose 40 can easily travel and be transported to the olfactory bulb 82 through sub perineural epithelial space surrounding the 20 olfactory nerves, and transported through cerebro spinal fluid around the olfactory bulb to the rest of brain to infect. The olfactory mucosa also transports many therapeutic agents directedly to the brain bypassing the blood brain barrier through the sub perineural epithelial space (Shantha TR Bypassing the BBB: Drug Delivery from the Olfactory Mucosa to the CNS. Drug development and delivery. 2017, Vol 17, # 1, 32-37.).

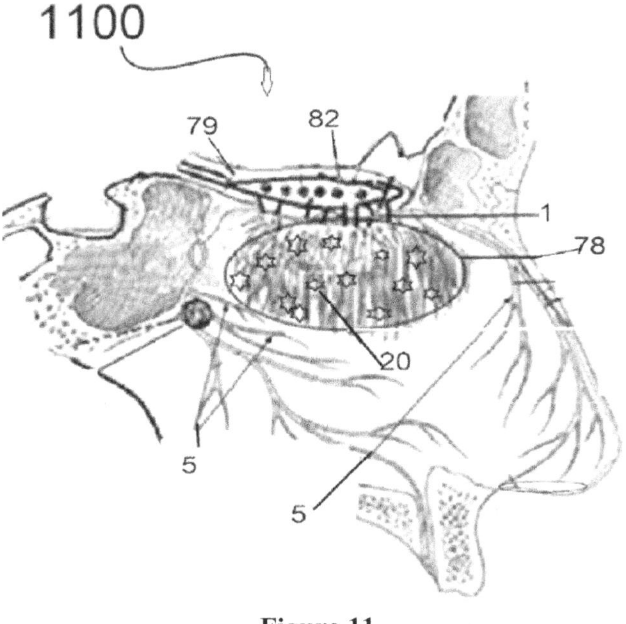

Figure 11

Figure 12, below, is the diagram 1200 showing the olfactory mucosa of the nose 78 and its histological structure. It shows how easy it is for coronavirus from olfactory mucosal 78 contact to be transported to the olfactory bulb from the olfactory mucosal lining 82 of the nose. Note the coronavirus 40 on the mucosa, which then passes between the receptor apical junctions 83, then passes on to sub perineural epithelial space 81 made of perineural epithelium 80 that surround each olfactory nerves fasciculi and allow the coronavirus to enter olfactory bulb and sub arachnoid space that surrounds the olfactory bulb with cerebro-spinal fluid. From here coronavirus has free accesses to spread through the cerebro-spinal fluid to any part of the brain through the Virchow-Robin space, paravascular route and aquaporins. Maybe the coronavirus binds to the chemoreceptors instead of angiotensin receptors. It does not need angiotensin receptors to latch on once inside the cerebro-spinal fluid pool.

There is constant movement of the cerebro-spinal fluid from olfactory mucosa and sub arachnoid space, hence, the coronaviruses are easily carried to the CNS and the olfactory bulb through this cerebro-spinal fluid waterway. Some of the coronavirus may enter the axon and the inter-axonal spaces that are carried to the olfactory bulb glomeruli, and multiply and cause inflammation ***resulting in loss of smell***. Note that the receptor cells are constantly dying and being in the processes of replacement with new cells, thus creating inter-cellular space between the olfactory mucosa receptor cells for easy entrance and transport of coronavirus to the central nervous system (CNS) to cause inflammation of neurological structures.

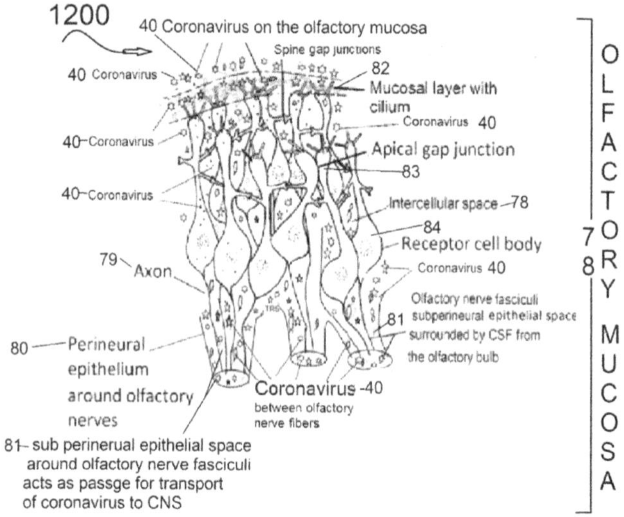

Figure 12

Figure 13, below, is the diagram 1300 showing the olfactory mucosa of the nose 78 on the nasal roof and upper walls with olfactory mucosa. It shows cranial nerves CN1 entering the cribriform plate of the ethmoid bone connecting and entering the olfactory bulb 82 which is surrounded by sub arachnoid space 79 and perineural epithelium 80 and sub perineural epithelial space 81 that communicates with the sub arachnoid space of the CNS, and the cerebro-spinal fluid communicated with fluid around the olfactory nerves CN-1 and sub arachnoid space around the olfactory bulb.

Hence, the coronavirus 40 attached to the olfactory mucosa's sticky mucus can eastly travel to the olfactory bulb 82 and infect the glomeruli 88, then then coronavirus can infect olfactory tracts and the rest of the brain through the cerebro-spinal fluid system through Virchow-Robin space and aquaporins, from the sub arachnoid space. The inflammation of the olfactory receptor cell 84 and glomeruli 88 contributes to loss of smell and this infection can be prevented by using nasal sprays described in this invention, thus save the neurological affliction of the brain such as encephalopathy.

Olfactory mucosal may or may not have angiotensin receptors and the infection of coronavirus spreads between the intercellular spaces, and the sub arachnoid space around the olfactory bulb then reach the neuronal component of the olfactory bulb and other parts of the brain to cause various signs and symptoms of CNS affliction. Loss of smell indicates that the coronavirus is already seeded in the olfactory mucosa, olfactory bulb and edging to travel to the rest of the nervous system (CNS) to cause various symptoms. It can also be due to cytokine affecting the CNS function.

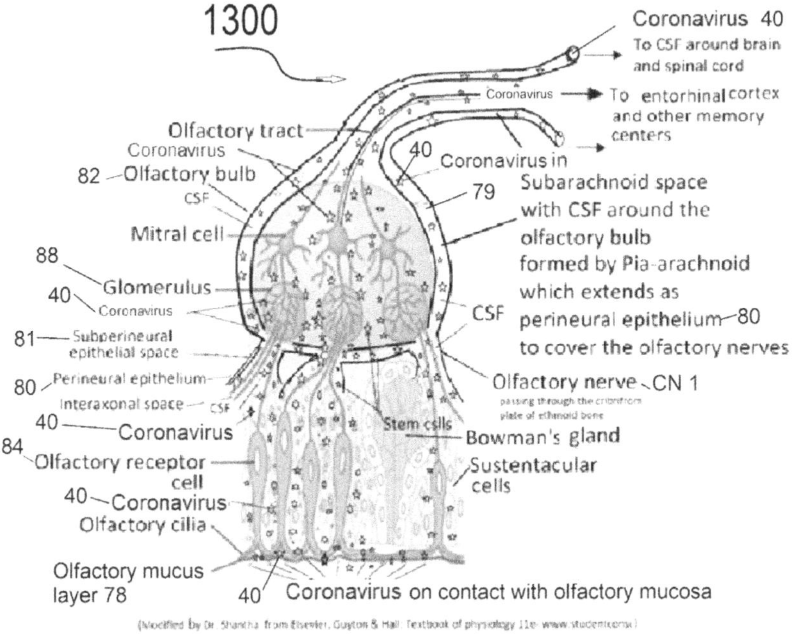

Figure 13

Figures 11, 12, 13 show the coronavirus and how easy for it to attack the nervous system besides lungs though the nose olfactory mucosa. Hence it is important to keep the inspired and expired air free of any coronaviruses and keep it away from the olfactory mucosal contact.

Figure 14, below, shows the effect of cytokine and the organs affected by 1400 due to coronavirus infection producing COVID-19. The figure shows that almost every organ in the body can be affected from minor to major degree depending upon the amount of coronavirus infection and the type of cytokines insult and the age of the patient. The figure shows the effect of coronavirus attack of the respiratory system 76, especially in the elderly with comorbidities resulting in high morbidity and mortality. Breathed in coronavirus in the lung alveoli attacks the blood vessels 19, 19a that supply the alveoli 18 that can result in producing SARS and ARDS.

Cytokines produced as a result of the coronavirus attack of the immune system and the general body 81 produce fever, fatigue, lack of appetite, headache, and pain all over the musculoskeletal system 80. As cytokine production by immune system and other cells to fight the virus continues, the cytokines affect the lungs 76, brain and the spinal cord 70, olfactory mucosa 71a, blood vessels 71, micro-vasculature of the alveoli 18, 19, 19a, heart 72, liver 73, kidneys 79, gastro-intestinal track 74, 75, stomach 78, spleen 77, producing organ related symptoms as described. Gallbladder and the pancreas are also affected by this coronavirus attack and cytokine produced by the immune system.

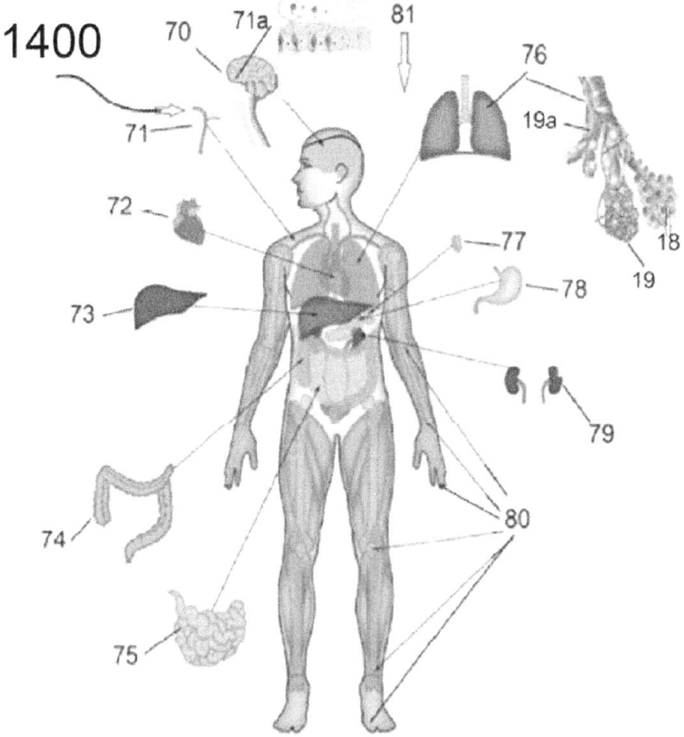

Figure 14

Chapter 22

"Cytokine release syndrome" (CRS) "cytokine storm" (CSS) and its effects on the lungs and rest of the body

Cytokines are the general category of messenger molecules (ILs, INFs, TNFs, CSFs and such) made up of small proteins which cannot cross the cell membrane to enter the cell cytoplasm. They are produced by broad range of cells, including immune cells like macrophages, B lymphocytes, T lymphocytes and mast cells, as well as endothelial cells, fibroblasts, and various stromal cells that act on cell surface receptors produced specifically by host immune responses due to infection, inflammation, trauma, sepsis, cancer, and reproduction.

Chemokines are a special type of cytokine that directs the migration of white blood cells to infected or damaged tissues as it happens in severe cases of COVID-19. On the other hand, Prostaglandins are lipid compounds-fatty acid derivatives, found in every tissue in the human body. They can cause alveolar capillary dilations and prevent platelet aggregation, thus allow more inflammatory fluid to leak into alveoli from leaking alveolar blood vessels during COVID-19.

Cytokine release syndrome (CRS) or cytokine storm syndrome (CSS) simply labelled as cytokine storm is a systemic inflammatory response syndrome (SIRS) that can be triggered by a variety of factors such as coronavirus infections, in addition to adoptive T-cell therapies to treat cancers. Severe CRS or cytokine reactions can occur in a number of infectious and non-infectious diseases including graft-versus-host disease (GVHD), acute respiratory distress syndrome (ARDS), sepsis, Ebola virus infection, avian influenza, smallpox, systemic inflammatory response syndrome (SIRS) and infusion of antibody based therapies, non-protein-based cancer drugs such as oxaliplatin, and as effect of some monoclonal antibody drugs infusion, also known as infusion reaction [Shimabukuro-Vornhagen et al. Cytokine release syndrome. Journal for ImmunoTherapy of Cancer, (2018) 6:56]. Severe cases have been called cytokine storms, and when occurring as a result of drug administration, it is also known as an infusion reaction.

In COVID-19 in the respiratory system, the storms occur when large numbers of white blood cells are activated by the coronavirus, or the white blood cells try to thwart the massive virus infection, with release of inflammatory cytokines, which in turn activate yet more white blood cells with production of more cytokines, and the vicious cycle sets in. The cytokines produced excessively start attacking the body besides viruses. As the battle between the immune system and coronavirus rages inside our body, it produces the following symptoms as labelled in Chapter 21, Figure 14:

1. 70 CNS: moderately severe headache, delirium, confusion, hallucinations, aphasia, epileptic seizures.
2. 71a olfactory mucosa and olfactory bulb coronavirus infection leasing to the loss of smell, a diagnostic symptom of COVID-19.
3. 71 Blood: cytopenia, coagulopathy (PTT up, INR down), elevated ESR
4. 72 Heart: tachycardia, hypotension, troponin elevation, arrhythmias, QT prolongation, cardiomyopathy, acute heart failure
5. 73 Liver: enlarged, elevated liver enzymes, hyperfibrinogenemia, finally liver failure,
6. 74 Colon: pain, diarrhea
7. 75 Small intestines: colic, pain, bloating, diarrhea
8. 76 Lungs: tachypnea, hypoxia, hypercarbia, pulmonary edema, respiratory failure.
9. 77: spleen: enlarged, tender to touch
10. 78 Stomach: nausea, vomiting
11. 79 Kidney: Acute kidney injury, renal failure
12. 80 Musculo-skeletal system: muscle and joint pain, rigors, rash, swelling, blue toe, lips and finger tips
13. 81 Unspecific symptoms: fever, fatigue, anorexia, headache (due to air sinus infections), ear ache, loss of smell and taste

Cytokine storm is graded and treatment varies on the grade:

1. Grade 1: Fever, constitutional symptoms, treated like other ailments of cold or flu, low grade CRS is treated symptomatically with antihistamines, antipyretics, antibiotics, and fluids.
2. Grade 2: Hypotension treatable by fluids intravenous, vasopressors
3. Grade 3: Shock requiring high dose multiple vasopressors, hypoxia requiring supplemental oxygen 40%, organ toxicities, elevated transaminases in the blood, treated in ICU.

4. Grade 4: life-threatening situation requires prompt and aggressive treatment. Mechanical ventilation, organ failure with high transaminases, needs ICU management. This is noticed mostly in the elderly with co-morbidities in the nursing homes or extended care facilities high morbidity and mortality.

COVID-19 of the lungs is due to the coronavirus antigen biding to the immune system cells receptor and its subsequent activation which also activates bystander immune cells and non-immune cells, such as endothelial cells of the blood vessels that surround the lung alveoli results in the massive release of a wide range of cytokines which enter the pulmonary veins and the right heart distributed all over the body. The results are catastrophic as shown in Chapter 21, Figure 14. IL-6, IL-10, and interferon (IFN)-Y are among the core cytokines that are consistently found to be elevated in serum of patients with CRS. In additional cytokines such as IL-18, IL8, IP10, MCP1, MIG, and MIP1β are also elevated (Teachey DT, Lacey SF, Shaw PA, Melenhorst JJ, Maude SL, Frey N, et al. Identification of predictive biomarkers for cytokine release syndrome after chimeric antigen receptor T cell therapy for acute lymphoblastic leukemia. Cancer Discov. 2016). Inhibiting the massive immune system white blood cell response due to coronavirus infection, by administration of *rapamycin* and maybe even administration of ivermectin or Ebselen, may prevent the development of ARDS and cytokine storm if given early as the disease is taking a turn.

Ebselen is a potent protease inhibitor and anti-oxidant. It should be prescribed for those in nursing homes, extended care facilities, and the aged living independently to prevent coronavirus infection and to tame down or inhibit the development of comorbidities in the aging population, that can lead to increased morbidity and mortality. Ebselen prevents cell death (apoptosis) caused by injury or toxic exposure by blocking proteins involved in apoptosis (thiol groups).

Amazingly, Ebselen crosses the blood-brain barrier, that has an immense effect on the health of the CNS. It blocks an enzyme in the brain involved in the breakdown of acetylcholine (acetylcholinesterase), thus improving cognitive impairment and memory loss. It also decreases the production of proteins involved in Alzheimer's disease (amyloid precursor protein and β-secretase) and reduces toxic protein accumulation (amyloid plaques) in the brain. Ebselen should be prescribed for all neurodegenerative disease patients including Alzheimer's and Parkinson's. Increases serotonin production, which acts as an antidepressant especially in the aging population with comorbidities.

We want to use Ebselen in nasal and oral sprays, mouth wash along with other therapeutic agents for prophylaxis described in the invention examples. It also acts against biofilm of oral and nasal cavities, the rest of respiratory system, esophageal and gastro-intestinal track.

We advise it be taken orally between 400-600 mg twice daily once diagnosed with COVID-19. It could be a life saver, a chemical compound with anti-viral, anti-inflammatory, anti-oxidative, bactericidal, and cell-protective properties.

To make things worse, after the first infection, these patients develop a secondary infection that occurs in patients with CSS, predominantly of bacterial (could be fungal) origin, primarily involving the respiratory tract, treated with azithromycin. A plausible explanation for infection could be that the massive release of cytokines in CRS induces a form of immune paralysis, which predisposes the patients to an increased risk of infection.

The treatment for CSS is described below in the claims. Studies show that administration of monoclonal antibodies against IL-6 (siltuximab) and its receptor (tocilizumab) led to rapid resolution of CRS symptoms [Lee DW, Gardner R, Porter DL, Louis CU, Ahmed N, Jensen M, et al. Current concepts in the diagnosis and management of cytokine release syndrome. Blood. 2014; 124:188–195.]. Due to 69% positive treatment response of CRS, tocilizumab has quickly become the gold standard for the initial treatment of severe CRS. FDA has approved Tocilizumab for the treatment of CRS in patients 2 years of age or older.

It is the intent of this invention to provide high doses of **vitamin C, vitamin D3, vitamin B1 (thiamine hydrochloride in the form of Benfotiamine), dexamethasone, and zinc (may be Ebselen, auranofin) to reduce viral infection pathological effects including cytokine storm.** These supplements improve immune system function, reduce the capillary leak around the alveoli and decrease the probabilities that an infected person will have a catastrophic outcome (SARS-ARDS) and at the same time decrease the COVID-19 related morbidity and mortality. During viral pandemics and flu seasons, these supplements are a must to beef up the immune system to all the inhabitants of the nursing homes, extended care facilities, and the aged living independently as a prophylactic against microbial infections, and to provide protection from the coronavirus aggressive action and catastrophic outcome. These lifesaving vitamin supplements (vitamin C, thiamine hydrochloride, vitamin D3, zinc) are overlooked by most care providers for the COVID-19 afflicted. They should be part of the protocol for these patients.

Chapter 23

Method of prophylaxis and treatment for coronavirus

Our method of prophylaxis and treatment for coronavirus infection

Coronaviruses are highly contagious and virulent. So far, the coronavirus has shown no mutation, but with lapse of time it will develop mutation. We do not know if the mutated coronavirus is less virulent or more virulent. The elderly past age 60-70-80 with prevailing health conditions are at the greatest risk of increased morbidity and mortality compared to young. As of now there is not a vaccine or therapeutic agents to prevent COVID-19 and to treat if a person develops symptoms. The recent pandemic has shown that knowledge about this virus is lacking regarding clinical management and treatment.

Remdesivir, oseltamivir, foscarnet, intravenous immunoglobulin, antibodies from patient suffered from this COVID-19, and other drugs such as dihydrochloroquin and chloroquine phosphate, telmisartan and losartan, vitamin C and such have been used to treat patients. Now the dihydrochloroquin and chloroquine phosphate have appeared as treatments also. Unfortunately, no effort has been made to test vitamin C for treating mild to severe cases of COVID-19 so also EDTA. In the past since 1948, vitamin C has been used to treat various viral infections including flu, measles, mumps, herpes, polio and so on. We also have used it to treat infections to cancers. In the invention we want to explore and describe our methods of prophylaxis and treatment of patients afflicted with this coronavirus using various therapeutic agents including vitamin C. This method we describe can be easily adopted for other virus epidemics.

Instead of finding a single drug to treat the coronavirus, I would recommend a cocktail of drugs that targets different stages of coronavirus life cycle starting from initial attachment of the coronavirus to healthy cell to is replication and budding for propagation from the host cells. This virus is probably going to be like a number of other viruses: it will undergo mutation and selection, so if you use only one antiviral drug, you are going to ultimately select another for resistance developed virus. The treatment will be most effective when given to a patient early on, perhaps even before symptoms develop. Taken very early in the course of exposure, the antiviral drugs could have a real impact on this virus infection and its resulting disease—production of signs and symptoms. After someone is already in the hospital in respiratory distress and a high fever, we propose high dose vitamin C, thiamine hydrochloride, dihydrochloroquin/chloroquine phosphate, protease inhibitor remdesivir, zinc, angiotensin receptor blockers telmisartan and losartan, and hydrocortisone with or without azithromycin. Maybe even chelation using ethylenediaminetetraacetic acid (EDTA).

Treating COVID-19 means lessening the symptoms by providing respiratory support, maintaining proper oxygen levels in the blood, lowering the fever and making sure patients are hydrated while at the same time maintaining the vital signs in ICU among other things such as intubation, ventilation, nor-rebreathing oxygen masks use, and as well as kill the coronavirus that is causing the COVID-19 as they are budding from the infected cell surface creating lung and capillary bed injury by viral load and bringing cytokine storm. The vast majority of coronavirus patients today are receiving some of the treatment like patch work, but not all the treatment we discuss in this invention.

The aim of the invention is to provide combination of antiviral therapeutic agents that exhibit anti–SARS-CoV activity to prevent COVID-19:

1. to have direct antiviral effects as they enter the nasal, oral, respiratory passages to prevent their attachment to healthy cells with aerosolized medicaments such as superoxide dismutase and angiotensin receptor blockers, viricidal mouth washes and sprays, antibody nanoparticles, and antibody serum from the COVID-19 affected people.
2. to inhibit coronavirus entry by attaching to healthy cells, and replication within the host cells, and budding of virions from the host cells to spread to other healthy cells.
3. to provide proper care in the hospital and in the ICU settings to maintain vital signs, prevent hypoxia, prevent and / or inhibit the cytokine storm, viral and secondary bacterial attack of lungs responsible for lung affliction that increases morbidity and mortality.
4. to provide post COVID-19 care to maintain good health without post coronavirus and microbial infections related complications, and at the same time treat other underlying health conditions that might have led to the severity of COVID-19 including post COVID-19 depression, weariness, and exhaustion of the patients and care givers.

Now we describe and discuss the inventive methods to achieve the above goals.

Chapter 24

Ebselen, remdesivir and other protease inhibitors

Ebselen: Potent main protease inhibitor (Mpro) of coronavirus

Ebselen is a research substance that contains selenium and is touted as one of the therapeutic agents against coronavirus, other viruses and multiple systemic diseases. Ebselen is a Glutathione peroxidase mimic; peroxynitrite scavenger (as well as Disulfiram, Tideglusib, Carmofure, Shikonin, PX-12.). It also inhibits ferroptosis, lipoxygenase, cyclooxygenase, nitric oxide synthase, protein kinase C and H+/K+-ATPase activity. It inhibits the hepatic carcinogenic effects of aflatoxin B1. It is an antioxidant and anti-inflammatory as well as acts against biofilm coating of the oropharynx. It selectively inhibits Gq protein signaling and enhances differentiation of brown adipocytes. Most of all, it inhibits SARS-CoV-2 Mpro (main protease) in vitro (IC50 = 0.67 μM in FRET assay). It has the brightest future to become one of the most sought after and widely prescribed therapeutic agents/supplement to treat comorbidities and extend longevity.

Mechanism of action

Oxidants (hydroxyl radicals and hydrogen peroxide) in our body ravage 24/7/365 and can damage proteins, fats, and DNA contributing to many acute and chronic diseases. This activity ***speeds up aging*** and leads to age-related diseases, including stroke, heart disease, Parkinson's, Alzheimer's disease, cancers and such. Ebselen blocks the activity of oxidants and reduces cell damage via multiple mechanisms:

1. It mimics the activity of an antioxidant enzyme (glutathione peroxidase) in our body: Ebselen converts harmful products made by oxidants back to harmless products (such as converting selenic acid to selenol). It also converts oxidants, such as hydrogen peroxide to water.
2. It prevents cell death (apoptosis) caused by injury or toxic exposure by blocking proteins involved in apoptosis (thiol groups).
3. It crosses the blood brain barrier and blocks an enzyme in the brain involved in the breakdown of acetylcholine (acetylcholinesterase),

which might improve memory. It also decreases the production of proteins involved in Alzheimer's disease (amyloid precursor protein and β-secretase) and reduces toxic protein accumulation (amyloid plaques) in the brain, Ebselen should be prescribed for all neurodegenerative disease patients including Alzheimer's and Parkinson's. It increases serotonin production, acting as an antidepressant.

4. It irreversibly blocks an enzyme involved in cell communication and growth (Inositol Monophosphatase) in a similar way to lithium (approved treatment for bipolar disorder).
5. It prevents the spread of brain cancer cells (glioblastoma cells).
6. It decreases the production of pro-inflammatory compounds (cytokines IL-6, IL-8)
7. It prevents iron overload in the heart, a major cause of heart failure.
8. Membrane protases (Mpro) facilitate the virus' ability to make proteins from its genetic material—RNA—and enable the pathogen to replicate within the host cell, a process blocked by Ebselen.
9. Ebselen is able to decrease Mpro's activity in two different ways: "In addition to binding at the 1. catalytic site of the enzyme, 2. Ebselen also binds strongly to a distant site, which interferes with the enzyme's catalytic function by relying on a mechanism in which information is carried from one region of a large molecule to another region far away from it through subtle structural reorganizations." (Dr. Drag of Chicago)

Potential uses of Ebselen

This list of its benefits keeps on growing as it becomes more and more popular. It is used for stroke, bipolar patients without toxicity like lithium in normal doses, noise-induced hearing loss, decreased impulsive actions, both prevention and treatment of Alzheimer's, arthritis, for decreasing PMN leukocytes and T lymphocytes, osteoporosis (reduces bone breakdown), rescues complications of diabetes (arteriosclerotic vascular disease and kidney damage), cancers (58% reduction in pancreatic cancers), reduces chemotherapy side effects, reduce inflammation of lungs, and its fluid buildup as seen in COVID-19 ARDS, reduces alcohol damage of the liver, reduces the manganese and cadmium induced testicular damage, effective in treating drug resistance bacterial infection (MRSA) better than present antibiotics and many such therapeutic effects to prevent, curtail, stop the progression of and / or cure the diseases will be added to the list. Besides, it is an effective anti-aging supplement, along with rapamycin, lithium, metformin, bioactive curcumin and such.

Now we have Ebselen as an anti-coronavirus agent and treatment for COVID-19

Coronavirus's main protease enzyme Mpro is a key enzyme that plays a central role in its life cycle. The enzyme cuts through proteins and allows the pathogen to survive, multiply and spread. Professor Drag of Chicago University said: "It allows it to survive. Stopping the enzyme immediately causes the virus to die." The enzyme studied by Professor Drag and his team is a so-called protease, dubbed SARS-CoV-2 Mpro.

Study co-author Juan de Pablo from the University of Chicago explained this activity: Protease enzymes inhibitor Ebselen is responsible for cutting down proteins into smaller particles and allowing the virus to perish. Ebselen in addition to binding at the catalytic site of the enzyme, also binds strongly to a distant site, which interferes with the enzyme's catalytic function by relying on a mechanism in which information is carried from one region of a large molecule to another region far away from it through subtle structural reorganizations. Ebselen is a chemical compound with anti-viral, anti-inflammatory, anti-oxidative, bactericidal, and cell-protective properties. It is used to treat multiple diseases, including bipolar disorders and hearing loss.

We want to use Ebselen in nasal and oral sprays, and mouth wash for a prophylaxis. We advise it be taken orally between 400-600 mg twice daily once diagnosed with COVID-19. ***It could be a life saver, a chemical compound with anti-viral, anti-inflammatory, anti-oxidative, bactericidal, and cell-protective properties.***

Remdesivir and other protease inhibitors to prevent biding of COVID to host cells

Remdesivir (RDV) is a type of broad-spectrum antiviral medication called a nucleotide analog. It is currently a drug being tested for COVID-19, now extensively used in hospital settings. Coronavirus is an RNA virus. (RNA is the molecular transcription tool organisms use to build proteins using DNA instructions). RNA viruses are dependent on an RNA polymerase enzyme to grow the RNA chain. Remdesivir substitutes this RNA polymerase enzyme, meaning the RNA can't develop, so the virus cannot replicate itself, spread and become infectious (Chapter 21, Figures 5, 6).

Protease inhibitors are synthetic drugs that inhibit the action of HIV-1 protease, an enzyme that cleaves two precursor proteins into smaller fragments. These fragments are needed for viral growth, infectivity and replication by proteases. Protease inhibitors bind to the active site of the protease enzyme and prevent the maturation of the newly produced virions so that they remain non-infectious. Now with COVID-19, protease inhibitors are going to play a therapeutic agents' role to prevent the multiplication of RNA virus into small fragments for viral growth, infectivity and replications. There are many protease inhibitors in the market already; they can be used as a spray to prevent inhaled virus attaching and preventing their multiplication besides inhibiting the viral multiplication inside the infected host cells. They also play a role in preventing the virus attaching to the angiotensin receptors on the host cells due to structural change as they come in contact with the protease inhibitor.

We know that combinations of protease and other inhibitors in one pill have worked against HIV and made the patients to zero load of viruses in the blood test, with restoration of T4 immune cells, though it still cannot eliminate the hidden virus in the lymph nodes, spleen, liver and such (hiding places). As coronavirus is an RNA virus like HIV, similar viral clearance/inhibition of multiplication coronavirus also takes place. Unlike HIV, coronavirus virus ***has no hiding place*** to produce lifelong chronic disease. We propose using HIV drug remdesivir or Kaletra®, which is a combination of antivirals lopinavir and ritonavir which could be effective in treating COVID-19 along with dihydrochloroquin and chloroquine, high dose vitamin C, thiamine hydrochloride and dexamethasone as well as angiotensin receptor blockers and potent anti-inflammatory auranofin.

This inventive new protocol will stop the coronavirus multiplication, production of various cytokines (produced by immune system to kill the coronavirus) that damage the lungs and blood vessels and making them leaky, stabilize the endothelium of blood vessels surrounding the alveoli and prevent their leaking of the plasma and white cells inside the alveoli. It is this pathology that is responsible for death due to COVID-19 by creating severe acute respiratory distress syndrome (SARS, ARDS).

This method of treatment with other therapeutic agents discussed below will reduce the hypoxia, and kill the coronavirus inside the host cell as they are liberated by budding to infect another cell. That is why high dose vitamin C is very effective in treatment of all phases of coronavirus infection acting as an antioxidant and producing hydrogen peroxide which is viricidal, thus, reducing severe morbidity and mortality associated with COVID-19. The combination of angiotensin receptor II blockers telmisartan and losartan, protease inhibitor remdesivir, high dose vitamin C, thiamine hydrochloride and hydrocortisone, vitamin D3, auranofin, and bioactive curcumin act as antioxidants and reduce the free radicals generated by the coronavirus infection that damage the lung parenchyma, thus acting as an anti-inflammatory and viricidal.

Proteases enzymes are needed for viruses to enter and to replicate, and acid media is needed for virus particle to form and assemble in the Golgi-Lysosome complex inside the cell (Chapter 21, Figure 6) which are inhibited with changing of pH in the Golgi-lysosome complex by protease inhibitor and dihydrochloroquin / chloroquine phosphate.

A team of academic and industry researchers has reported new findings on how exactly an investigational antiviral drug stops coronavirus from replicating. The paper was published the same day that the US National Institutes of Health (NIH) announced that the drug in question, remdesivir, is being used in the country's first clinical trial of an experimental treatment for COVID-19, the illness caused by the SARS-CoV-2 virus. Now it is approved by FDA to be used on all needed COVID-19 patients.

Previous research in cell cultures and animal models has shown that remdesivir can block replication of a variety of coronaviruses, but until now it has not been clear how it does so. The researchers, from the University of Alberta, US, and Gilead, studied the drug's effects on the coronavirus that causes Middle East Respiratory Syndrome (MERS). They found that *remdesivir blocks a particular enzyme that is required for viral replication.* Coronaviruses replicate by copying their genetic material using an enzyme known as the *RNA-dependent RNA polymerase* (Chapter 21, Figure 6 #84, 82, 93, 98).

Matthias Götte, a virologist and professor at the University of Alberta, found that the enzymes can incorporate remdesivir, which resembles an RNA building block, into new RNA strands. Shortly after adding remdesivir, the enzyme stops being able to add more RNA subunits. This halts genome replication. RNA containing remdesivir takes on a strange shape that does not fit into the enzyme—hence viral replications ceases. They suggest the viral RNA polymerase of coronaviruses as a target in finding therapeutic agents to prevent the coronavirus infection. The study was published in the Journal of Biological Chemistry. I believe a similar mode of action with other protease inhibitors with slight variation will be successful.

Protease inhibitors are widely used in the treatment of human immunodeficiency virus (HIV infection) and acquired immune deficiency syndrome (AIDS). Remdesivir was developed to treat Ebola virus infection. It is an ***adenosine analogue,*** which incorporates into nascent viral RNA chains and causes their pre-mature termination of their reproduction factory. It was developed by Gilead Sciences as a treatment for Ebola virus disease and Marburg virus infections [Warren TK, Jordan R, Lo MK, Ray AS, Mackman RL, Soloveva V, et al. (March 2016). "Therapeutic efficacy of the small molecule GS-5734 against Ebola virus in rhesus monkeys". Nature. 531 (7594): 381–5.] though it subsequently was found to show antiviral activity against other single stranded RNA viruses such as respiratory syncytial virus, Junin virus, Lassa fever virus, Nipah virus, Hendra virus, and the coronaviruses (including MERS and SARS viruses).

It is being studied for SARS-CoV-2 and Henipa virus infections (Brunk D. "Remdesivir Under Study as Treatment for Novel Coronavirus". Medscape. Retrieved 11 February 2020. Holshue ML, DeBolt C, Lindquist S, Lofy KH, Wiesman J, Bruce H, et al. (Washington State 2019-nCoV Case Investigation Team) (March 2020). "First Case of 2019 Novel Coronavirus in the United States". The New England Journal of Medicine. 382 (10): 929–936). Based on success against other coronavirus infections, Gilead provided remdesivir to physicians who treated an American patient in Snohomish County, Washington in 2020, who was infected with SARS-CoV-2, and is providing the compound to China to conduct a pair of trials in infected individuals with and without severe symptoms.

Yuen Kwok-Yung, a microbiologist at the University of Hong Kong agrees that *remdesivir is the most promising drug for Covid-19 and MERS.* He is now calling for the drug to be available in China and would also like to test lopinavir and ritonavir in combination with interferon beta-1b in randomized, controlled studies. We want to add remdesivir and other *newly formulated super-monoclonal antibodies* to block the protease and prevent binding of the spike to angiotensin receptors on the cell and thus prevent the virus entry into the healthy cell as well besides inhibiting the viral replication inside the host cell.

Protease inhibitors are already in widespread use in the treatment of HIV infection and they are FDA approved. Saudi Arabia now has a study under way in which patients with the Middle East respiratory syndrome (MERS), a distant relative of COVID-19, receive either the *lopinavir and ritonavir* combination plus interferon beta-1b, which boosts immune responses. They have also tested patients with experimental drug remdesivir, that interferes with the viral polymerase enzyme, and have recorded positive results.

Studies show that Kinase inhibitors such as Gilteritinib and Ralimetinib, required lower concentrations than remdesivir in order to kill off 50% of the virus which may be used to treat COVID-19 and need further studies.

Initial spike (Chapter 21, Figures 2, 3, 4, 6) protein priming by transmembrane protease, serine 2(TMPRSS2) is essential for entry of SARS-CoV-2 [Hoffman M, et al. April 2020. "SARS-CoV-2 Cell Entry Depends on ACE2 and TMPRSS2 and Is Blocked by a Clinically Proven Protease Inhibitor" [Cell. 181: 1–10.]. After a SARS-CoV-2 virion attaches to a target cell, the cell's protease TMPRSS2 cuts open the spike protein of the virus, exposing a fusion peptide. Remdesivir can block spike protein binding to cell membrane and the entry of coronavirus inside the host cells to multiply. The virion then releases RNA into the cell, forcing the cell to produce copies of the virus that are disseminated to infect more cells ["Anatomy of a Killer: Understanding SARS-CoV-2 and the drugs that might lessen its power". The Economist. March 12 2020. Archived from the original on March 14 2020. Retrieved March 14 2020.]. SARS-CoV-2 produces at least three virulence factors that promote shedding of new virions from host cells and inhibit immune response [Wu C, et al. February 2020). "Analysis of therapeutic targets for SARS-CoV-2 and discovery of potential drugs by computational methods". Acta Pharmaceutica Sinica B. doi:10.1016/j. apsb.2020.02.008].

In late April 2020, Dr. Fauci revealed preliminary results from the NIH trial showing remdesivir shortened the time to recovery to 31%, 11 days on average versus 15 days for those just given usual care.

A viral entryway facilitated by protease of the host cell membrane blocked by Camostat, a serene protease inhibitor

For the coronavirus to infect a human host, the virus must be able to gain entry into individual human cells. They use these host cells' machinery to produce copies of themselves, which then spill out and spread to new cells. On Feb. 19, 2020 in the journal Science, a research team led by scientists at the University of Texas at Austin described the tiny molecular key on SARS-CoV-2 that gives the virus entry into the cell. This key is called a spike protein, or S-protein. Zhou and his team in 2020 described the rest of the puzzle: the structure of the ACE2 receptor protein (which is on the surfaces of respiratory cells) and how it and the spike protein interact. The researchers published their findings in the journal Science on March 4, 2020.

Employing a pathogenic animal model of SARS-CoV infection, researchers demonstrated that viral spread and pathogenesis of SARS-CoV is driven by serine rather than cysteine proteases and can be effectively prevented by camostat, used to treat chronic pancreatitis, and thus represents an exciting potential therapeutic modality for COVID-19. Serine protease enzymes have a variety of functions in the body, and so camostat has a diverse range of uses. Camostat is, in Japan, approved for treatment of postoperative reflux esophagitis. It is used in the treatment of some cancers and against viral infections, inhibiting fibrosis in liver or kidney disease or pancreatitis.

It is an inhibitor of the enzyme transmembrane protease (Chapter 21, Figures 4-6), serine 2 (TMPRSS2). Camostat mesylate is an orally active protease inhibitor. It is known to inhibit trypsin and various inflammatory proteases including plasmin, kallikrein and thrombin. Inhibition of TMPRSS2 partially blocked infection by SARS-CoV and human coronavirus NL63 in HeLa cells. Another in vitro study showed that Camostat reduces significantly the infection of Calu-3 lung cells by SARS-CoV-2, the virus responsible for COVID-19 (Hoffman, Markus (2020-03-05). "SARS-CoV-2 Cell Entry Depends on ACE2 and TMPRSS2 and Is Blocked by a Clinically Proven Protease Inhibitor". Cell. Retrieved 2020-03-05.). For chronic pancreatitis, Camostat's typical dose is 600 mg daily, for postoperative reflux esophagitis 300 mg are taken. The daily dose is split in 3 doses and taken after each meal. Its application for COVID-19 is still not determined.

Chapter 25

Monoclonal antibodies and COVID-19

Monoclonal antibodies and COVID-19: for prophylaxis and treatment of coronavirus

Now attempts have been made to block the virus by using monoclonal antibodies that binds to coronavirus to prevent the coronavirus activity to the healthy cells and maybe even within our bodies making it ineffective. Andre Brandli (17 March, 2020, Derek report), describes a human monoclonal antibody known as 47D11 that was found to bind to SARS-CoV-2 and SARS-CoV and potently inhibit the virus' infection of Vero cells. Erasmus MC of Utrecht University, First Report of Human Monoclonal Antibody That Blocks SARS-CoV—they found an antibody against COVID-19. The scientific publication of the group of ten scientists is ready for assessment by the leading journal (Nature https://www.erasmusmagazine.nl/en/2020/03/14/unique-discovery-in-erasmus-mc-antibody-against-corona/ Reference: Wang et al. (2020). A human monoclonal antibody blocking SARS-CoV-2 infection. bioRxiv).

Antibody-based treatments with monoclonal antibodies makes them very valuable tools for designing specific antibody treatments to infectious agents. These attributes have already caused a revolution in new antibody-based treatments in oncology and inflammatory diseases, with many approved products on the way. The high cost of monoclonal antibody therapies, the need for parallel development of diagnostics, and the relatively small markets are major barriers for their development in the presence of cheap antibiotics. So far, only one monoclonal antibody, ***palivizumab,*** for the prevention and treatment of respiratory syncytial virus, is approved for infectious diseases. With lapse of time there will be an anti-coronavirus monoclonal antibody for the treatment of COVID-19.

Biocon Biologics pharmaceuticals has received approval from DCGI for use of its biologic psoriasis drug ***Itolizumab***, repurposing its use to treat moderate to severe COVID-19 patients. Its use as part of the treatment had no mortalities, compared to the control group with many fatalities, according to the producer of these therapeutic agents. This monoclonal antibody most likely acts against the hyperactive immune system and downgrades their excessive cytokine production, thus reducing the cytokine release syndrome effects; saving the lungs and heart injury thus allowing the patients to recover form ill effects of coronavirus infection precipitously, with least morbidity and no mortality.

In severe cases, with massive inflammation of the blood and tissue due to virus and cytokines, Biocon Biologics CytoSorb blood cleaning method may be tried to lower the viral and cytokine load as described below.

Active coronavirus specific antibody moiety extracted from convalescent plasma, attached to the monoclonal antibody and delivered to attack coronavirus as prophylaxis and treatment

It has been shown that the antibody containing serum/plasma taken from the coronavirus recovered/convalescent plasma has reversed and saved many lives threatened by COVID-19 in the ICU and hospital. It is passive immune therapy and effects do not last longer. This therapy is not new and was used by Behring for diphtheria in the year 1890, and has been used throughout medical practice on and off. Now we have COVID-19 for which it touted to be used in severe cases (Behring EA, Kitasato S. 1890. Ueber das zustandekommen der diptherie-immunität und der tetanus-immunität bei thieren. Deutch Med Woch 49:1113–1114. Casadevall A. 1996. Antibody-based therapies for emerging infectious diseases. Emerg Infect Dis 2:200–208. Hey A. History and Practice: Antibodies in Infectious Diseases. Microbiol Spectr. 2015 Apr;3(2): AID-0026-2014. doi: 10.1128/ microbiolspec. AID-0026-2014.).

Antibodies and passive antibody therapy in the treatment of infectious diseases is the story of a treatment concept which dates back more than 130 years, to the 1890s, when the use of serum from immunized animals provided the first effective treatment options against infections with Clostridium tetani and Corynebacterium diphtheriae. Convalescent plasma has previously been used against viral illnesses such as rabies, hepatitis B, polio, measles, influenza and Ebola. It was also used in the outbreaks of MERS and SARS-1, where faster viral clearance following convalescent plasma therapy was observed. Important aspect of convalescent plasma therapy is that the antibodies from convalescent plasma will suppress viremia (the presence of viruses in the blood) by mopping up the inciting viral antigens.

Now (2020) intravenous antibody infusion of plasma from coronavirus recovered patients past 28 days who have tested positive for antibodies, is used in ICU and to those patients who are fighting to recover in many hospitals all over US and the world. FDA approved the use of convalescent plasma for treatment of COVID-19 on August 23, 2020. "I don't want you to gloss over this number," said Dr. Stephen M. Hahn, commissioner of the FDA, who insisted that 35 out of very 100 hospitalized COVID-19 patients "would have been saved because of the administration of plasma."

We believe that isolating the most active antibody moiety from the plasma from the coronavirus recovered, then isolating the specific antibody components, attach it to the monoclonal body, make it a specific *monoclonal coronavirus antibody*, and administer it parenterally will prevent the attack of this virus specifically and kill the coronavirus during outbreak season specially to the vulnerable population. Such a preparation can be administered to lower the coronavirus load, who are already infected, as a treatment. By this method, we can eliminate the infection or convert the infection specially in the elderly to a minor flu like disease.

I believe such a biologic preparation can be used as a prophylactic treatment during the flu-coronavirus season in the future. The immunity and protection which it provides may not be long lasting, maybe just months. Once the coronavirus season is gone, it has served its purpose and saved many lives especially aged with comorbidities from severe morbidity and mortality. It also can prevent hospitalization with escalating costs, and post intensive care – intubations and complication. It may also provide protection against blood clotting and its related embolic phenomenon that affects COVID-19 patients as well as lung scarring, kidney failure, stroke, and cardio-vascular episodes. If this method is successful, it can be prepared to treat many other diseases, which produce antibodies besides COVID-19, such as rabies, genetic disorders, viral and bacterial infections, autoimmune diseases including type I diabetes, neurodegenerative diseases and such. *This method of treatment can be developed to any and all antibody producing diseases and such. The antibody spray can be used to coat the lungs to prevent the effect of cytokines and cytokines storm, can be a prophylactic against coronavirus infection.*

Nasal and mouth sprays, and mouth washes to prevent coronavirus seeding in the oral cavity, gums, cheek, fauces, and oropharynx, and be used in the nose wash also using Navage nasal irritation devices

There are multiple antimicrobials vs. antiseptic vs. antibacterial mouthwashes in the market and none of them prevent the infection of the coronavirus inside the healthy cells. Each one is labeled slightly differently, promising different benefits such as "cavity-fighting," "antibacterial," "antiseptic," "breath-freshening" and so on. The Centers for Disease Control and Prevention defines antimicrobial products as those that are designed to kill or inactivate various kinds of microbes, which include fungi, bacteria, parasites and viruses. Antibacterial agents (also known as antibiotics) kill, slow down or inactivate bacteria specifically. As Merck Manuals explains, the terms "antibacterial" and "antibiotic" are often used interchangeably. In contrast, antiseptic products are typically spread over a specific area of the body to reduce the risk of infection, according to the Microbiology Society.

Most of the mouth washes contain Chlorhexidine gluconate germicidal, alcohol, fluoride, hydrogen peroxide, methyl salicylate, antifungal nystatin, povidone -iodine, toxic alkaloids such as sanguinarine, sodium bicarbonate, salt, sodium lauryl sulfate a foaming agent, antibiotics, the list is endless, and they decrease bacteria in the mouth. None of these mouth washes or sprays contain antiviral products that prevent virus from invading nasal and oral cavity during a viral epidemic or pandemic. They cannot be used as prophylactic against coronavirus or flu or cold virus laden droplets infect us mostly through the nose breathing.

Harald Loe, professor at the Royal Dental College in Aarhus, Denmark demonstrated that a chlorhexidine compound could prevent the build-up of dental plaque. The reason for chlorhexidine's effectiveness is that it *strongly adheres to surfaces in the mouth and thus remains present in effective concentrations for many hours*. Thus, it can be an effective spray with other therapeutic agents we describe for their effect to last longer in our oral and nasal passages. Chlorhexidine is included it in nasal, and oral sprays as well as mouth-nasal washes in our invention with EDTA, xylitol, for our antiviral agents to stick and expose the mucosa to prevent coronavirus seedings. This mouth wash can be used for the mouth 2-3 times a day also as nasal cleansing wash using nasal Navage® system during flu and coronavirus virus seasons. This coronavirus antivirals sprays and mouth washes include angiotensin receptor blockers, protease inhibitor, anti-oxidants, viricidal, antibacterial and such.

Superoxide dismutase (Orgotein, SOD, Super Dioxide Dismutase, Superóxido Dismutasa, Superoxydase Dismutase, Superoxyde Dismutase) as nasal, pharyngeal and respiratory system lung spray

Irwin Fridovich and Joe McCord at Duke University discovered the enzymatic activity of superoxide dismutase in 1968. Then it was found out that it is an anti-inflammatory drug "Orgotein" (McCord JM, Fridovich I (1988). "Superoxide dismutase: the first twenty years (1968-1988)". Free

Radical Biology & Medicine. 5 (5–6): 363–9.). Likewise, Brewer (Brewer GJ (Sep 1967). "Achromatic regions of tetrazolium stained starch gels: inherited electrophoretic variation". American Journal of Human Genetics. 19 (5): 674–80.) identified a protein that later became known as superoxide dismutase as an indophenol oxidase by protein analysis of starch gels using the phenazine-tetrazolium technique (Brewer GJ. 1967. "Achromatic regions of tetrazolium stained starch gels: inherited electrophoretic variation". American Journal of Human Genetics. 19 (5): 674–80.) (Brewer GJ (Sep 1967). "Achromatic regions of tetrazolium stained starch gels: inherited electrophoretic variation". American Journal of Human Genetics. 19 (5): 674–80.) Brewer GJ (Sep 1967).

Superoxide dismutase is an enzyme found in all living cells and in our body. An enzyme is a substance that speeds up certain chemical reactions in the body. The superoxide dismutase that is used as medicine comes from the melon, made in a lab from bovine red blood cell. SOD is an enzyme that alternately catalyzes the dismutation (or partitioning) of the superoxide (O2−) radical into ordinary molecular oxygen (O2) and hydrogen peroxide (H2O2).

SODs catalyze the disproportion Superoxides and reactive oxygen species free radicals in cells and tissue produced as byproduct of oxygen metabolism: $2\ HO_2 \rightarrow O_2 + H_2O_2$. In this way, O_2 is converted into two less damaging species.

Thus, SOD is an important antioxidant defense in all living cells exposed to oxygen, especially the lungs. Superoxides are produced as a by-product of oxygen metabolism, and if not regulated, cause many types of cell damage (Hayyan M, Hashim MA, Al Nashef IM (2016). "Superoxide Ion: Generation and Chemical Implications". Chem. Rev. 116 (5): 3029–3085). Hydrogen peroxide is also damaging and is degraded by other enzymes catalase. In our present invention we use vitamin C to produce hydrogen peroxide to kill the coronavirus.

All eukaryotic cells in our body contain SOD enzyme in the cytoplasm with copper and zinc (Cu-Zn-SOD). Cu-Zn-SOD available commercially is normally purified from bovine red blood cells. Superoxide is one of the main reactive oxygen species in the cell. As a consequence, SOD serves as a key antioxidant role to protect the cells. Even some bacteria produce SOD to protect themselves. Oxidative stress is a major determinant of the rate of aging, SOD protects our cells from the oxidative damage. Diminished SOD activity has been linked to lung diseases such as Acute Respiratory Distress Syndrome (ARDS) or Chronic obstructive pulmonary disease (COPD) (Young RP, Hopkins R, Black PN, et al. 2006. "Functional variants of antioxidant genes in smokers with COPD and in those with normal lung function". Thorax. 61 (5): 394–9. Lob HE, Marvar PJ, Guzik TJ, Sharma S, et al. 2010. "Induction of hypertension and peripheral inflammation by reduction of extracellular superoxide dismutase in the central nervous system". Hypertension. 55 (2): 277–83,). Mutations in the first SOD enzyme (SOD1) can cause familial amyotrophic lateral sclerosis (ALS, a form of motor neuron disease) (Milani P, Gagliardi S, Cova E, Cereda C (2011). "SOD1 Transcriptional and Posttranscriptional Regulation and Its Potential Implications in ALS". Neurology Research International. 2011: 1–9).

SOD has powerful anti-inflammatory activity and has been used for the treatment of inflammatory bowel disease, to ameliorates cis-platinum-induced nephrotoxicity, is effective in the treatment of urinary tract inflammatory disease in humans, is used for radioprotection during radiation and to prevent radiation-induced dermatitis. SOD-mimetic nitroxides exhibit a multiplicity of actions in diseases involving oxidative stress, that is why it is important to use Superoxide dismutase to inhibit inflammation of the respiratory system by coronavirus resulting in COVID-19.

It is also given parenterally for treating pain and swelling (inflammation) caused by osteoarthritis and rheumatoid arthritis (RA). It has been promoted by alternative health providers as an anti-ageing agent and for treating scleroderma, radiation-induced cystitis, osteoarthritis, inflammation and urinary tract disorders. It is also taken by mouth, given as a shot, or applied directly to the eyes for inflammations. Now we want to use it applied to the respiratory system to prevent coronavirus infection and its associated inflammation and cytokine production. Superoxide dismutase is an enzyme that helps break down potentially harmful oxygen molecules in cells that are produced during coronavirus infection of the respiratory system. This might prevent damage to the lining mucous membrane of the respiratory system including the lungs, nasal and oral cavity. It is being researched to see if it can help conditions where harmful oxygen molecules are believed to play a role in disease as in severe acute respiratory distress syndrome. There are three major families of superoxide dismutase, depending on the protein fold and the metal cofactor: Cu/Zn type (which binds both copper and zinc), Fe and Mn types (which bind either iron or manganese), and the Ni type (which binds nickel).

We believe that aerosolized SOD coating of the respiratory system (nasal cavity, respiratory system larynx-traches-bronchioles- lung alveoli), sprays (nasal and oral sprays) and mouth washes we describe in this invention can coat the respiratory system airway mucous membrane lining of the lungs, and prevent the coronavirus attaching and attacking the lung parenchyma. It may even enhance the immune system to attack as the virus is attempting to get into lung parenchyma. *This superoxide dismutase aerosolized or other methods of sprays can be used to prevent cold-flu-coronavirus infection and other infections of the respiratory system that plague COPD patients, asthmatics, cystic fibrosis, idiopathic pulmonary fibrosis, and immunocompromised patients due to any number of reasons.*

We induce 350 million anesthetics every year all over the world, who wake up with sore throat and sore tongue due to physical trauma by oral airway (Shantha, Wieden. WO 2018/200063 A1.2018), endotracheal tubes, and laryngeal masks. All those who get these devices (mostly by anesthesiologists) to facilitate breathing should use our superoxide dismutase spray and superoxide dismutase aerosol coating to prevent infection and inflammation. It needs to be used in most cases of coronavirus infection affecting the lungs and to prevent severe acute respiratory distress syndrome and reduce or inhibit cytokine storm.

N-acetyl cysteine in the spray and mouthwash to soften the phlegm and mucus plug

N-Acetylcysteine Solution (USP) is for inhalation (mucolytic agent) or oral administration (acetaminophen antidote), available as a sterile, unpreserved solution. The solutions contain 20% (200 mg/mL) or 10% (100 mg/mL) acetylcysteine, with disodium edetate in purified water. This is added to nasal spray or mouth wash.

The viscosity of pulmonary mucous secretions and phlegm depends on the concentrations of mucoprotein and, to a lesser extent, deoxyribonucleic acid (DNA). The latter increases with increasing purulence owing to the presence of cellular debris mostly white blood cells and some mucosal lining cells of respiratory system. The mucolytic action of acetylcysteine is related to the sulfhydryl group in the molecule. This group "opens" disulfide linkages in mucus and phlegm thereby lowering the viscosity. In other words, it softens it to be expelled with cough with ease. The mucolytic activity of acetylcysteine increases the pH of mucus. It is indicated as adjuvant therapy for patients with abnormal, viscid, or inspissated mucous secretions in such conditions as: coronavirus affliction of the respiratory system including lungs, chronic bronchopulmonary disease (chronic emphysema, emphysema with bronchitis, chronic asthmatic bronchitis, tuberculosis, bronchiectasis and primary amyloidosis of the lung). We want to use it in nasal spray to prevent clogging the air passages during coronavirus infection.

Chapter 26

Oxygen supplement, extracorporeal membrane oxygenation (ECMO) and hyperbaric oxygen therapy (HBO)

Oxygen supplement for moderate to severe cases of COVID-19

The most important effects of COVID-19 symptoms are difficulty in breathing. Doctors will decide on the best treatment option to help the patient fight off the dyspnea with low blood oxygen saturation measured by oximeter. There are 4 modes of oxygen delivery to maintain vital function:

1. Basic oxygen therapy: COVID-19 patients become breathless, struggling to get oxygen, are fitted with a mask and oxygen-enriched air.
2. Pressurized oxygen therapy: Not getting enough oxygenation with simple mask, they will remain conscious, be fitted with an airtight mask and the oxygen-air gas delivered pressurized.
3. Mechanical ventilation: If the patient is still having difficulty breathing and not getting enough oxygen into their blood, intubate and put them on a ventilator in ICU to take the weight off the heart and lungs. Ventilators assist in the removal of carbon dioxide from the lungs, keep the patient alive and give the body time to fight off the virus.
4. Extracorporeal membrane oxygenation (ECMO): If the patient's lungs become too damaged due to any number of reasons including cytokine storm, and the mechanical ventilator does not adequately supply oxygen into the bloodstream, consider using an extracorporeal membrane oxygenation (ECMO) machine.

Extracorporeal membrane oxygenation (ECMO) for severe COVID-19 cases to maintain desired oxygenation of blood

ECMO, also known as extracorporeal life support (ECLS), is an out of the body technique of providing prolonged cardiac and respiratory support to persons whose heart and lungs are unable to provide an adequate amount of oxygen and eliminated carbon dioxide on their own. It is a gas exchange extracorporeal perfusion method to sustain life. The technology for ECMO is an off shoot from cardiopulmonary bypass, and the ECMO machine is the same as the heart-lung by-pass machine used in open-heart surgery. It pumps and oxygenates a patient's blood outside the body, allowing the heart and lungs to rest. Guidelines that describe the indications and practice of ECMO are published by the Extracorporeal Life Support Organization (ELSO "General Guidelines for all ECLS Cases" Extracorporeal Life Support Organization. Retrieved April 15, 2015).

ECMO works by temporarily drawing blood from the body to allow artificial oxygenation of the deoxygenated red blood cells and removal of carbon dioxide. Generally, it is used either post-cardiopulmonary bypass or in late-stage treatment of a person with profound heart and/or lung failure, although it is now being used as a treatment for cardiac arrest in certain centers, allowing treatment of the underlying cause of arrest while circulation and oxygenation are supported. ECMO is also used to support patients with the acute viral severe affliction of the lungs associated with COVID-19 in cases where artificial ventilation is not sufficient to sustain blood oxygenation levels. Usha Martin, Karim Jabr, and Bill Martin at Navicent Medical Center of Macon GA, are experts and experienced in this field. They have performed hundreds of ECMO procedures for many years and used it on severe COVID-19 patients also with 75% success. They state that ECMO cuts down the morbidity and mortality drastically and is a life saving measure. Everyone in ICU treating COVID-19 infection should consider using this modern innovative life saving measure. They use it for acute severe cardiac or pulmonary failure that is potentially irreversible and unresponsive to conventional management. Examples of clinical situations that may prompt the initiation of ECMO include severe cases COVID-19 infection and others such as:

a) Hypoxemic respiratory failure with a ratio of arterial oxygen tension to fraction of inspired oxygen (PaO2/FiO2) of <100 mmHg despite optimization of the ventilator settings, including the fraction of inspired oxygen (FiO2), positive end-expiratory pressure (PEEP), and inspiratory to expiratory (I:E) ratio, Hypercapnic respiratory failure with an arterial pH <7.20
b) Refractory cardiogenic shock, cardiac arrest and failure to wean from cardiopulmonary bypass after cardiac surgery. In those with cardiac arrest or cardiogenic shock, ECMO improves survival and good outcomes.
c) As a bridge to either heart or lung transplantation or placement of a ventricular assist device

d) Septic shock is a more controversial but increasingly studied use of ECMO
e) Hypothermia, with a core temperature between 28 and 24° C and cardiac instability, or with a core temperature below 24° C.

ECMO system can be used for rapid delivery and distribution of therapeutic agents all over the body including lungs, heart, brain and such to treat severe COVID-19 and other infections and disorders.

Extracorporeal blood purification (EBP) to clear viruses and cytokines

Many of very ill COVID-19 patients though intubated and cared for in ICU, cannot shed the effect of coronavirus and cytokines pathological effects and may continue to deteriorate and succumb, if they are cleared from the blood circulation. In such cases, it is also worth consider using extracorporeal blood purification (EBP) method by means of a device developed by Bicon Biologics of India named ***CytoSorb*** therapy. It has been approved by the Drugs Controller General of India (DCGI is akin to US FDA) and FDA cleared for emergency compassionate use. CytoSorb device reduces cytokine and viral load in the circulating blood. It saves the patient from the deadly inflammatory response through blood purification, by reducing the viral and cytokine load in the blood; thus the coronavirus and cytokine injury may be mitigated or prevented, which accelerates the recovery.

CytoSorb is a plug-and-play compatible with the most commonly used blood purification machines or pumps in the ICU used to treat COVID-19 patients, including hemoperfusion, hemodialysis, continuous renal replacement therapy (CRRT), and extracorporeal membrane oxygenation (ECMO) machines. In April, the USFDA granted Emergency Use Authorization (EUA) of CytoSorb for use in patients with COVID-19 infection.

Hyperbaric oxygen therapy—HBO for those who have severe acute respiratory distress syndrome (SARS) and adult acute respiratory distress syndrome (ARDS)

The English scientist Joseph Priestley discovered oxygen in 1775. In 1937, Behnke and Shaw first used it in the treatment of decompression sickness (Sharkey S (April 2000). "Current indications for hyperbaric oxygen therapy". ADF Health). In 1955 and 1956 Churchill-Davidson, in the UK, used hyperbaric oxygen to enhance the radio sensitivity of tumors. I have used HBO during chemotherapy and radiation treatment of cancers to increase the free radicals in cancers in the body to kill the multiplying cancers cells and augment the effect of radiation and chemotherapy.

The Undersea Medical Society (now Undersea and Hyperbaric Medical Society) formed a Committee on Hyperbaric Oxygenation which has become recognized as the authority on indications for hyperbaric oxygen treatment. Hyperbaric medicine and HBO therapy is medical treatment in which an ambient pressure greater than sea level atmospheric pressure is used. The treatment comprises hyperbaric oxygen therapy (HBOT), the medical use of oxygen at an ambient pressure higher than atmospheric pressure, and therapeutic recompression for decompression illness, intended to reduce the injurious effects of systemic gas bubbles by physically reducing their size and providing improved conditions for elimination of bubbles and excess dissolved gas. Now there are numerous HBO centers spread all over USA, easily accessible and used for a host of other diseases other than decompression illness.

The United States Undersea and Hyperbaric Medical Society, known as UHMS, lists approvals for reimbursement, in certain treatment in hospitals and clinics. The following Hyperbaric Oxygen Therapy Indications (The Hyperbaric Oxygen Therapy Committee Report, 12th ed. "Indications for hyperbaric oxygen therapy". Undersea & Hyperbaric Medical Society) are approved uses of hyperbaric oxygen therapy as defined by the UHMS Hyperbaric Oxygen Therapy Committee:

a) Air or gas embolism
b) Carbon monoxide poisoning
c) Carbon monoxide poisoning complicated by cyanide poisoning
d) Central retinal artery occlusion
e) Clostridial myositis and myonecrosis (gas gangrene)
f) Crush injury, compartment syndrome, and other acute traumatic ischemias
g) Decompression sickness
h) Enhancement of healing in selected problem wounds
i) Diabetically derived illness, such as short-term relief of diabetic foot diabetic, retinopathy, diabetic nephropathy
j) Exceptional blood loss (anemia)
k) Idiopathic sudden sensorineural hearing loss
l) Intracranial abscess
m) Muco-mycosis, especially rhino-cerebral disease in the setting of diabetes mellitus

n) Necrotizing soft tissue infections (necrotizing fasciitis), Osteomyelitis (refractory)
o) Delayed radiation injury (soft tissue and bony necrosis)
p) Skin grafts and flaps (compromised), Thermal burns

I add to the list the neurodegenerative diseases, multiple sclerosis, neuropathies, sympathetic dystrophy, PTSD, autism, Lyme disease mental fog, alcohol syndromes, fibromyalgia, chronic fatigue syndrome, senile and many other dementias and such. I have used it for autism, cancer, diabetes neuropathy and ulcers, HIV/AIDS, Alzheimer's disease, during radiation and chemotherapy of cancers, Bell's palsy, cerebral palsy, depression, arteriosclerotic vascular disease and other heart diseases, migraines, multiple sclerosis, Parkinson's disease, spinal cord injury, sports injuries during participation (head trauma including concussions) and after retiring specially in football, and boxing (contact sports, traumatic brain injuries), or stroke and even used it for jet lag and hangover and such. One needs to remember the ear drum damage in the young. *Now that we have coronavirus disease with severe acute respiratory distress syndrome, HBO should be considered.*

Chapter 27

Immune senescence: aging of the immune system, and supplements to prevent or reduce its effects

Immune senescence: aging of the immune system—a comorbidity in the aged who give way to coronavirus infection more often. How to prevent or inhibit immune system aging, to counter coronavirus and other viral attack, enhance the personal immunity to ward off diseases and other infections at the same time increase longevity

Immune senescence is a ***comorbidity*** associated with aging; nobody talks about it, and *it is an aging illness.* I call it a **comorbidity or illness** in the aged, though no doctor diagnoses and label it. Doctors don't tell you about it, billions of people don't even know they have it, people start developing it as they pass the age of 50s. People do not even notice that ***immune senescence*** is the aging disease of the immune system going on within the body, which may determine what diseases people get, and how long or short people live. A more insidious impact of age-related immune system impairment (immunosenescence) leads to a higher rate of cancer, severe influenza, pneumonia, neurodegenerative diseases, age related lung afflictions, and the list is endless - now COVID-19. Immune senescence leads to many chronic inflammations called inflamm-aging which contribute to most of our age-related diseases. In the present-day coronavirus pandemic, it is deadly to the aged because they have ***immune senescence***, besides other diseases called comorbidities that accompany aging that develop due to immunosenescence related inflamm-aging.

Children below the age of 8 may be resistant to coronavirus infection, probably due to some special protein in the lung that will not allow the coronavirus to infect and thrive, and paucity of angiotensin receptors in the nasal mucosa. Maybe they have not developed well developed angiotensin receptors that are needed to virus to latch on to enter the healthy cell, even if they have them, they have fewer of them.

Free radicals are unstable molecules with an unpaired electron produced 24/7 in the body, which are countered by endogenous anti-oxidants described below. These instable molecules lead them to steal electrons from other healthy molecules that inflict damage to delicate cellular structures known as oxidative stress as seen in coronavirus storm. It is also responsible for *senescence.* Free radical oxidants can be deadly in COVID-19.

Without going into details, the following are the measures people including the aged can take to maintain a healthy full-bodied, robust immune system, an immune regenerative protocol to fight viral infections such as coronavirus. They are as follows:

a) Zinc acetate lozenges have shown a direct effect of inhibiting the ability of certain viruses enter cells, latch on to the cells in the back of the throat where they multiply and can descend into the lungs to potentially cause pneumonia.

b) Selenium boosts the immune system and can provide protection against colds-flu and coronaviruses infections. Deficiency increases susceptibility to viral infection, and may be associated with higher mortality.

c) Melatonin—a high dose of 10 mg-25 mg at bedtime induces a potent immune response, besides sleep, anti-aging, antiviral. Enhancing the responses of antibodies that "tag" specific viruses, bacteria, and other invaders to be attacked by different components of the immune system. Reducing chronic inflammation, a cause of nearly all age-related chronic diseases, and Enhancing the activity of T cells, helping to more efficiently destroy pathogens. I used it in all my cancer patients as part of anti-cancer therapy.

d) Lithium prevents cognitive decline, extends longevity and health health span. I advise taking 300 micrograms of lithium after the age of sixty, and I drink lithium water myself. It helps maintain longer telomeres at the end of chromosomes, thus maintains healthy DNA structure. The longer the telomere, the longer you live and have less disease afflictions. In those taking lithium, the brain is thicker and memory centers are well maintained, indicating it effectiveness in treatment of many dementias especially in bipolar disorder. It also regulates and inhibits the overactivity of GSK-3 which raises the risk of many chronic diseases of older age including Alzheimer's, type II diabetes, mood disorders, cancer, and others. Telomeres are stretches of repetitive DNA strands that cap the ends of chromosomes. Like the burning of a fuse, telomeres at the ends of our chromosomes steadily shorten every time a cell replicates itself. Once telomeres reach a critically short length, cell renewal virtually stops, leading to accelerated aging, diseases (senescence), or death of the cell. Combining with rapamycin and other supplements such as bioactive curcumin, Nicotinamide mononucleotide (NMN), and melatonin may make even more effective as anti-aging brain function maintaining cognition and immune system function.

e) Aged garlic extractm - 3,600 mg a day are unique immune-boosting antiviral allicin compounds
f) Dehydroepiandrosterone (DHEA) and its metabolites have demonstrated immune-enhancing and antiviral effects by increasing B and natural killer immune cells, increase interleukin 2.
g) Vitamin D3, 5000 IU twice a day, anti-inflammatory, enhances immune system.
h) Vitamin C 1000 mg orally twice a day (see details in the following chapters), antioxidant, and antiviral by producing hydrogen peroxide.
i) Vitamin E: vitamin E is critical for maintaining efficient immune function. Clinical evidence has shown vitamin E supplementation can increase resistance to infection, especially in older individuals. In a study in elderly men and women, supplementation with 200 mg per day vitamin E significantly enhanced immune parameters including neutrophil, T-cell, B-cell, and NK-cell function, bringing their values close to those of younger healthy adults (De la Fuente M, Hernanz A, Guayerbas N, Victor VM, Arnalich F. Vitamin E ingestion improves several immune functions in elderly men and women. Free radical research. Mar 2008;42(3):272-280.). Increased vitamin E intake has been shown to restore the decline in T-cell function associated with aging. This improvement in T-cell function results from vitamin E's direct impact on T cells as well as inhibition of prostaglandin E2, a mediator of inflammation and a T-cell suppressor (Wu 2014; Wu 2008; Han 2006). In a mouse model, vitamin E supplementation reversed the age-associated decline in naïve T-cell function (Adolfsson 2001). This supplement is of great help in the aged to inhibit the morbidity and mortality after coronavirus and such infections.
j) Bioactive curcumin 1000-2000 mg a day acts as an anti-oxidant, anti-inflammatory, anti-septic, anti-cancer, anti-cardio-vascular diseases, anti-neurodegenerative diseases, anti-scarring, and has anti-fat-protein clump accumulation effects.

k) NAD+ is the acronym for nicotinamide adenine dinucleotide found in every cell, co-factor in cell ***energy*** transfer, energy using, plays a critical role in regulating aging processes. includes mitigating chemical stress, inflammation, DNA damage, and failing mitochondria. At the same time, NAD promotes longevity by facilitating DNA repair and providing cellular benefits associated with caloric restriction and exercise. By age 80s, NADlevels drop to only **1%** to **10%** expressed in youth. Supplementation with nicotinamide riboside offers a way of supporting essential body systems. NAD+ is unstable and cannot be used as a supplement, but nicotinamide riboside is a useful precursor to NAD+ that is capable of restoring cellular NAD+ levels.

l) Lactoferrin is a protein found in highest amounts in colostrum, the first type of mother's milk produced after a baby is born. It stimulates macrophages, which in turn may help to induce cell-mediated immunity, inhibits viral infection by interfering with the ability of certain viruses to bind to cell receptor sites. It defends against a wide range of viruses such as cold/flu, herpes, hepatitis, Respiratory syncytial virus (RSV), Epstein Barr virus (EBV), Enteroviruses and such. As we know that the viral infections are contracted in the nasal and oral passage mucous membrane, they move to throat and lungs due to rapid proliferation, hence lactoferrin intake is an effective antiviral and is effective in the aged. Lactoferrin has an antiviral effect in two different ways. First it binds to the virus directly, blocking the viruses' surface proteins' ability to recognize binding sites on the surface of cell membranes. Second, it binds to surface sites on the cell's outer membrane that are targeted by viruses. There is a compound found on cell surfaces, called heparan sulfate, which is a common target for various viruses. Several studies have shown that lactoferrin binds to structures containing heparan sulfate, which can prevent viruses from recognizing and entering (Lang J, Yang N, Deng J, et al. Inhibition of SARS pseudo-virus cell entry by lactoferrin binding to heparan sulfate proteoglycans. PLoS One. 2011 ;6(8): e23710. Chen JM, Fan YC, Lin JW, et al. Bovine Lactoferrin Inhibits Dengue Virus Infectivity by Interacting with Heparan Sulfate, Low-Density Lipoprotein Receptor, and DC-SIGN. Int J Mol Sci. 2017 Sep 12;18(9).

m) Cimetidine (Tagamet®): 800 mg - 1,200 mg each day, sold over the counter, acts by boosting immune function by activating natural killer cells and reducing T-suppressor cells. I have used it for cancers to prevent metastasis, before and after surgery. Those who take cimetidine suffer less from colds and flu during flu season. **Don't use cimetidine if cytokine storm suspected in COVID-19**.
n) Famotidine anti-ulcer drug in high doses is showing superior promise as a potential inhibition/cure for COVID-19
o) Favipiravir (Avigan®) is approved to fight influenza by interfering with the virus replication process by selectively inhibiting RNA (ribonucleic acid) polymerase necessary for influenza virus replication, can be used during coronavirus and cold-flu seasons.
p) Whey protein prevents muscle wasting specially during periods of inactivity. It is a high-quality protein for aging people, especially those in nursing homes, extended care facilities, and the aged living independently, or inactive hospital patients. It is an excellent supplement and it promotes glutathione synthesis, which is an important endogenous anti-oxidant. Whey protein is a mixture of proteins isolated from whey; the liquid material created as a by-product of cheese production. The proteins consist of α-lactalbumin, β-lactoglobulin, serum albumin and immunoglobulins. Used by body builders along with creatine and glutamate extensively, whey protein has been shown to be better compared to other types of protein, such as casein or soy (Tang, Jason E. et.al. 2009. "Ingestion of whey hydrolysate, casein, or soy protein isolate: effects on mixed muscle protein synthesis at rest and following resistance exercise in young men". Journal of Applied Physiology. 107 (3): 987–99).
q) Take magnesium L-threonate which is absorbed and passes blood brain barrier to the brain with ease. It is needed for 80% of the body metabolic function and acts as a secondary activator of NMDA receptors that is vital for synaptic plasticity improving the mental function with increased short term and long-term memory. The gastro-intestinal track has a large network of nerve cells and processes, that is why it is called the ***second brain***. By taking 1000 mg it will improve the brain and intestinal function, that also improves health and makes old people more attentive to their health, thus less likely to develop coronavirus and such infections, decreasing the morbidity and mortality in the aged.

r) Endogenous antioxidants are formed within the body include ***superoxide dismutase (SOD), glutathione peroxidase, catalase, and glutathione.*** These compounds got their name because they fight against oxidative stress. They find free radicals and neutralize them. These decline with aging hence, taking supplements of SOD is helpful in fighting the coronavirus infection.

s) Exogenous antioxidants are formed outside the body but can be absorbed and used by our cells if we consume them orally. Examples include vitamins C and E, carotenoids, curcumin, turmeric, ginger, flavonoids, polyphenols and other plant extracts including green tea. Endogenous antioxidants (like SOD) are the first line of defense, particularly against the free radicals that are formed within our cells due to normal metabolic processes.

t) I would like to add anti-aging metformin 250-500 mg after every meal, which has countless anti-aging effects and has effect in reducing all age-related diseases. Metformin as the first-ever anti-aging medication. It is taken by millions of type II diabetics. Studies show that metformin acts by boosting the activity of AMPK, a master metabolic regulator that favors fat- and sugar-burning and prevents their accumulation, and facilitates DNA repair, I am not diabetic, but I have been taking it for more than a decade. AMPK activity decline with age, and it plays a major role in preventing age-related disorders including cancer, cardiovascular disease, obesity, and neurocognitive decline, Alzheimer's and Parkinson's and age-related diseases including immune senescence.

u) Of course ***life style changes*** need to be included to keep the immune system at par besides above supplements such as: daily exercise, not smoking, curtail alcohol, maintain proper body weight, stress free job and life, control blood sugar if diabetic and blood pressure (a must), weekly sauna (whole body hyperthermia), Mediterranean diet rich in omega 3 rich fish, whey protein, Reishi mushroom extract, enzymatically modified rice bran, beta glucan, herbal green tea, *8 hours of sleep* and such will enhance the immune system and immune response to coronavirus and ward off other diseases.

v) Those with recent history of heart attack in nursing homes, extended care facilities, and the aged living independently, consider taking 0.5 mg of Colchicine orally intermittently to prevent stroke, maybe even another heart attack besides reducing inflammatory cytokines needed if coronavirus infection.

Chapter 28

Our body's low-level vitamin D3, "senescent cells" (called "zombie cells"), immune system decline, and chronic inflammation in the aged

Low vitamin D3 levels in those with COVID-19 and the relation to morbidity and mortality

Patients from countries with high COVID-19 mortality rates, such as Italy, Spain and the United Kingdom, had lower levels of vitamin D compared to patients in countries that were not as severely affected, according to the study. The researchers also found a strong correlation between vitamin D levels and cytokine storm, which is a hyperinflammatory condition caused by an overactive immune system (https://www.foxnews.com/science/vitamin-d-levels-covid-19-mortality-rates). People after surgery recover faster when supplemented with vitamin D3. All nursing home residents, extended care patients, the aged living independently, and coronavirus infected patients in the hospital and ICU should be on vitamin D3, 5000 IU 2-3 times a day *to inhibit the onslaught of coronavirus affliction and its associated cytokine storm* with destruction of lung and its function.

Our body "senescent cells" also called "zombie cells" accumulation as we age and their effect on our health and COVID-19

Our body normally gets rid of damaged, abnormal old cells as part of its daily functioning. These damaged cells are programmed to die off, through a process called *apoptosis*, preventing the buildup of dysfunctional senescent cells in tissues. Senescent cells have been found in virtually every human tissue (Calcinotto A, et al. Cellular Senescence: Aging, Cancer, and Injury. Physiol Rev. 2019; 99(2):1047-78). Besides immune senescence, these senescent cells accumulations can be labelled ***senescent comorbidity***, another disease process like other age-related afflictions which is not diagnosed or talked about that affects those in nursing homes, extended care facilities, and the aged living independently. These cells have reached the end of their natural lifespans. But instead of dying off in the normal process called apoptosis, they accumulate within tissues, where they secrete protein-degrading enzymes that damage other normal healthy cells and their function.

As we get older, senescent cells numbers build up, become aged, damaged cells that, rather than self-destructing, survive to increase inflammation and death of healthy functioning cells. As senescent cells accumulate, they damage cellular DNA and mitochondria, the genetic and metabolic factory within all our cells that make our body, while causing telomere dysfunction (shortening of telomeres). Telomeres are nucleotide sequences at the ends of chromosomes. Loss of telomere structure accelerates degenerative aging processes, dysfunctional DNA and predisposition to age related diseases and death.

In the elderly, aged cells (senescent) cease functioning properly, fail to accomplish their normal tasks, and lose their ability to divide. But instead of dying off, these cells linger and accumulate in various tissues. This is known as ***cellular senescence and I label it as a comorbidity in the aged akin to chronic inflammation in the body***. Scientists have discovered that these dysfunctional, older cells are a major contributor to many of the diseases and co-morbidities that come with advancing age due to their secretion of high levels of toxic compounds collectively called as SASP (senescence—associated secretory phenotype) that contain protein degrading enzymes that induce chronic inflammation. A major factor in old-age decline is the accumulation of senescent cells that:

a) Impede organ function
b) Create chronic inflammation
c) Emit protein-destroying enzymes
d) Shorten healthy lifespan

One of the ways senescent cells contribute to chronic disease is by spewing out hundreds of toxic proteins that degrade healthy tissues. In addition to this effect, a newly published study has discovered that many of these secreted proteins also play roles in the control of blood clotting, that is why we have thromboembolic events aged afflicted with COVID-19. The study showed that these secretions promote clotting, potentially contributing to deep venous thrombosis and other clotting-related conditions as well as heart attack, stroke, arthritis, neurodegenerative diseases including Alzheimer's and Parkinson's.

Compounds that circumvent this aging mechanism are called ***senolytics (Life Extension, June 2020).*** "Seno" refers to old cells that secrete toxic chemicals, "lytic" refers to their destruction. Senolytic compounds selectively destroy these old (senescent) dysfunctional cells by inducing ***apoptosis (death)*** in senescent cells, triggering their ability to die off. Fruits, vegetables, onions, apples and berries contain small amount of senolytics. For the aged, those in nursing and extended care facilities should have a senolytics regimen that includes:

a) 1,250 mg/day of quercetin with or without phytosome enhancer
b) 100 mg/day of dasatinib (an anticancer drug low dose so not toxic, so also navitoclax)
c) theaflavins derived from black tea, apigenin and strawberry flavonoid called fisetin
d) I add bioactive curcumin, vitamin D3, micro-doses of rapamycin, lithium, mini doses of metformin, Ebselen, and weekly whole-body hyperthermia to enhance autophagy of senescent cells

The aged living in nursing homes, extended care facilities, and the aged living independently are more susceptible to COVID-19, and should be treated with the above anti-senescent protocol to reduce the morbidity and mortality in them if they develop COVID-19 and viral infections.

How aging accelerates immune system decline that predisposes to increased morbidity and mortality in the aged with comorbidities who have the coronavirus

a) Decreased numbers of naïve T cells, or cells that can easily be activated to recognize and respond to diseases like the flu, coronavirus, cancers, and such
b) Increased numbers of senescent memory T cells that have lost their ability to divide and function properly and instead release pro-inflammatory cytokine molecules
c) Decreased natural killer (NK) cell activity, which is linked to increased illness and death from infections, atherosclerosis, cancers, and diminished response to the flu and other vaccines
d) Cytomegalovirus infection may be associated with immune senescence, cardiovascular disease, frailty, and mortality

Age-related chronic inflammation and the effect of coronavirus inhabitants of the aged in the nursing home and the use of rapamycin

"Inflamm-aging" is a pro-inflammatory phenotype which is observed in aging mammals including humans. Cytokines, chemokines, prostaglandins and such are the major players of the inflammatory response, allow communication between immune and somatic cells. Their chronic response might be partially responsible for frailty, development of age-related diseases, and functional decline, with increased morbidity and mortality. That is why I consider it as comorbidity in the aging.

Multiple causes of inflamm-aging are examined: accumulation of proinflammatory tissue damage over time, secretion of proinflammatory cytokines by senescent cells, dysfunctional immune system failing in efficiently clearing pathogens and dysfunctional host cells, mismanagement of apoptosis due to mitochondrial dysfunction, defective autophagy response and accumulation of damaged proteins triggering inflammatory response, change in the levels of hormones that regulate cytokine production (including testosterone), enhanced activation of NF-κB transcription factor. These alterations result in an enhanced production of interleukin 1β (IL-1β), a tumor necrosis factor that induces the release of TNF-α. Interleukin-6 (IL-6) is released by TNF-α and IL-1β which attack healthy organs and cells- and these cytokines are reduced by use of rapamycin.

Nuclear factor-κB (NF-κB) is implicated in immune response and response to stress such as coronavirus. It controls several genes associated with inflammation and is considered as a molecular culprit for inflamm-aging. Remarkably, SIRT-1 expression inhibits NF-κB. Sirtuins could also influence inflamm-aging. Studies have revealed that, by deacetylating histones and components of inflammatory signaling pathways such as NF-κB, SIRT1 downregulates inflammation-related genes. Reduction of SIRT-1 levels correlates with the development and progression of many inflammatory diseases, and pharmacological activation of SIRT1 may prevent inflammatory responses in mice.

Once again, the age-associated inflammation can be interpreted as a defense mechanism that ultimately becomes deleterious overtime and may even put on end to our healthy life longevity. Low levels of inflammatory response may be beneficial to tissue repair and regeneration by activating the immune system, but higher levels aggravate damage as it happens in cytokine storm of the COVID-19 as seen after BCG and other vaccination. Age-related inflammation could also severely undermine stem cell function. By using rapamycin, mTOR is inhibited, acting on various inflammation pathways as described and hence has anti-aging and anti-cytokine effect. It may be a time to put the elderly on micro-dose rapamycin as I advise my friends to maintain youthful activity and reduce or prevent cancers. It should be used with other therapeutics to maintain health and fight coronavirus and other age-related diseases. Does rapamycin help to calm or prevent cytokine storm in the COVID-19, YES IT DOES. When combined with ivermectin, it may be very effective to prevent, curtail, stop the progression of and/or cure COVID-19.

Chapter 29

Rapamycin: to protect from age related increased morbidity, mortality and to inhibit cytokine storm with extreme immune system reactions

Age-related chronic inflammation and the effect of coronavirus in the aged in the nursing homes and the use of rapamycin (sirolimus), and rapamycin analogues (rapalogue), everolimus, biolumus, temsirolimus and ridaforolimus (RAD001) or specific cephalosporin such as cefixime

"Inflamm-aging" is a pro-inflammatory phenotype is observed in aging mammals including humans. Cytokines, the major players of the inflammatory response, allow communication between immune and somatic cells. Their chronic response might be partially responsible for development of age-related diseases and functional decline, with increased morbidity and mortality. Multiple causes show accumulation of proinflammatory tissue damage over time, secretion of proinflammatory cytokines by senescent cells, dysfunctional immune system failing in efficiently clearing pathogens and dysfunctional host cells, mismanagement of apoptosis due to mitochondrial dysfunction, defective autophagy (house cleaning) response and accumulation of damaged proteins triggering inflammatory response, change in the levels of hormones that regulate cytokine production (including testosterone), enhanced activation of NF-κB transcription factor. These alterations result in an enhanced production of interleukin 1β (IL-1β), a tumor necrosis factor that induces the release of TNF-α. Interleukin-6 (IL-6) is released by TNF-α and IL-1β which attack healthy organs and cells- and these cytokines are reduced by use of rapamycin or its analogues.

Nuclear factor-κB (NF-κB) is implicated in immune response and response to stress such as coronavirus. It controls several genes associated with inflammation and is considered as a molecular culprit for inflamm-aging. Remarkably, SIRT-1 expression inhibits NF-κB. Sirtuins could also influence inflamm-aging. Studies have revealed that by deacetylating histones and components of inflammatory signaling pathways such as NF-κB, SIRT1 downregulates inflammation-related genes. Reduction of SIRT-1 levels correlates with the development and progression of many inflammatory diseases, and pharmacological activation of SIRT1 may prevent inflammatory responses in mice.

Rapamycin (Sirolimus) or rapamycin analogues (rapalogue), everolimus, biolumus, Temsirolimus and ridaforolimus): to prolong health span and longevity to protect from increased morbidity and mortality in the aged and from coronavirus infection (COVID-19)

I am one of the most passionate advocates for taking micro-doses of rapamycin as an anti-aging drug against age-related diseases including cancers, arteriosclerotic vascular disease, diabetes, elevated blood pressure, Alzheimer's and Parkinson's, obesity and such. The drug is researched extensively by US based scientist Mikhail Blagosklonny, (with others) who now works at the Roswell Park Cancer Institute in Buffalo. A native of St. Petersburg, he was working on cancer treatments in the early 2000s when he realized the same qualities that made rapamycin effective at slowing tumor growth (used by rapamycin discoverer Dr. Sehgal himself) might also help it slow the aging process. Acceptance as an anti-aging drug was supported by the NIH study that mice on rapamycin lived 9-14% longer. I advocate for rapamycin to all cancer patients and senior citizens as adjuvant therapy along with chemotherapy and those with metastasis also, besides using it in micro-mini-doses to inhibit the ravages of aging and aging related diseases.

Studies identified rapamycin as ***the first pharmacological agent to extend maximal lifespan*** in a mammalian species, with effects in both males and females (Blagosklonny MV. Aging and immortality: quasi-programmed senescence and its pharmacologic inhibition. Cell Cycle. 2006;5(18):2087–2102. Harrison DE, Strong R, Sharp ZD, et al., Rapamycin fed late in life extends lifespan in genetically heterogeneous mice. Nature. 2009;460(7253):392–395.). Remember the quote by Ashley Montagu (1905–1999) ***"The idea is to die young as late as possible."*** **Rapamycin, metformin, lithium, Ebselen and bioactive curcumin are the answers for such an outcome.**

Even today, there is no cure for aging and age-related neurodegenerative diseases, Alzheimer's and Parkinson's, cardio-vascular diseases, cancers, 152 autoimmune diseases and such that plague our lives besides self-inflicting drug addiction, alcoholism, obesity, gun violence, suicide, accidents and such. Rapamycin is the answer for attacking the comorbidities in the aged, a specific inhibitor of the mTOR signaling complex, the central regulator of cell nutrient sensing and energy metabolism (Zoncu R, Efeyan A, Sabatini DM.

mTOR: from growth signal integration to cancer, diabetes and ageing. Nat Rev Mol Cell Biol. 2011;12: 21–35).

Ehninger D, et al., and Blagosklonny describes in *"longevity and aging"* (Dan Ehninger, Frauke Neff, and Kan Xie. Cell Mol Life Sci. 2014; 71(22): 4325–4346. Blagosklonny MV, 2006.) the use of rapamycin to block the aging and aging related diseases. The rapamycin decreases a subset of heart dimensional measures, such as diastolic left ventricular internal diameter (LVIDd), and also decreases overall heart mass, improves heart function. Immunoglobulin plasma concentrations are often robustly elevated in aged mice. Rapamycin tended to decrease plasma immunoglobulin concentrations in several cases that may reduce the body inflammation.

One of the well-documented age-related alterations is the strong decline in adult hippocampal neurogenesis of the brain (main memory center). Neff F. et al., determined 1-year rapamycin treatment initiated at 4 months, *prevents the age-related decline in hippocampal neurogenesis* [Neff F. et al., Rapamycin extends murine lifespan but has limited effects on aging. J Clin Invest. 2013;123(8):3272–3291.]. This is an important effect of rapamycin to restore all levels of memory and recall. I believe that rapamycin should be part of protocol for all dementias including Alzheimer's and Parkinson's, senile dementia, and such. Intermittent or pulsed frequent micro-doses of rapamycin should be given to all old nursing home and extended care patients, and those who show signs and symptoms of dementia and other comorbidities. Normal population after the age of 60s should consider taking it for longevity health span.

Another important cellular process regulated by mTOR signaling that is controlled by Rapamycin is autophagy. Autophagy, a process by which the cell recycles macromolecules and organelles, allows for the removal of damaged cellular constituents and enables the cell to mobilize substrate under nutrient-poor conditions. mTORC1 regulates autophagy by phosphorylating and inhibiting the autophagy-initiating kinase Ulk1. In C. elegans, autophagy has been reported to be required for the *lifespan extension caused by genetic mTOR inhibition. Lifespan extension, caused by rapamycin, is dependent on autophagy effects also.* It can also be true in humans.

Detailed cause-of-death analyses in rapamycin-treated mice and controls indicated that both groups die primarily (i.e., in >80 % of cases) due to cancers, but rapamycin-*treated animals do so later in life than controls.* Rapamycin has well-***known anti-cancer properties***, including inhibitory effects on de novo cancer formation, as well as suppression of established tumors via inhibition of cancer growth, promotion of apoptosis of neoplastic cells and/or a modification of the host response to the tumor (for example, inhibiting new blood vessel formation—angiogenesis) (Kopelovich L, Fay JR, Sigman CC, Crowell JA. The mammalian target of rapamycin pathway as a potential target for cancer chemoprevention. Cancer Epidemiol Biomark Prev. 2007;16(7):1330–1340). My advice is to take rapamycin if one had cancer as part of the treatment or to prevent cancer development.

Aging is associated with a number of neurobehavioral and neurological changes, such as cognitive decline, alterations in motor coordination, balance, reduced muscle strength and such. Rapamycin has an effect on improving all these conditions with lowering of brain inflammation and reducing the neurodegeneration. *We believe that injections of rapamycin in micro doses directly to cerebro-spinal fluid pool will arrest the progression of Alzheimer's and Parkinson's and bring life back to the brains of the 5.8 million patients in US (Dementia: how to prevent and treat. Under preparation for publication by Dr. T. R. Shantha).*

Neff et al. also used the virtual drum vision test to examine if rapamycin can prevent the aging-associate decline in visual acuity which is seen in aged animals. Rapamycin treatment significantly elevated red blood cell counts in aged animals and tended to have similar effects on red blood cell counts in young animals, indicating that rapamycin effects were likely linked to aging-independent effects on erythrocyte production and/or turnover.

One of the well-documented age-related alterations is the strong decline in adult hippocampal neurogenesis (important memory center complex). Neff et al. determined if a 1-year rapamycin treatment, initiated at 4 months, *prevents the age-related decline in hippocampal neurogenesis*. The data showed the expected substantial age-related increase in thyroid follicle sizes. Rapamycin treatment significantly decreased thyroid follicle size in aged mice and restore its normal function.

Shorter-term rapamycin treatment (for 6 months) yielded expression alterations of 100 genes (32 up-regulated and 68 down-regulated) in males and 1,427 genes (675 up-regulated and 752 down-regulated) in females. In summary, both studies provide initial insights of rapamycin associated gene regulation in a single organ (the liver). The ***aging traits found to be ameliorated by rapamycin*** were either related to immune system changes (e.g., plasma immunoglobulin concentrations, frequency of specific T cell subsets, cytokine concentrations in blood and heart, response to vaccination), age-related alterations in body mass, organ size and dimensions (body weight, fat mass, lean mass, thyroid follicle size, cardiac dimension, heart weight), tumors and pre-cancerous lesions, as well as neurobehavioral changes (motor activity, learning, memory and recall).

According to Blagosklonny (2006), anti-aging formula includes rapamycin combined with commonly available drugs such as *metformin, aspirin, propranolol, angiotensin II receptor blockers and angiotensin-converting enzyme inhibitors.* He states that calorie restriction deactivates mTOR and increases life span, Rapamycin prevents obesity and extends life span in similar fashion.

Glucose, amino acids and fatty acids, insulin, insulin-like growth factor 1 (IGF-1), tumor necrosis factor (TNF) activate the mammalian target of rapamycin (mTOR) signaling pathway in cells. Overactivation of the mTOR pathway causes insulin resistance. mTOR is involved in diabetic complications. mTOR is involved in aging and age-related diseases. These are counteracted by rapamycin and the it extends life span and health span in all species tested (M V Blagosklonny, TOR-centric view on insulin resistance and diabetic complications: perspective for endocrinologists and gerontologists. Cell Death Dis. 2013 Dec; 4(12): e964). Due to coronavirus infection, there is overactivation of immune system cells, that is inhibited by simple method of treating with rapamycin.

A study has investigated the effect of rapamycin feeding on immune protection in old mice (Hinojosa et al., 2012) and demonstrated that rapamycin-treated old mice exhibited improved survival following pneumococcal infection which would die within 4 days without rapamycin. Now we have the coronavirus and its related lung infection. Those who are old can protect their immune system by use of micro-domes of rapamycin to protect increased morbidity and mortality from coronavirus infection. If one with cancer wants to live long life, add rapamycin to chemo not matter what stage the cancer is. It is already used to treat renal cancers (Abraham RT, Gibbons JJ. The mammalian target of rapamycin signaling pathway: twists and turns in the road to cancer therapy. Clin Cancer Res. 2007;13: 3109-3114.) and in coronary stents to prevent clots around them. I have had a prostate cancer patient with extensive metastasis on rapamycin whose alkaline phosphatase and PSA markers have come down, and his metastasis was not growing as they did before the therapy. One of the sarcoidosis and other autoimmune diseases patients was on rapamycin with reduced symptoms, so also a bladder cancer patient.

For anti-aging and prevention of neurodegenerative diseases including Alzheimer's and Parkinson's and arteriosclerotic vascular diseases, as anticancer, obesity and such in nursing homes, use micro doses of rapamycin.

That is about 4-6 mg (depending on the body weight) taken orally once a week. We want to introduce micro doses of rapamycin by direct delivery to cerebro-spinal fluid (CSF) as described in another chapter to treat Alzheimer's and other neurodegenerative diseases as well as brain cancers, as soon as it is diagnosed. It will arrest the disease and reduce or eliminate brain inflammation that is responsible for Alzheimer's Aβ plaques formation and other toxic substance in the brain around the neurons in neurodegenerative diseases and their processes and connection (synapses). I have had cancer and autoimmune diseases patients on micro-dose rapamycin. It also increases our youthful activity which we lose as we age, hence it is a must for the elderly.

Micro-doses of rapamycin have hardly any complications as seen when used in megadose to prevent kidney and other transplant rejection. If they occur, they are milder such as mouth ulcers (canker sores,), skin infection, impaired wound healing and mild Broncho-pneumonia which can all be treated with antibiotics and discontinuing the rapamycin if need be. According to Buck Institute Kennedy, it reduces bone loss, reverses cardiac aging, reduces chronic inflammation (Bloomberg, Time, Feb 2015). If you want to remain healthy for a long time, rapamycin has a role, especially for those in nursing homes, extended care facilities, and the aged living independently, who succumb to coronavirus and other comorbidities.

I describe rapamycin administration dose into 3 categories as described in Part III of the book:

1. **Mega-doses** rapamycin: Given 15-40 mgs loading dose with reduced doses daily to prevent organ transplant rejection, fraught with many complications.
2. **Mini-doses** rapamycin: Take 6 mg loading dose and 2 mg every day for a limited time. Given to counteract cytokine and viral storm of COVID-19 as described above or for severe flu or in septic shock (maybe), till the symptoms and diseases state dissipate—that is for a week or two.
3. **Micro-doses**: Intermittent (pulsed doses) intake of rapamycin: take 3-6 mg depending upon the comorbidities, weight, age once every 7-10 days to prolong remaining life, to bring back some youth, prevent or inhibit or counteract or decrease expansion of comorbidities that escalate our morbidity and mortality as we age. Take rapamycin 6-8 weeks then take one- or two-weeks ***drug holiday*** or stop taking if complication develop such as bronchitis, oral ulcers, and infection. Then restart the ragmen. I take it to prevent age related diseases from neurodegenerative diseases to cancers, cardio-vascular diseases and such, and so far, have not observed any complication.

Chapter 30

Xylitol, DMSO, ozone, vitamin D3, bioactive curcumin, zinc, EDTA

Xylitol as anti-viral and bacterial in the nasal-oral-pharyngeal-laryngeal mucosa sprays and mouth washes

Numerous studies indicate that the addition of xylitol to oral and nasal spray can considerably enhance the effect of antibacterial agents locally in the nasal and oral cavity. Also, the study listed below demonstrated that xylitol in the mouth affects the composition of oral flora by a decrease in S. mutans and L. acidophilus. Xylitol inhibits the growth of bacteria via the inducible fructose phosphotransferases system and formation xylitol-5-phosphate. Xylitol can disturb the protein synthesis and causes ultrastructural changes bacteria (Kontiokari T, Uhari M, Koskela M. Effect of xylitol on growth of nasopharyngeal bacteria in vitro. Antimicrobe Agents Chemother 1995;39: 1820-3). This study confirms that S. mutans and L. acidophilus also possess the fructose pathway and could thus be inhibited by xylitol as indicated in the previous report on L. paracasei, L. brevis and L. fermentum.

Xylitol has been shown to prevent the adherence of Pneumococci and Hemophilus influenzae to nasopharyngeal cells due to a fructose phosphotransferase system-mediated uptake and phosphorylation of xylitol in the cell. Streptococcus mutans has been implicated as a principal etiological agent of dental caries in humans because this oral bacterium has the ability to adhere to and produce acid at the tooth surface when carbohydrates, such as sucrose, serve as a substrate. Studies show the reduction of tooth decay and plaque buildup by use of xylitol tooth paste and chewing gum due to inhibition of bacteria and their multiplication.

Our study of using xylitol spray of the nose and mouth showed that there was reduction of bacterial and flu affliction in children attending kindergarten and nurseries. Adding angiotensin receptor blockers made it even more effective. Hence, we add the xylitol to inhibit the COVID-19 virus sticking to nasal-oral-pharyngeal-laryngeal mucosa, the S spike on the virus gets coated with therapeutic agents due to sticky xylitol and inhibit its infection for multiplication within these nasal and pharyngeal mucosal cells.

Dimethyl sulfoxide as a cell membrane modifier for nasal or oral spray and to enhance the uptake of virus stat and viricidal therapeutic agents inside the afflicted cells.

Dimethyl sulfoxide (DMSO) is a small amphiphilic molecule which is widely employed in cell biology as an effective penetration enhancer, cell fusogen, and cryoprotectant. I have used this since 1964 in my medical practice with successes and no complications. A recent study has suggested that DMSO induces pores in the membrane (Norman, R.; Noro, M.; O'Malley, B.; Anwar, J. J. Am. Chem. Soc. 2006, 128, 13982-13983) for enhanced activity of therapeutic agents on the surface and inside the cell wall and stand alone has healing effects by reducing the inflammation..

DMSO is observed to exhibit three distinct modes of action, each over a different concentration range.

1. At low concentrations, DMSO induces cell membrane thinning and increases fluidity of the membrane's hydrophobic core.
2. At higher concentrations, DMSO induces transient water pores into the membrane allowing more therapeutic agents inside the cell.
3. At still higher concentrations, individual lipid molecules of the cell membrane are desorbed from the membrane followed by disintegration of the bilayer structure, thus not allowing the time for the coronavirus to multiply inside the coronavirus infected cells.

The study provides further evidence that a key aspect of DMSO's mechanism of action is pore formation, which explains the significant enhancement in permeability of membranes to hydrophilic molecules by DMSO. The reduction in the rigidity and the general disruption of the membrane induced by DMSO are considered to be prerequisites for membrane allowing therapeutic agents inside the cell and for fusion processes. The findings also indicate that the choice of DMSO concentration for a given application is critical, as the concentration defines the specific mode of the solvent's action. Knowledge of the distinct modes of action of DMSO and associated concentration dependency should enable optimization of current application protocols on a rational basis and also promote new applications for DMSO in the prophylactic and therapeutic spray to prevent coronavirus attachment to healthy cells, spread and its entrance to the healthy cell for multiplication, and killing them with many anti-coronavirus therapeutic agents described herein within the cell cytoplasm (Gurtovenko Andrey A. and Jamshed Anwar. Modulating the Structure and Properties of Cell Membranes: The Molecular Mechanism of Action of Dimethyl Sulfoxide, J. Phys. Chem.B2007, 111, 10453-1060). *DMSO may interrupt the*

S-spike attaching to the angiotensin receptors on the healthy cells due to changes it brings about in the configuration due to water holes, and change in the lipid molecule.

Ozone autohemotherapy, IV ozonized saline as coronavirus viricidal

Ozone, with extra molecule of oxygen, is known to be viricidal and bactericidal. Intravenous (IV) ozone therapy has been used for decades treating systemic conditions including cancer, autoimmune diseases, lime disease co-infections, flu, Methicillin-resistant Staphylococcus aureus (MRSA), cardiovascular conditions, and neurodegenerative diseases like Multiple Sclerosis, Alzheimer's, ALS, and Parkinson's by some traditional and alternative medical practice providers. These intravenous therapies appear to aid in speeding up the detox process, help eradicate bacterial and viral infections, and modulate the immune system, leading to healing. There are a few ways to administer ozone intravenously: major-autohemotherapy, and ozonated saline. *Never use direct ozone gas intravenously which is fraught with deadly complications. The physician who uses ozone therapy needs to have knowledge, training and experience using ozone therapy methods.*

Mode of action: as a treatment for coronavirus, ozone attacks the envelope of coronaviruses, according to Dr. Papadakos (director of anesthesia and critical care at Ibiza, Spain, where ozone can be legally administered). "By affecting cysteine, ozone disrupts viral proteins, lipoproteins, lipids and glycolipids in the actual virus. As a result, ozone creates a dysfunctional virus, which cannot replicate." Ozone therapy also affects certain coagulation parameters.

Method of preparation: To prepare ozone for IV, the patient's blood is saturated with ozone from a medically approved ozone generator, then the patient's blood containing the ozone molecule is infused through the peripheral venous access into the patient as described above. The ozone/oxygen mixture is adjusted to a 1:1 ratio, with a starting dose of 25 mic/mL of blood. "Although ozone has a long track record in Europe, where it is approved, ozone is not FDA approved in the United States, but can be used as "off label, compassionate use and repurposing old drug." Both ozone and vitamin C with thiamine hydrochloride and hydrocortisone/ dexamethasone of Keller (1948) and Marick (2017) can mitigate COVID-19, save the life of the patient, and prevent the patient being intubated for ventilatory support.

Ozone autohemotherapy (through blood) protocol: For COVID-19 infection, obtain exclusive venous access for administration. Use central line for ICU patients, and remove 100 ml blood into the sterile glass IV bottle (preferred) or bag. Add to the bottle, anticoagulant such as 1) heparin sodium: 1,000 IU for each 100 ml of blood removed or 2) sodium citrate 3.13%; 10 mL for each 100 ml of blood removed, or 3) instead, one can use ACD-A (Anticoagulant Citrate Dextrose A); 14 ml for each 100 ml of blood. Add initial dose, 40 mcg/ml of ozone. Increasing dose is acceptable in ensuing days to a maximum of 70 mcg/m. Take 100 ml volume of gas O2/O3, add to the blood, shake the bottle or bag gently once ozone is mixed with the blood, and shake it every 5 minutes during administration. Do 2 sessions per day in noncritically ill patients and ±3-4 sessions per day in critically ill patients until there is improvement.

Ozone through saline: Use pure oxygen to generate ozone using an FDA certified or approved ozone generator, at 1/8 LPM as the oxygen flow rate for 20 minutes generates between 33-40 gamma which will produce 2 PPM ozone concentration in the saline. This is for the 500 ml IV saline drip system, bubbled for 20 minutes or 100-200 ml of ozone can be drawn in a glass syringe from the ozone generator and added to 250-500 ml saline drip in a glass infusion bottle through a stopcock, shake, and slowly infuse over 90 minutes. It can be an effective COVID-19 viricidal in the blood, blood vessels, respiratory system and in the microvasculature that surrounds the alveoli and other part of the body.

Watch the infusions, and as soon as the infusate is coming to an end, stop infusion to prevent the air/ozone enter the systemic circulation. Never inject pure ozone gas intravenously, though it can be used rectally. The physician who uses ozone therapy needs to have knowledge, training and experience using ozone therapy methods.

Ozone through the mask and endotracheal tube: Ozone can be piped in through the endotracheal tube or breathing mask or through nasal canula as the patient takes deep breath 3-6-times, breaths every 1-3 hour to neutralize the free floating and mucosal sticking coronavirus. Be aware that prolonged ozone administration can damage air passages and lung parenchyma.

CDC recommended antiviral drugs for flu. Can they block COVID-19?

The CDC recommends the following antiviral drugs for the common flu. They are baloxavir, marboxil (Xofluza), oseltamivir (Tamiflu), and zanamivir (Relenza) for both flu prevention and treatment. Flu drugs are most effective when taken within the first 48 hours of flu symptoms, although they may help prevent severe disease when taken later. Oseltamivir, a commonly prescribed anti-flu drug is a neuraminidase inhibitor, a competitive inhibitor of influenza's neuraminidase enzyme. The enzyme cleaves the sialic acid which is found on glycoproteins on the surface of human cells that helps new virions to exit the cell. Thus, oseltamivir prevents new viral particles from being released. We believe that some of these therapeutic agents, such as zanamivir, work by binding to the active site of the neuraminidase protein, rendering the influenza virus unable to escape its host cell and infect others (Cyranoski D. 2005. "Threat of pandemic brings flu drug back to life". Nature Medicine. 11 (9): 909.). It is also an inhibitor of influenza virus replication in vitro and in vivo. In clinical trials, zanamivir was found to reduce the time-to-symptom resolution by 1.5 days if therapy was started within 48 hours of the onset of symptoms. We do not know how effective these anti-flu and cold medications would be for the virulent highly infective coronavirus, why not take it, because there are hardly any harmful effects. Our anti-coronavirus spray invention will be more effective to prevent flu as well as coronavirus infection.

Vitamin D3 supplement as an anti-inflammatory and immune system enhancer

Vitamin D binds to cell receptors all over the body, causing changes in cellular function, Duggan explains. A study published in Frontiers of Immunology in 2013 found that the vitamin D receptor in combination with hormonally active vitamin D serves to regulate more than 900 genes that influence many physiological functions. Vitamin D also helps the body absorb calcium. The body can make vitamin D from exposure to the sun, and a few foods have naturally occurring vitamin D. Salmon, tuna, cod liver oil, and egg yolks are some of the best. Mounting evidence indicates that vitamin D3 regulates the renin-angiotensin system. These results suggested that decreased plasma 25(OH)D levels were associated with increased activity of the renin angiotensin system (Chapter 21, Figure 8). This is why vitamin D affects blood pressure, diabetes, and many other maladies related to the renin-angiotensin system.

Studies in the Morehouse School of Medicine have shown that vitamin D3, in high doses, augments surgical healing and reduces inflammation. (Matthews LR, et al. Worsening severity of vitamin D deficiency is associated with increased length of stay, surgical intensive care unit cost, and mortality rate in surgical intensive care unit patients. Am J Surg. 2012 Jul;204(1):37-43.). COVID-19 is nothing but inflammation of the respiratory system and the rest of the body with *intense response from the immune system*. Hence, simply taking vitamin D to correct deficiency (over 30ng/ml 25(OH)D serum level) has proven positive effects on human health. As an anti-inflammatory it may reduce the cytokine storm, and inflammation of infected cells due to coronavirus infection of the respiratory system. I believe that many of the elderly patients who are severely affected by COVID-19 may have severe vitamin D deficiency along with co-morbidities, hence we advocate vitamin D3, as an immune-supporting effect against the coronavirus for those in nursing homes, extended care facilities, and the aged living independently.

Bioactive curcumin as an anti-oxidant, anti-inflammatory, anti-septic, anti-coagulant and such to prevent, curtail, stop the progression of and/or cure the disease

There are hundreds of research papers on the benefits of curcumin (Goel A, Jhurani S, Aggarwal BB. Multi-targeted therapy by curcumin: how spicy is it? Mol Nutr Food Res. 2008;52(9):1010-30. Aggarwal BB, Sundaram C, Malani N, Ichikawa H. Curcumin: the Indian solid gold. Adv Exp Med Biol. 2007; 595:1-75.). Curcumin is one of the most important supplements and has the most anti-oxidants, is an anti-inflammatory, anti-septic, anti-cancer, anti-cardio- vascular diseases, anti-Alzheimer's, anti-neurodegenerative diseases, anti-scarring, anti-coagulant, and anti-fat protein clump accumulation and such effects. Patients with COVID-19 affliction should take 1000-3000 mg of bioactive curcumin orally daily.

One of the reasons that curcumin works so well on such a wide range of diseases is because it is such a powerful anti-oxidant and anti-inflammatory. We know that almost all chronic diseases—from diabetes to heart disease to arthritis to Alzheimer's disease—have something in common: unchecked, destructive inflammation. Unlike synthetic drugs, which typically work against only a single inflammation pathway, natural curcumin reduces inflammation through its effects on multiple inflammation targets. In technical terms, curcumin has been found to:

a) Suppress the activation of the NF-kB, which regulates the expression of pro-inflammatory gene products, thus reducing the cytokine productions.
b) Downregulate COX-2, the enzyme linked to most inflammation and reduces the alveolar capillary leakage of alveolar blood vessels.
c) Inhibit 5-LOX, another pro-inflammatory enzyme that augments cytokine production.

d) Downregulate the expression of cell surface adhesion molecules linked to inflammation, thus prevent the coronavirus binding to the new cells and respiratory system to cause infection.
e) Inhibit the activity of TNF, one of the most pro-inflammatory cytokines (cell-signaling protein molecules), thus suppressing the lung inflammation.

Severe COVID-19 is due to massive inflammation of the respiratory system lining cells, and it is a physio-pathological immune system response with production of free radicals that is triggered when the body begins to attack the invading virus. However, inflammation should be limited, with a definite beginning and end, but not with coronavirus infection. It should not continue, day after day, but it does. Hence the bioactive curcumin is a must to treat coronavirus positive patients as an anti-inflammation anti-oxidant supplement. It thins the blood like anti-coagulants and aspirin, thus prevents thromboembolic phenomenon in the aged coronavirus infected.

On-going, persistent inflammation is destructive, not restorative. One of the keys to improving chronic diseases (heart disease, diabetes, arthritis, asthma, neurodegenerative diseases etc.) is stopping the cycle of chronic inflammation. As discussed earlier, curcumin, unlike synthetic drugs, works on multiple inflammation pathways to help return the body to a normal inflammation balance. Additionally, curcumin has specific, unique mechanisms of action that make it invaluable in treating chronic diseases.

Because of this anti-inflammatory activity, as well as its ability to kill/suppress coronavirus and tumor cells, increase activity of protective antioxidants such as glutathione, and modulate tumor growth cell factors, curcumin is effective against hundreds of diseases. It has an ability to neutralize free radicals especially during coronavirus infection. In fact, curcumin has an antioxidant value of over 1,000,000 per 100 g—many times higher than that of blueberries (6,552), strawberries (3,577) or dark chocolate (powder) (40,200)—well known food-based antioxidants.

Curcumin should be used as a nutritional supplement as well as a prophylactic and therapeutic agent in all coronavirus infections so that it can prevent the lung complications needing oxygen supplementation with intubation and ventilation. Every COVID-19 patient should take 1000-3000 mgs bioactive curcumin. It is not toxic up to the dose of 8,000 mg and it is also inexpensive (from the forthcoming book on Indian Gold by Dr. T. R. Shantha 2021).

Zinc and its health benefits in fighting flu and COVID-19 (Chapter 21, Figure 7)

A randomized trial showed that the administration of zinc and other micronutrients together were significantly more efficacious for cell-mediated immunity. The presence of zinc in the diet affects various aspects of cell-mediated immunity, including expression of interleukin-2 and interferon-γ. Interleukin-2 stimulates generation of natural killer and cytolytic T cells that kill viruses including coronavirus, bacteria, and tumor cells. Interferon-γ and interleukin-2 together activate macrophage monocytes that kill parasites. Zinc also suppresses or inhibits the protease from HIV type viruses and probably has the same effect on the coronavirus also. The elderly may be deficient in zinc, hence coronavirus attack augmented leading to increase in morbidity and mortality. The latest studies combining dihydrochloroquin and chloroquine, azotomycin and zinc resulted in rapid recovery without entering to ICU or on ventilatory support from COVID-19.

Impaired immune function in people with zinc deficiency can lead to the development of respiratory, gastrointestinal, or other infections, e.g., pneumonia, frequent cold, flu and such. I believe that the elderly in nursing homes, and extended care facilities and those suffering from chronic diseases are zinc deficient, hence prone to severe coronavirus affliction and its adverse health effects. The levels of inflammatory cytokines (e.g., IL-1β, IL-2, IL-6, and TNF-α) in blood plasma are affected by zinc deficiency and zinc supplementation produces a dose-dependent response in reduction in the level of these cytokines. Every elderly person should be on the zinc oral supplement. During inflammation, there is an increased cellular demand for zinc and impaired zinc homeostasis from zinc deficiency is associated with chronic inflammation (Chapter 21, Figure 7). Zinc supplements used for the treatment of the common cold at doses of 75 mg/day within 24 hours of the onset of symptoms have been shown to reduce the duration of cold symptoms by about 1 day in adults. The intranasal use of zinc-containing nasal sprays has been associated with the loss of the sense of smell.

The human rhinovirus – the most common viral pathogen in humans – is the predominant cause of the common cold. The hypothesized mechanism of action by which zinc reduces the severity and/or duration of cold symptoms is the suppression of nasal inflammation and the direct inhibition of rhinoviral receptor binding and rhinoviral replication in the nasal mucosa.

Zinc increases immunity and fights colds taken as lozenges, tablets, or in syrup. It has been proposed as an antiviral supplement for decades. Zinc is an effective anti-inflammatory and antioxidant agent. When taken over months, zinc may reduce a person's risk of becoming sick with the common cold, flu including coronavirus. We advocate taking 50-75 mg of zinc and use of zinc lozenges in those affected with COVID-19, and add zinc nasal and oral spray to inhibit the virus biding to healthy cells and suppress the cytokine production. Injectable zinc sulfate is available, and can be used with high dose vitamin C to treat COVID-19 in moderate to severe cases. Chloroquine also acts as a zinc ionophore, thereby allowing extra cellular zinc to enter inside the cell and inhibit viral RNA dependent RNA polymerase. This could be the potential mechanism of action of chloroquine and zinc against COVID-19.

Ethylenediaminetetraacetic acid (EDTA) as antibacterial and antibiofilm to prevent coronavirus sticking to nasal and oral mucosa

Ethylenediaminetetraacetic acid (EDTA) is a well-known metal-chelating agent, extensively used for the treatment of patients who have been poisoned with heavy metal ions such as mercury and lead as chelation therapy to prevent or reduce arteriosclerotic vascular diseases of the coronary blood vessels. I have used it personally and in my medical practice. It is used by hundreds of physicians all over the world as cardio-vascular diseases and cancer protractor and detoxifying agent.

More recently EDTA has been used as a permeating and sensitizing agent for treating *biofilm-associated conditions in the oral cavity in dental care*, on medical devices, and in veterinary medicine. In human medicine EDTA is presently formulated into commercially available wound dressings that are used to modulate matrix metalloproteinases (MMP) and manage wound infections. EDTA compositions are also being developed and employed for reducing biofilms in intravascular and urinary catheters and therefore represents an antibiofilm agent, which can significantly help to reduce catheter-related bloodstream infections (Thomsen TR, Hall-Stoodley L, Moser C, Stoodley P. The role of bacterial biofilms in infections of catheters and shunts. In: Bjarnsholt Thomas, Jensen Peter Østrup. Moser Claus, Høiby Niels, eds. Biofilm Infections. New York: Springer, 2011:91–109).

EDTA has been utilized for the control of microorganisms and biofilms biguanide (PHMB) quaternary ammonium compounds, silver, iodine, surfactants, and other antiseptics (Simon Finnegan1 and Steven L. Percival. EDTA: An Antimicrobial and Antibiofilm Agent for Use in Wound Care. Adv Wound Care (New Rochelle). 2015 Jul 1; 4(7): 415–421. Chang Y, Gu W, McLandsborough L. Low concentration of ethylene diamine tetra acetic acid (EDTA) affects biofilm formation of Listeria monocytogenes by inhibiting its initial adherence. Food Microbiol 2012;29: 10–17. Ayres HM, Payne DN,

Furr JR, Russell AD. Effect of permeabilizing agents on antibacterial activity against a simple Pseudomonas aeruginosa biofilm. Lett Appl Microbiol 1998; 2:79–82. Percival SL, Bowler P, Parson D. (2005) Antimicrobial composition. US/WO2007068938A2?cl=en).

The effect of EDTA on bacteria was reported 50 years ago (Rown MR, Richards RM. Effect of ethylenediamine tetraacetate on the resistance of Pseudomonas aeruginosa to antibacterial agents. Nature 1965;207: 1391–1393). Numerous EDTA compositions and combinations provide powerful antiseptic activities and function as antimicrobial agents against bacteria and pathogenic yeast and viruses.

Environments can increase our risk to infections caused by an array of different microorganisms including bacteria, protozoa, fungi, yeasts, and viruses. such as coronavirus. They can attack surfaces of objects (internally and externally), fluids and fluid conduits of nasal and oral cavities and respiratory passages. These infections lead to longer hospital stays for patients and increased hospital costs, worse, a large number of patient deaths are attributed to infections. Our anti- coronavirus spray with EDTA will prevent and or inhibit the coronavirus binding to healthy cell surfaces.

There is growing evidence that EDTA, specifically tetrasodium EDTA (tEDTA), is an antimicrobial and antibiofilm agent. EDTA compositions are highly effective in eliminating existing biofilms, and preventing biofilm formation. Based on the characteristics of EDTA, it is the aim of this invention to use in nasal and oral sprays as well as mouth washes and intravenously to act as coronavirus viricidal, prevent its attachment to healthy cells with its spike protein and allow the therapeutic agents in the sprays and intravenous fluids administration to clear the microbial infections and expose the respiratory track system to prevent the entry of coronavirus, flu and cold viral particles and other microbes.

These mouth washes, nasal and mouth sprays with EDTA of our invention should be provided in the hospital bed side table, bath room counters, and every house should have them next to the bath room sink. They are a must in nursing homes, and long care facilities. These can be formulated to suit the need using the below described formulations in this invention. The EDTA can be formulated with chlorohexidine, and with any of anti-microbial therapeutic agents as described in this invention. For example, a 5% solution of disodium EDTA has a pH of 4.0–5.5, trisodium EDTA a pH range of 7.0–8.0, and tEDTA 8.50–10 and above as specified in the British Pharmacopoeia. We select trisodium EDTA which is closer to our blood pH. There are alternative to EDTA type chelator available such as S, S isomer of EDTA, ethylenediamine-N, N′-disuccinic acid (EDDS), deferoxamine, Iminodisuccinic acid (IDS), Polyaspartic acid, Trisodium dicarboxymethyl alaninate, also known as methylglycinediacetic acid (MGDA), and such.

At low concentrations EDTA has been shown to prevent biofilms by inhibiting the adhesion of bacteria. Furthermore, it has also been shown to reduce biofilm colonization and proliferation. The effect of EDTA and its ability to chelate and expose the cell walls of bacteria and viruses including coronavirus and its ability to destabilize a biofilm by sequestering calcium, magnesium, zinc, and iron makes it a suitable agent for use in the prevention and management of biofilms in the mouth and nose and allow our antiviral to ***prevent the adhesions and entry of coronavirus into healthy cells***.

The action of EDTA may enhance the therapeutic effect of other antimicrobials and coronavirus viricidal by disrupting the biofilm structure in which the target microorganisms are encased. Use of our sprays and mouth washes reduces or eliminates the possibility of microorganism including coronavirus sticking to healthy nasal and oral mucosa and spreading to the rest of the respiratory system. Furthermore, when combined with different antimicrobials its *synergistic ability* for enhancing the antimicrobial and anti-viricidal efficacy is also evident.

In view of its low cost, effectiveness as an anticoagulant, antibiofilm, antiviral, and antimicrobial activity tEDTA should be choice of mouth wash as well as intravenous administration (measured amount, less than chelation infusion) with other therapeutic agents. Selected amount of EDTA intravenous infusion is therapeutic to prevent the thrombo-embolic phenomenon that is prevalent as one of the outcomes of COVID-19 reduce or eliminate vital load of the infected.

Chapter 31

Vitamin C and its effect in treatment of COVID-19 infection

Vitamin C for the treatment of COVID-19 infection

Our body does not synthesize vitamin C. It is obtained through food and as a supplement. It plays a major role as an antioxidant, builds collagen and maintains integrity between cells (prevents leaking from blood and lymph vessels and alveolar lining cells due to the coronavirus infection). In our body, collagen is the most abundant protein. If a person weighs 75 kg, 12.5 - 15 kg of that is protein. Of this weight again 6 – 7.5 kg (15-18 lbs.) are collagen and vitamin C plays a major role in its production and function.

Vitamin C acts as a *potent barrier of the coronavirus* as well as other microbes that try to enter the nose, mouth, and the rest of the respiratory system, as well as the gastro-intestinal track and the skin. It restores the barrier function by producing collagen and repairing the junction between cells lining the nose, pharynx, the rest of respiratory system, and the microvasculature endothelial lining surrounding the alveoli to prevent inflammatory exudates entering the lungs and thus protect the respiratory system onslaught by the coronavirus, its inflammatory reaction, and cytokines production as a result. At the same time, it enhances immune system activity.

Vitamin C is directly involved in response to coronavirus and other infections. It acts by enhancing immune system cell migration such as neutrophils and augmenting the function of the lymphocyte (second most common cells nest to neutrophils) of the immune system (B cell, T cells, and NK natural killer cells) at the affected site by multiplying and amplifying their killing effect by hydrogen peroxide production by neutrophils and antibody production by lymphocytes. It is a powerful antioxidant (bioactive curcumin is more powerful anti-oxidant compared to vitamin C) and viricidal, hence should be part of the protocol in the treatment of COVID-19. It should also be taken 1000 mg 2-3 times a day orally as a prophylactic to prevent coronavirus infections and obtund its effects.

Vitamin C is one of the most important vitamins available for oral and intravenous administration and even local application. Many old people infected with coronavirus develop classic respiratory distress syndrome

(RSD) due to accumulation of inflammatory fluid in the alveolar sacs and respiratory passages as phlegm-respiratory-inflammatory exudate due to leaking between the endothelial and alveolar lining cells. Vitamin C blocks it.

If the leaking continues, they become hypoxic and need to be intubated and ventilated and patients are in a state of shock associated high mortality with the present management. We have treated similar patients with high dose vitamin C and corticosteroids and antibiotics along with N-acetyl cysteine spray to liquify phlegm. This was the method of treatment by Dr. Klenner 7 decades ago, and it was used to treat every viral infection including poliomyelitis (Klenner, F. R.: Virus pneumonia and its treatment with vitamin C. Southern Med. Surg., Feb. 1948, Klenner, F. R.: Encephalitis as a sequela of the pneumonias. Tri-State Med. J., Feb. 1960. Klenner, F. R.: An insidious virus. Tri-State Med. J., June 1957 Massive doses of vitamin C and the virus diseases. Southern Med. Surg., 1951. Pauling, L.: Vitamin C and the Common Cold. W. F. Freeman & Co., 1970. US Patent Pub. No. 2016/0074480 A1, 2016, TR Shantha).

Marik et al. 2017 used high dose intravenous (IV) with thiamine and hydrocortisone with great success in treating shock and severe sepsis of various etiologist akin to Klenner method. (Marik PE, Khangoora V, Rivera R, Hooper MH, Catravas J. Hydrocortisone, vitamin C, and thiamine for the treatment of severe sepsis and septic shock: a retrospective before-after study. Chest. 2017;151: 1229–1238. Life Extension Journal, by William Faloon, September, 2018, As We See It, Sepsis: Is There a Cure? pages 7-13). Cytokine storm of COVID-19 is a septic shock and this vitamin C, thiamine hydrochloride and hydrocortisone with zinc should be used against it as part of treatment in ICU and in the hospital settings as soon as the coronavirus infection is diagnosed. This therapy is a must in children who suffer from ***"Pediatric multisystem inflammation syndrome"*** akin to Kawasaki disease. The study in UK shows improvement in outcome coronavirus infected seriously ill and ventilator using dexamethasone. It is said to inhibit over active immune system, that produce cytokines and cytokine storm which is a modified approach of Marik, Klenner and our work.

Now we have a pandemic due to coronavirus infection of the lung putting the patients in a state of shock due to development of severe acute respiratory distress syndrome (SARS) due to respiratory failure. As described in various studies, COVID-19 affects the elderly who are immunocompromised with underlying health conditions with leaking endothelial cells around the aveoli. Ventilating with 100% is a life saving measure, so also use of extracorporeal membrane oxygenation (ECMO). We advise all the intensive care units and physicians to use high dose of vitamin C, thiamine hydrochloride and hydrocortisone or dexamethasone in combination as soon as the coronavirus lung affliction diagnosed and those in ARDS with or without ventilatory support. This protocol was used by Klenner decades ago, and Marik further expanded the protocol to treat sepsis and septic shock with dramatic recovery of the afflicted. It is also time tested, why not use it to treat COVID-19?

Cytokine storm of COVID-19 is a forerunner of septic shock. By using the protocol for COVID-19 we can save most of these COVID-19 respiratory distress syndrome (RSD) afflicted thus reducing the morbidity and mortality (Mohadeseh Hosseini Zabet, et al. Effect of high-dose Ascorbic acid on vasopressor's requirement in septic shock. J Res Pharm Pract. 2016 Apr-Jun; 5(2): 94–100. Khalili H. Ascorbic acid in septic shock J Res Pharm Pract. 2016 Oct-Dec; 5(4): 301–302.).

In the latest study, using vitamin C on the critically ill patients with COVID-19 in ICU, they found an average reduction in ventilator time of 25% among patients who received 1-6 grams of intravenous or oral vitamin C per day (Nobuaki Shime, Save the ICU and save lives during the COVID-19 pandemic. *Journal of Intensive Care* volume 8, Article number: 40 (June 2020).

Further, in the process of fighting infection, immune cells rapidly use up vitamin C, hence we need give vitamin C to all ICU and hospitalized COVID-19 patients to boost the immune system function. Administer 1.5-3 grams of vitamin C intravenously and repeat once or twice a day.

Vitamin C maintains the barrier function repairing junctions between cell lining the respiratory system and microvasculature that surrounds the alveoli, thus reduces the alveoli drowned in inflammatory phlegm due to these microvasculature leak into alveoli and trachea-bronchial tree. It reduces the coronavirus burden by activating the immune system as explained above.

Follow below protocol to treat COVID-19

a) Vitamin C: 1.5 grams (administered as an infusion over 30 to 60 minutes) every 8-16-24 hours four days or until ICU discharge and patient shows improvement. We have used higher doses depending upon the severity of infection and the level of sepsis and septic shock. Infuse IV in 250 ml normal saline drip run over 30 minutes. Vitamin C kills microbes by production of hydrogen peroxide which may reduce the inflammation and pathologic deposits in the brain. Vitamin C is a logical approach to reducing the impact of the inflammatory changes due to coronavirus infection.

b) Thiamine (vitamin B1): 200 mg every 12 hours for four days or until ICU discharge or other symptoms improve. Vitamin C with thiamine protects against capillary inflammatory damage, capillary leak, and endothelial dysfunction of inner lining of the blood vessels of the alveoli and rest of the body microvasculature by coronavirus. It reverses the major site of damage in mitochondria due to inflammation by coronavirus and restores its function by preventing inactivation of the 2-oxoglutarate dehydrogenase complex by coronavirus cytokines to restore the mitochondrial function and prevent the capillary leak into lung alveoli.

c) Hydrocortisone 50 mg or dexamethasone 10-20 mg every 12 hours for weeks or until ICU discharge and / or symptoms of the disease improve, followed by a tapering dose over three days. Vitamin C, thiamine hydrochloride and hydrocortisone will have a dramatic reversal of loss of vascular barrier function, i.e., capillary permeability, prevent capillary leakage of intravascular fluids into lung alveoli and stop the progression of the pathology in the lungs due to COVID-19 that can result in ARDS.

d) Combining this therapy with intravenous remdesivir, zinc, oral angiotensin receptor blockers such as telmisartan and losartan and antioxidant like auranofin will be very effective in moderate to severe cases of COVID-19, may save the lung from the onslaught of coronavirus infection associated with cytokine fury. It may also prevent post infection scarring of the lungs. May even consider ivermectin one dose therapy early in the COVID-19 described in this invention.

The above doses of therapeutic agents can be modified depending upon the response, severity of sepsis and shock, signs and symptoms of the diseases other than sepsis, age and weight of the patient and other COVID-19 associated comorbidities.

We have used oral bioactive curcumin 100-2000 mg orally and vitamin D3, 5000-10,000 IU in high doses to maintain the integrity of gastrointestinal track in suspected sepsis, severe infection and septic shock, after chemotherapy, to remove free radicals and reduce inflammation due to hypoxic damage in 35 feet of gut wall inside lining which is teeming with trillions of pathogenic and nonpathogenic microbiomes. This is needed to protect the gut from amplifying and expanding the sepsis and septic shock produced by ischemic gut wall and the trillions of bacteria it lodges that produce exo and endo toxins, by removing the free radicals and reducing the inflammation of the gut wall. The effect spreads all over the body if untreated. I believe adding EDTA infusion before or after vitamin C infusion, can have a dramatic effect on therapeutic effectiveness of vitamin C, thiamine hydrochloride and hydrocortisone or dexamethasone and other therapeutic agents against the coronavirus infection.

Mode of action of vitamin C, corticosteroids and thiamine therapeutic agents on the lungs subjected to cytokine attack produced by coronavirus

A large body of experimental data has demonstrated that both corticosteroids and intravenous vitamin C act synergistically to reduce activation of nuclear factor kB (NF-kB), thus attenuating the release of pro-inflammatory mediators (cytokines, chemokines, prostaglandins), reducing the endothelial injury of micro-blood vessels and alveolar lining characteristic of severe coronavirus infection sepsis of the lungs, thereby reducing endothelial permeability and improving micro-circulatory flow, augmenting the release of endogenous catecholamines and enhancing vasopressor responsiveness. This dose of vitamin C is very safe with no recorded complications. Vitamin C with thiamine and hydrocortisone protects against capillary inflammatory damage, capillary leak, and endothelial dysfunction of inner lining of the blood vessels and the alveolar lining thus reducing the pathology in the blood vessels as well on the cell membranes, making them less leaky. It can be used in any and all shocks including COVID-19 induced shock due to cytokine storm which affects the alveoli and its vasculature.

Mechanisms of action of high dose vitamin C, thiamine hydrochloride and hydrocortisone with auranofin for the treatment of coronavirus induced lung pathology that leads to pulmonary edema associated cytokine storm

Corticosteroids and intravenous vitamin C reduces activation of nuclear factor kB (NF-kB) attenuating the release of pro-inflammatory mediators (cytokines, chemokines, and prostaglandins), reducing the endothelial and alveolar injury characteristic of coronavirus sepsis, thereby reducing endothelial permeability and improving microcirculatory flow of the alveoli, augmenting the release of endogenous catecholamines and enhance vasopressor responsiveness and reducing the fluid in the lungs, and allowing for proper oxygen exchange in the lung.

Low dose corticosteroids have proven to be safe, act synergistically to reduce activation of nuclear factor kB (NF-kB), thus attenuating the release of pro-inflammatory mediators (cytokines, chemokines, prostaglandins), reducing the endothelial injury of micro-blood vessels characteristic of coronavirus infection sepsis of the lungs, thereby reducing endothelial permeability and improving micro-circulatory flow, augmenting the release of endogenous catecholamines and enhancing vasopressor responsiveness. Corticosteroids decrease vasopressor dependency. This dose of vitamin C is very safe with no recorded complications.

Vitamin C with thiamine along with corticosteroids protects against capillary inflammatory damage, capillary leak, and endothelial dysfunction of inner lining of the blood vessels and the lining cells of the alveoli reducing the pathology in the blood vessels as well on the alveolar cell membranes. Similarly, intravenous vitamin C has been evaluated in ICU patients. Klenner used it 7 decades ago for many years with successes to every kind of viral infection. In vitro data has suggested that vitamin C and hydrocortisone with vitamin B1 may act synergistically and protect the vascular endothelium from damage by cytokine. One of the important aspects of Klenner's protocol to treat multiple sclerosis uses high dose vitamin B1, which I have used successfully. Based on these clinical and experimental data initiate treatment protocol for patients with severe sepsis and septic shock condition of the lungs due to coronavirus infection.

Klenner and Marik et al. and other studies have demonstrated reproducibly and consistently that this therapeutic cocktail of vitamin C, thiamine hydrochloride and hydrocortisone reverse the lung dysfunction with a marked reduction in mortality. This protocol with modifications has now been or is being adopted at multiple centers across the world with remarkably consistent and reproducible results.

Furthermore, treatment with the vitamin C cocktail we describe here in this invention reduces the duration of mechanical ventilation, duration of vasopressor support and duration of ICU stay. We postulate that this intervention will limit muscle loss thereby limiting critical illness myopathy (CIM) characteristic of critical illness and COVID-19 is no exception.

A large body of experimental data has demonstrated that both corticosteroids and high doses of intravenous vitamin C reduce activation of nuclear factor kB (NF-kB) attenuating the release of pro-inflammatory mediators by the immune system. Thus, it prevents the endothelial and alveolar injury thereby reducing endothelial permeability and improving microcirculatory flow to clear the inflammatory cytokines, and inhibiting their release and improving the alveolar function. It prevents the massive accumulation fluids and cells in the alveoli due to leaky pulmonary blood vessels, micro-vasculature and alveolar lining leakage and allows the patients to recover much faster with least morbidity and mortality.

The lungs get 2.5 liters of blood every minute. Hence even the small leaking alveolar blood vessels can spill plasma and other cellular contents of the blood to create massive lung pathology. On top of that, the infection is augmenting the production and spilling of the cytokines from the immune system cells making the COVID-19 disease even more critical due to difficulty in getting the oxygen through the lungs to supply the oxygen needs of the body.

The air that is inhaled has 20 percent oxygen, and the air that is exhaled is about 15 percent oxygen, so about 5 percent of the volume of air is consumed in each breath and converted to carbon dioxide. Therefore, a human being uses about 550 liters of pure oxygen (19 cubic feet) per day. With COVID-19 lung affliction, this is cut down drastically, because oxygen cannot enter the arterial blood due to severe acute respiratory distress syndrome (ARDS) resulting hypoxia needing 100% supplemental oxygen delivered to the lungs through an oxygen mask, endotracheal tube ventilation with or without Positive End-Expiratory Pressure (PEEP). In COVID-19 patients on spontaneous ventilation using non-invasive ventilation (NIV), use CPAP (continuous positive airway pressure) which is analogous to PEEP, but the pressure applied is maintained throughout the respiratory cycle (during both inspiration and expiration) can also increase and improve oxygenation of blood.

All the COVID-19 patients on the ventilatory oxygen supplement support should get 1.5-3 grams of vitamin C intravenous to shorten the duration of ventilatory support according to J Intensive Care. 2020 Feb 7;8: 15 publication. Some of these patients may need ***veno-venous extracorporeal membrane oxygenation (ECMO)*** to maintain the oxygen if all other measures such as intubation and PEEP ventilation.

The protocol to follow to treat the respiratory distress syndrome (RSD) of COVID-19 is as follows.

Start this therapy when the COVID-19 patients show signs and symptoms of difficulty in breathing with low oxygen saturation. This can be instituted as soon as the diagnosis is made and follow our protocol described below.

a) Vitamin C: Vitamin C is provided by the manufacturer as a 50 ml vial at a concentration of 500 mg/ml. Three (3) ml of vitamin C is placed in a 50 ml bag of Normal Saline (1500 mg vitamin C in 50 ml bag) which will then be infused over 1 hour. The dosing schedule is 1500 mg every 6-12-24 hours depending upon the patient condition / until the signs and symptoms of the disease abates.

b) Hydrocortisone: Patients are treated with hydrocortisone 50 mg IV q 12-24 hourly for until ICU discharge. Optional dosing strategy: Hydrocortisone 50 mg bolus (or dexamethasone), followed by a 24-hour continuous infusion of 200 mg for one day or for 7-15 days or as needed.

c) Thiamine hydrochloride: Patients will receive intravenous thiamine 200 mg q 12 -24 hourly for 7-15 days or until ICU discharge or as needed. It acts as metabolic resuscitator of blood vessels.

d) Administer ivermectin one dose could help to bring about the end to coronavirus infection.

e) We administer azithromycin 500 mg once a day or appropriate antibiotics in addition to prevent any secondary infection of the lungs and other systems, then 200 mg once a day.

f) We also add angiotensin receptor blockers telmisartan 10-20 mg immaterial of blood pressure to prevent the shed new viral particles attaching and attacking the healthy cells. They block the entrance of coronavirus on the alveoli and their blood vessels lining cells and inhibit their leaking to alveolar sacs, thus reduce the cytokine shock and maintain pulmonary function, oxygenation and elimination of carbon dioxide.

g) Administer dihydrochloroquin 200 mg twice a day to prevent the coronavirus multiplication and their spread.

h) Protease inhibitors such as remdesivir can be added depending upon the response to above therapy.

i) One dose ivermectin may be tried as soon as the person has tested positive for coronavirus.

Do not overhydrate, and maintain proper ICU protocol used to treat patent in shock. We can also add well-established anti-inflammatory such as auranofin, vitamin D3, curcumin also. EDTA with 1-2% DMSO infusion can be initiated depending on the condition of the patients.

Chapter 32

Antimalarial Dihydrochloroquin and chloroquine phosphate chloroquine against coronavirus infection

Anti-malarial drug controversy as coronavirus antiviral therapy

I firmly believe that Dihydrochloroquin and chloroquine work in the treatment of COVID-19. Given at early stages of coronavirus infection, it is an effectual valuable treatment, and ***I reject the political and pseudoscientist ideas promulgated by some in the media that it should not be used.*** I have used it for treating cancers, Parkinson's, autoimmune diseases, flu and such.

Antimalarial as prophylactic against coronavirus and flu virus infection

I believe that chloroquine acts as a prophylactic if one takes it once or twice a week if that is needed because it has a long plasma life. This micro-dose has no effect on the heart. It is one of the most important measures you can take besides wearing the mask and social distancing. That is why we have added Dihydrochloroquin and chloroquine phosphate in our nasal spray and mouth washes. Further we noticed those cancer patients who were treated with antimalarial hardly developed flu, akin to angiotensin receptor blockers.

It has been proposed that dihydrochloroquin and chloroquine phosphate (names used interchangeably), used as an antimalarial drug for more than many decades, can be helpful in fighting against coronavirus outbreak. Now it is no longer experimental or compassionated use drug. I advocate every COVID-19 positive should take it till the alternate specific coronavirus therapy is discovered. In small doses, they can be taken as prophylactics also by those working in high risk areas such as in the hospitals, ICU and at gatherings. ***It should be prescribed by qualified physician and one needs to watch for heart irregularities due to the conduction block effect it has.*** I do not believe that using it in sprays has any such adverse effects.

Reports that Dihydrochloroquin and chloroquine phosphate do not work for treatment of coronavirus infection are misleading, and it is due to using these therapeutic agents in cases far advanced with COVID-19 where the immune system and coronavirus are literally at war. If it is given early, it will act as an effective anti-coronavirus therapeutic agent. It can be an effective prophylactic agent also in the spray and taken orally as we describe below. It can be a life saver when remdesivir is not available.

It is important to remember that research studies show chloroquine has both prophylactic and therapeutic effects. chloroquine has strong antiviral effects on SARS-CoV infection of primate cells. These inhibitory effects are observed when the cells are treated with the drug either before or after exposure to the virus, suggesting both prophylactic and therapeutic advantage (Virol J. 2005; 2: 69. Published online 2005 Aug 22. doi: 10.1186/1743-422X-2-69). In addition to the well-known functions of chloroquine such as elevations of endosomal pH, which blocks coronavirus assembly that needs an acid media. This drug appears to interfere with terminal glycosylation of the cellular receptor, angiotensin-converting enzyme 2. This may negatively influence the virus-receptor binding and abrogate the infection, with further ramifications by the elevation of vesicular pH, resulting in the inhibition of infection and spread of SARS CoV at clinically admissible concentrations.

Chloroquine can inhibit a pre-entry step of the viral cycle by interfering with viral particles binding to their cellular cell surface receptor which could be due to an antiadhesive effect. Chloroquine also interferes with ACE2 receptor glycosylation thus preventing SARS-CoV-2 binding to target cells, hence, we add in our prophylactic spray. Angiotensin receptor blocker can augment this effect, it is included in the spray to prevent entry of coronavirus into healthy cells. Preliminary data indicate that chloroquine interferes with SARS-CoV-2 attempts to acidify the lysosomes and presumably inhibits cathepsins, which require a low pH for optimal cleavage of SARS-CoV-2 spike protein, a prerequisite to the formation of the autophagosome. Chloroquine could also interfere with proteolytic processing of the M protein and alter virion assembly and budding (Fig. 6). Finally, in COVID-19 this drug could act indirectly through reducing the production of pro-inflammatory cytokines and/or by activating anti-SARS-CoV-2 CD8+ T-cells.

Further studies have shown that chloroquine appears as a versatile bioactive agent reported to possess antiviral activity against RNA viruses as diverse as rabies virus, polio virus, HIV, hepatitis A virus, hepatitis C virus, influenza A and B viruses, influenza A H5N1 virus, Chikungunya virus, Dengue virus, Zika virus, Lassa virus, Hendra and Nipah viruses, Crimean–Congo hemorrhagic fever virus and Ebola virus, as well as various DNA viruses such as hepatitis B virus and herpes simplex virus (Devaux et al., New insights on the antiviral effects of chloroquine against coronavirus: what to expect for COVID-19? International Journal of Antimicrobial Agents online

12 March 2020, 105938 https://doi.org/10.1016/j.ijantimicag.2020.). According to preliminary reports from the Chinese authorities suggesting that approximately 100 infected patients treated with chloroquine experienced a more rapid decline in fever and improvement of lung computed tomography (CT) images and required a shorter time to recover compared with control groups, with no obvious serious adverse effects, the Chinese medical advisory board has suggested chloroquine inclusion in the SARS-CoV-2 treatment guidelines.

Excitement has grown about the use of the drug to fight COVID-19, with research studies showing chloroquine is effective at preventing and treating the virus that causes severe acute respiratory syndrome. SARS is in the same coronavirus family as COVID-19, Dr. Len Horovitz, a pulmonologist and internist at Lenox Hill Hospital in New York City, told ABC News. The theory of the experiment with primate cells was that chloroquine could be for preventing viral infection or as a treatment for viral infection after it had occurred. Our invention includes Dihydrochloroquin and chloroquine phosphate in the nasal and oral sprays and mouth was as prophylactic against coronavirus attacking the healthy cells. Now we have protease inhibitor such as remdesivir developed for Ebola as therapeutic agents to prevent coronavirus.

Dihydrochloroquin and chloroquine phosphate are antimalarial drugs which are weak bases that are known to affect lysosomal (may be even Golgi vesicle) acid vesicles inside the cells to more alkaline milieu with coronavirus leading to dysfunction of several enzymes and stop the viral reproduction and final assembly. These antimalarial drugs cross the cell membrane and have direct molecular effects on endosomal-lysosomal activity, autophagy and signaling pathways. It is thus not surprising that chloroquine/hydroxychloroquine is concentrated within acidic organelles such as the endosome, Golgi vesicles, and the lysosomes, where the pH is low, meaning makes them less acidic, the viral proteins inside are ruined—they unfold because they are no longer at the correct acid pH for the virus to assemble. What this means for coronavirus is that its viral proteins are rendered nonfunctional because chloroquine changed the pH of the endosomes carrying them.

In addition to ruining the viral invader's proteins, chloroquine can distort the shape of proteins called cytokines, which can reside in the endosomes of immune cells that fight infection. Sometimes, the immune system gets too excited and creates what is commonly called a "cytokine storm," which has been a major complication of COVID-19. By quelling this cytokine storm, chloroquine provides a dual advantage in helping the body combat coronavirus infection. Positively charged chloroquine exerts direct antiviral effects, inhibiting pH-dependent steps of the replication of several viruses including members of the flaviviruses, retroviruses, and coronaviruses (Chapter 21, Figure 5, 6).

Chloroquine is a cationic drug, accumulates in acidic cellular compartments and binds to phospholids with a consequent increase in lysosomal, maybe even in Golgi pH, and induces phospholipidosis. It appears that impairment of lysosomal function by chloroquine leads to anti-inflammatory effects by inhibition of arachidonic acid release and prostaglandin E2 synthesis thus reducing the inflammation in the COVID-19 and in coronavirus storm.

Moreover, the anti-inflammatory properties of chloroquine may temper the noxious immune hyperactivation and cytokine storm that results in severe acute respiratory distress syndrome. Chloroquine/hydroxychloroquine, due to their wide range of antiviral and anti-inflammatory properties, have an effect on SARS-COVID-19 treatment, and the treatment is no longer speculative hypothesis or a compassionated use of the drug.

It is well established that weak bases, by increasing the pH of lysosomal and trans-Golgi network (TGN) vesicles, disrupt several enzymes including acid hydrolases and inhibit the post-translational modification of newly synthesized proteins such as coronavirus assembly to from new viruses, as well as delays the movement of virion within the ER—Golgi—Lysosomal organelles (Chapter 21, Figure 6). The chloroquine-mediated rise in endosomal pH modulates iron metabolism within human cells by *impairing the endosomal release of iron* from ferrated transferrin, thus decreasing the intracellular concentration of iron. This decrease in turn affects the function of several cellular enzymes involved in pathways leading to replication of cellular DNA and RNA for expression of different genes in the coronavirus replication.

For enveloped viruses such as coronavirus, post-translational modification of the envelope glycoproteins occurs within the endoplasmic and trans-Golgi network (TGN) vesicles. This process involves proteases and glycosyl-transferases, some of which require a low pH (acid). In line with the pH-dependence of these events, Dihydrochloroquin and chloroquine phosphate raises pH of the lysosomal - Golgi complex (Chapter 21, Figure 6) to inhibit budding of virus particles, and to induce accumulation of non-infectious virus particles in the TGN. Chloroquine also inhibits the replication of members of the Flaviviridae family by affecting the normal proteolytic processing of the flavivirus prM protein to M. As a result, viral infectivity is impaired and the mechanism of inhibition seems to be inhibition of glycosylation of the envelope glycoproteins of the virus. Due to this effect the coronavirus cannot reproduce and this reduces their infectivity as they are shed. Dihydrochloroquin and chloroquine phosphate accumulates in lysosomes (Chapter 21, Figure 6) inside the cell and inhibits intra lysosomal catabolism, on the synthesis, transport, and degradation of cell-coat glycoproteins in absorptive cells, thus they have an effect on coronavirus replication and assembly.

Dihydrochloroquin and chloroquine phosphate indirectly reduce the production of anti-inflammatory cytokines by various cell types. In vitro, they inhibit the production of IL-1, IL-6, TNF and IFNγ by mononuclear cells. Furthermore, treatment with hydroxychloroquine inhibits the production of TNF, IFNα, IL-6 and CCL4 (also known as MIP1β) in pDC and natural killer cells. It is these effect that reduce the cytokine storm syndrome and the lung complications seen in the coronavirus infected elderly with underlying health conditions. This may also cause uncontrolled immunodepression resulting in pulmonary superinfection, that is why the use of z-pack antibiotic (azithromycin) or other appropriated antibiotic therapy should be incorporated for the effective of severe acute respiratory distress syndrome (SARS) of coronavirus infection.

The World Health Organization announced in March of 2020, that it is launching a multi-country clinical trial to test four drugs as potential COVID-19 therapies. Chloroquine is one of the drugs. The other drugs are the antiviral drug remdesivir; a combination of two HIV drugs, lopinavir and ritonavir; and lopinavir and ritonavir plus interferon beta. All four of the drugs have shown potential to be effective against COVID-19, either in animal studies or in studies done in laboratories and on many humans. Ten countries have already indicated they will take part in the trial: Argentina, Bahrain, Canada, France, Iran, Norway, South Africa, Spain, Switzerland and Thailand.

People become infected with both severe acute respiratory distress syndrome and COVID-19 when protein spikes as explained above on the surfaces (Chapter 21, Figures 2-6) of those viruses bind to special receptors on the outside of human cells called angiotensin receptor (Chapter 21, Figure 5, 6 # 49) to enter the inside the healthy cell to multiply and spread to other healthy cells. Dihydrochloroquin and chloroquine phosphate as well as remdesivir with or without telmisartan worked against SARS by interfering with those receptors, thereby interfering with the virus's ability to bind to cells. Researchers believe there is a chance Chloroquine can do the same against COVID-19. "The way that it worked against SARS was by preventing of the attachment of the virus to the cells. Chloroquine interfered with the attachment to that receptor on the cell membrane surface," Dr. Len Horovitz, a pulmonologist and internist at Lenox Hill Hospital in New York City, said. "So, it's disrupting a lock and key kind of mechanism of attachment."

One of our inventive methods is to include chloroquine and angiotensin receptor blockers in health care workers and those exposed or that are at risk to use our nasal-oral-pharyngeal-laryngeal mucosa to see if it can help prevent coronavirus from getting inside the healthy cells to multiply and spread to the rest of the body and also as oral prophylactic and therapeutic agents.

The state-owned Xinhua news agency reported in February that chloroquine was effective against COVID-19 in China. In addition to China, South Korea and Belgium are using chloroquine to treat patients with COVID-19, according to a story from Business Insider. According to Medscape, chloroquine combined with the antibiotic azithromycin has shown to reduce the amount of virus in patients with COVID-19. Studies also show that protease inhibitor Remdesivir and antimalarials (dihydrochloroquin and chloroquine phosphate) are very effective against the coronavirus preventing the coronavirus infection. The result was that Remdesivir stopped the virus from replicating in a lab dish. Further, the group found that Chloroquine was effective in preventing the infection from spreading in human cells in the lab. This could be because of its anti-adhesion effect owing to its inhibition of inflammation owing to inhibition of production of production of IL-1 and IL-6 in human monocytes and T cells and down regulation of (TNF)-α production and signaling in macrophages. This is the same as using nasal-oral-pharyngeal-laryngeal mucosa spray to block the coronavirus, where the mucosal lining of these areas acts as a petri dish used to grow these viruses in the vitro setting.

Chloroquine has antiviral effects, which work by increasing endosomal pH required for the virus/cell fusion process. Chloroquine also seems to act as ***a zinc ionophore***, thereby allowing extra cellular zinc to enter inside the cell and inhibit viral RNA dependent RNA polymerase. This could be one of the potential mechanisms of action of chloroquine against COVID-19 along with zinc.

Neither chloroquine nor hydroxychloroquine underwent conventional drug development, but their use has become a part of current treatment guidelines for rheumatoid arthritis (RA), systemic lupus erythematosus (SLE), antiphospholipid syndrome (APS) and primary Sjögren syndrome (as repurposing off label use). I have used it in the treatment of many different kinds of cancers and Parkinson's as adjuvant therapy with successes. ***I have talked to a number of these patients on this dihydrochloroquin and chloroquine phosphate antimalarial and they tell me that they rarely develop flu and colds.*** In addition to having direct immunomodulatory effects, chloroquine and hydro chloroquine can reduce rates of atherosclerosis, improve hyperglycemia and hyperlipidemia and protect against infections in patients with inflammatory rheumatic diseases and other 152 or so autoimmune diseases.

Initial promising results continue to hold up, hence the chloroquine, which is easily available, offers a *cheap and safe treatment* for those suffering from COVID-19, compared to expensive remdesivir and such therapeutic agents.

Caution: Chloroquine has been used for decades and has a reputation for being extremely safe, although it is not used in patients with epilepsy, myasthenia gravis, or glucose-6-phosphate dehydrogenase deficiency and it can also cause severe reactions in patients with psoriasis. It can cause abnormal heart rhythms such as QT interval prolongation and a dangerously rapid heart rate called ventricular tachycardia. They can cause a complete heart block due to disturbance in the in-conduction system such as AV block, hence this is contraindicated in patients with preexisting cardiac conduction disturbances (Chatre C et al. 2018).

Summary of probable mechanism how dihydrochloroquin and chloroquine phosphate act against corona and other viruses

Chloroquine has multiple mechanisms of action (Chapter 21, Figures 5, 6) that may differ according to the pathogen studied.

a) Chloroquine inhibited budding with accumulation of non-infectious particles in the trans-Golgi network.
b) The intracellular site of coronavirus budding is determined by the localization of its membrane M proteins that accumulate in the Golgi complex beyond the site of virion budding inhibited by chloroquine.
c) Besides affecting the virus maturation process, pH modulation by chloroquine can impair the proper maturation of viral protein. Chloroquine makes the milieu in the endosome-lysosomes-Golgi complexes inside the virus infected cell to alkaline from acid environment.
d) Chloroquine on the immune system includes increasing the export of soluble antigens into the cytosol of dendritic cells and the

enhancement of human cytotoxic CD8+ T-cell responses against viral antigens.

e) Chloroquine improves the cross-presentation of non-replicating virus antigens by dendritic cells to CD8+ T-cells recruited to lymph nodes draining the site of infection, eliciting a broadly protective immune response.
f) In the model of HCoV-229 coronavirus, chloroquine-induced virus inhibition occurs through inhibition of p38 MAPK.
g) Chloroquine is a well-known immunomodulatory agent capable of mediating an anti-inflammatory response of the immune system.
h) Dihydrochloroquin and chloroquine phosphate increases antiviral interferon production to inhibit the coronavirus.
i) Chloroquine with iontophoresis allows more Zinc inside the afflicted cells to wreck the coronavirus reproduction machinery.
j) Chloroquine was found to inhibit interferon-alpha (IFNα), IFNβ, IFNγ, TNFα, IL-6 and IL-12 gene expression and their release activated by the virus. With addition of vitamin C, the effects will be more pronounced.
k) Chloroquine induces post-translational modifications, which involve proteases and glycosyltransferases, occur within the endoplasmic reticulum or the trans-Golgi network vesicles and may require a low pH. Chloroquine increases the pH and prevent the cellular processes of assembling viruses in the lysosomes-Golgi complex.
l) In the absence of antiviral drug such as dihydrochloroquin and chloroquine phosphate, the virus is targeted to the lysosomal compartment where the low pH, along with the action of enzymes, liberates the infectious nucleic acid for formation of viruses which is disrupted/misfolded/unfolded by chloroquine.
m) Chloroquine disrupts the viral particle, enzymes necessary for its replication. Chloroquine-mediated inhibition of hepatitis A virus was found to be associated with uncoating, thus blocking its entire replication cycle with similar effect on the coronavirus.
n) Chloroquine along with vitamin C and zinc inhibits interleukin-1 beta (IL-1β) mRNA expression in THP-1 cells and reduces IL-1β release. Chloroquine-induced reduction of IL-1 and IL-6 cytokines was also found in monocytes/macrophages. This reduces the production and effect of cytokines.
o) Chloroquine-induced inhibition of tumor necrosis factor-alpha (TNFα) production by immune cells was reported to occur either

through disruption of cellular iron metabolism, blockade of the conversion of pro-TNF into soluble mature TNFα molecules and/or inhibition of TNFα mRNA expression in coronavirus. Vitamin C with zinc will facilitate this process.

In a recent publication by multicenter, a randomized study of antimalaria for coronavirus has disappointing results. In patients hospitalized with COVID-19, hydroxychloroquine was not associated with reductions in 28-day mortality but was associated with an increased length of hospital stay and increased risk of progressing to invasive mechanical ventilation or death (doi: https://doi.org/10.1101/2020.07.15.20151852). It may give disappointing results in very severe cases of COVID-19 affecting the aged with comorbidities administered late in the disease. I believe that it still has an antiviral effect, given at an appropriate time with combination of other therapeutic agents such as remdesivir, ivermectin, vitamin C, thiamine hydrochloride and hydrocortisone/dexamethasone.

Chapter 33

Ivermectin, Niclosamide, auranofin (Ridaura®), colchicine, cimetidine against the coronavirus infection COVID-19

Anti-parasitic drug ivermectin for COVID-19 to lower morbidity and mortality and as a possible cure

Another treatment option that appears to be safe and effective is the use of the anti-parasitic (anti-helminthic) drug ivermectin (sold under brand names: Heart Gard, Sklice and Stromectol in US). It is a macrocyclic lactone derived from the bacterium Streptomyces avermitilis, now being tested for chikungunya and yellow fever as well as Nonalcoholic Fatty Liver Disease.

Ivermectin kills parasites by interfering with nervous system and muscle function, in particular by enhancing inhibitory neurotransmission. It works by causing the parasite's cell membrane to increase in permeability, resulting in paralysis and death. Increased swelling of the virus due to increased permeability make it unable to bind to the angiotensin receptors on the healthy cells and multiply. It is on the World Health Organization's List of Essential Medicines, used for river-blindness (onchocerciasis), pinworm, whipworm, ascariasis, lymphatic filariasis; scabies, lice, rosacea (local application). ivermectin has been widely used on the continent of Africa for many years as an anti-parasitic (river blindness) and is believed to be a primary reason that COVID 19 has not severely impacted the African population. (https://www.worldhealth.net/news/ivermectin-displaying-promising-results/).

Now ivermectin is being tested and has been used also for coronavirus infection treatment. The Gates Foundation is funding a clinical trial of ivermectin in France, according to TrialSiteNews.com. Doctors familiar with clinical trials using it for COVID-19, described patients' results as dramatic with "statistically significant improvement in mortality," according to the source (Newsmax, July 15, 2020). "There's a common denominator here," said emergency medical physician Dr. Peter H. Hibberd, M.D., of Palm Beach County, Florida. "This drug is salvaging people from their death bed," he said. This tells us that this ivermectin can be used in any stage of coronavirus infection, early, middle or late stages of the disease.

A preprint published on 10 June 2020 reported on an observational retrospective study of COVID-19 patients at four Florida hospitals and found a significantly lower mortality in those who had received ivermectin [Rajter, Juliana Cepelowicz; Sherman, Michael; et al. (June 10, 2020). "ICON (Ivermectin in COVID Nineteen) study: Use of Ivermectin is Associated with Lower Mortality in Hospitalized Patients with COVID-19 (preprint)"]. As of June 4, ivermectin was being studied in 10 ongoing and 14 planned clinical trials. Just one ivermectin pill and then the course of antibiotics for ten days resulted in a 100% cure rate for COVID-19 patients according to the study.

In a separate trial, a team of doctors in Bangladesh reported that they administered a combination of ivermectin and doxycycline, a common antibiotic, to 60 coronavirus patients there. That team reported that within 72 hours after receiving the drug, their patients tested negative for the virus, and by the fourth day, they had recovered. Dr Tarek Alam, a senior doctor for the Bangladesh Medical College Hospital, told India's ZeeNews that the results were "astounding" and said all patients "experienced full recoveries."

The trials so far have shown ivermectin reduces the number of cell-associated viral DNA by 99.8 % in 24 hours. It is given as one dose with booster dose 7 days later if needed. Even if the FDA does not approve it for coronavirus infection, I would recommend it as *"off label, compassionate use and repurposing old drug."*

Dosage guidelines based on body weight for treating COVID-19:

a) 15 to 25 kg: 3 mg orally one time
b) 26 to 44 kg: 6 mg orally one time
c) 45 to 64 kg: 9 mg orally one time
d) 65 to 84 kg: 12 mg orally one time
e) 85 kg or more: 0.15 mg/kg orally one time

Ivermectin is taken as a single dose with a full glass (8 ounces) of water on an empty stomach 1 hour before breakfast.

Mode of action of ivermectin on coronavirus

a) Ivermectin is said to act by not allowing the multiplying viral particles to enter the nucleus, thus preventing or blocking the multiplication cycle of the virus inside the host cells.
b) According to health expert Dr Kaushal Kant Mishra, studies conducted at Monash University Australia and the Victorian Infectious Diseases Reference Lab found that ivermectin *eradicated the coronavirus within 48 hours*. The study also found that the use of the drug also weakened the RNA (genetic material) of

coronavirus by almost 93 per cent. Although the drug was not tried on humans in the study, the same was done by doctors of a private hospital in Bangladesh.

Because of its potent action on the coronavirus RNA and its effect on multiplication once inside the healthy cell, it may have application in nasal and oropharynx spray and mouth wash of our invention, hence we add it in our invention. I believe that it can act as a *chemical mask* in our nasal and oropharynx sprays that can be effectively used to prevent coronavirus entrance and seeding in the respiratory passages.

Mazzotti reaction by using ivermectin

There is an adverse reaction called Mazzotti reaction, first described in 1948. It is a symptom complex seen in patients after undergoing treatment of nematode infestation, particularly with the medication diethylcarbamazine (DEC). It can present when ivermectin is taken to treat nematode parasitic infestation, such as onchocerciasis (river blindness transmitted by black fly) infestation, but not seen when treating coronavirus infection, hence it is safe to use. Mazzotti reaction is characterized by fever, urticaria, swollen and tender lymph nodes, tachycardia, hypotension, arthralgias, edema, and abdominal pain that occur within seven days of treatment (appears like hypersensitive reaction to proteins from the dying nematodes). It is treated with intravenous methylprednisolone. COVID-19 patients are already on corticosteroids, and the chances of developing Mazzotti reaction are minimum to none.

Caution: Ivermectin passes on to breast milk. It is contraindicated in those with kidney diseases, liver diseases, cancers, HIV-AIDS, and other immune system weaknesses. Ivermectin should not be given to a child who weighs less than 33 pounds (15 kg), this may be a relative contraindication in coronavirus infection.

Niclosamide:

The recent pandemic outbreak of coronavirus disease 2019 (COVID-19) highlights an urgent need for therapeutics. Through a series of drug repurposing screening campaigns, Niclosamide, chlorinated salicylanilide, like above described ivermectin. an FDA-approved anthelminthic drug, was found to be effective against various viral infections with nanomolar to micromolar potency such as SARS-CoV, MERS-CoV, ZIKV, variety flavivirus, hepatitis C virus, Ebola virus, human rhinovirus, Chikungunya

Virus, Human Adenovirus, Epstein–Barr Virus known as human herpes and such, indicating its potential as an antiviral agent. In this brief review (Jimin Xu et al. ACS Infect. Dis. 2020, 6, 5, 909–915) summarize the broad antiviral activity of Niclosamide and highlight its potential clinical use in the treatment of COVID-19. It is used to treat broad or fish tapeworm, dwarf tapeworm, and beef tapeworm infections, taken 1.5 grams first day and then 1 gram for 6 days taken before meal. It will not work for other types of worm infections (for example, pinworms or roundworms). Niclosamide works by killing tapeworms on contact. Now it is being touted to be used on coronavirus and other viral infections, preventing the viral RNA multiplication at the endosomal level. With lapse of time, more such antiparasitic therapeutic agents such as suramin and pentamidine (Lomidine®) will emerge against coronavirus and other viruses.

Auranofin (Ridaura®) and its application to treat COVID-19 combined with other therapeutics

Auranofin was approved for the treatment of rheumatoid arthritis in 1985. Auranofin is [2,3,4,6-tetra-o-acetyl-L-thio-β-D-glycopyranp-sato-S-(triethyl-phosphine)-gold], has a well-known therapeutic application and its toxicity profile is considered safe for human use. After oral dosing (3-6 mgs), 15–25 % of the drug can be detected in the plasma, where it binds mostly to albumin. It is absorbed via the gastrointestinal tract within 20 minutes with peak plasma concentration of within 1-2 hours with drug plasma level of 6–9 μg/100 mL is reached. The beauty of this therapeutic agent is that the plasma ***half-life is 15–25 days with almost total body elimination after 55–80 days****, 85% excreted in the feces, and* 15 % appearing in the urine.

Auranofin inhibits inflammatory cytokine targets, thus suppressing the pro-inflammatory cascade associated with rheumatoid arthritis, neurodegenerative diseases, malaria, caners, so also in COVID-19 affliction. These drugs inhibit tumor necrosis factor (TNF)-α which is also inhibited by monoclonal antibodies infliximab, adalimumab, the TNF-α-receptor fusion protein etanercept, and the interleukin (IL)-1 inhibitor anakinra [Doan T, Massarotti E. Rheumatoid arthritis: an overview of new and emerging therapies. J Clin Pharmacol. 2005;45(7):751–762.].

The adverse effects of auranofin are gastrointestinal such as loose stools, abdominal cramping and watery diarrhea, which can develop after months of treatment. These symptoms are associated with changes in intestinal fluid movement and the net secretion of sodium, potassium, chloride and bicarbonate, as well as changes in the absorption of glucose and mannitol. But we are going to use auranofin for no more than 2-3 weeks, with other therapeutic agents such as vitamin C, dihydrochloroquin and chloroquine phosphate, remdesivir, thiamine hydrochloride, and such, hence we do not expect any gastro-intestinal-track problems. Auranofin has potential for new applications in the treatment of COVID-19 besides its use for rheumatoid arthritis, cancers, leukemias, parasitic infections, bacterial and viral infections, HIV, neurodegenerative disorders, Parkinson's disease, Alzheimer's and such.

Mechanism of action of auranofin

Mode of action of auranofin is through the inhibition of reduction/oxidation (redox) enzymes such as *thioredoxin reductase* (TrxR). These Redox enzymes are essential to many cellular processes, particularly in maintaining the intracellular levels *of reactive oxygen species (ROS)*. Controlling the level of ROS to prevent the resulting DNA damage and it is critical for the survival of all cell types, including cancer cells, parasites and memory T cells that shelter proviral HIV DNA so also in coronavirus infected respiratory tract cells. Diseases over-express redox enzymes, which increases the affinity of auranofin in the direction of these cells. Inhibition of redox enzymes alters the redox state of the cell and its milieu, leads to *increased production of hydrogen* peroxide and ROS that causes coronavirus cellular oxidative stress and ultimately intrinsic apoptosis of virus withing the infected cell. Auranofin also inhibits phagocytosis by macrophages and the release of lysosomal enzymes and antibodies involved in cytotoxicity reactions. Reducing the inflammation, decreasing the release of neurotoxins or increasing the release of neurotrophic factors could reduce neuronal loss, potentially slowing the progression of Alzheimer's and Parkinson's diseases as well. It is clear that auranofin inhibits cytokine targets, thus suppressing the pro-inflammatory cascade of cytokine storm associated with coronavirus infection as it does reduce the inflammation in rheumatoid arthritis and neurodegenerative diseases.

Auranofin acts against COVID-19, it acts as an anti-inflammatory, altering cytokine levels by increasing IL-8 and reducing IL-6 secretion from lipopolysaccharide-stimulating human monocytes, induces the anti-inflammatory enzyme heme oxygenase (HOX)-1 in the human monocyte cell line THP-1 cells, protecting the respiratory system cells of the alveoli and their blood vessels from oxidative stress and killing the viruses by hydrogen peroxide production.

Studies show that auranofin is effective against malaria which kills 600,000 people all round the world annually. Like cancer cells, parasites need to maintain cellular redox balance, and rely on redox enzymes such as TrxR, the thioredoxin–glutathione reductase system (TGR) and typanthione reductase (TypR) to prevent an accumulation of hydrogen peroxide and ROS in the cytosol and mitochondria. Auranofin is able to inhibit the action of the TrxR, TGR, thus leading death of cells containing coronavirus due to increased hydrogen peroxide and ROS, thus preventing coronavirus multiplication and spread.

For the treatment of coronavirus SARS, and ARDS, and inhibit the cytokine production and their storm to alveoli that results in SARS, we recommend 6 mg loading dose, then 3 mg every 12 hours, taper off the dose to 3 mg a day after symptoms abate for one week. Combine with high dose vitamin C which will augment the production of hydrogen peroxide, thus enhances the viricidal effect of auranofin which becomes an effective therapeutic agent to treat COVID-19 by killing the coronavirus. Auranofin inhibits redox enzyme, thus the hydrogen peroxide and ROS oxidants are elevated which are toxic to the virus within the cell where the coronaviruses are multiplying, and the result is their demise. That is how it works in coronavirus infected cells and it also blocks the cytokine ill effect on cells.

Colchicine: anti gout drug as coronavirus anti-inflammatory

Colchicine is a medication used to treat gout and Behçet's inflammatory disease mostly of oral cavity with mouth sores, prevention of pericarditis and familial Mediterranean fever. It is taken by mouth. Just like auranofin, it is used for gouty arthritis joint swelling. Colchicine is a compound originally extracted from the autumn crocus but also obtainable from the flame lily. It has been used since at least 1,500 BCE against rheumatism and joint swelling (Tardif JC, et al. Efficacy and Safety of Low-Dose Colchicine after Myocardial Infarction. *N Engl J Med.* 2019 Dec 26;381(26):2497-505), and also for pericardial inflammation. At 0.5mg one a twice a day it reduces inflammatory C reactive protein, and inflammatory signaling cytokines. It is also shown to reduce the risk of coronary artery disease, stroke, or acute coronary syndrome, hence it is worth considering in COVID-19 to prevent inflammation, cardio- vascular and CNS (stroke) event specially for those convalescing from COVID-19.

Cimetidine (Tagamet®) as antiviral and immune regulator

Cimetidine, approved by the FDA for inhibition of gastric acid secretion, is available over the counter. Several reviews have assessed the use of cimetidine in flu, warts, to prevent metastasis of colon and other cancers - I have used it in practice in the last decade. Cimetidine has immunomodulatory effects that include blocking suppressor T cells and facilitating cell-mediated immunity (CMI). The histamine-induced upregulation of IL-6 and IL-8 production is abrogated by a combination of pyrilamine and cimetidine. In patients with allergic rhinitis, cimetidine decreases the number of CD4+ and increases the number of CD8+ lymphocytes. Cimetidine inhibits nitric-oxide-associated nitrate production, and decreases interleukin 6 production which can block cell proliferation and c-fox gene transcription. Cimetidine can be use as a cold-flu and coronavirus infection prophylactic. Since cimetidine stimulates pro-inflammatory cytokines and inhibits regulatory T cells (Wang 2008), it may exacerbate the development of a cytokine storm and *should be avoided by individuals at risk for cytokine storm.*

I have prescribed cimetidine (available over the counter) for many patients in my practice to treat cancers (specially gastro-intestinal) patients before and after surgery, during and after chemotherapy and radiation to inhibit cancer growth and its metastasis. It is said to inhibit the attachment-seeding of free-floating cancer cell to the endothelium of blood vessels. It acts against adhesion molecule. If they don't get seeded, they cannot spread to distant parts of body and organs easily, and are picked up from the blood stream by the immune system as foreign body and eliminated.

Pentamidine aerosol spray

Pentamidine is life saver in the treatment of Pneumocystis jirovecii (carinii-**organism** is believed to be a protozoa; now it is identified as a fungus) pneumonia **in HIV patients.** It is also used to treat sleeping sickness due to leishmaniasis and trypanosomiasis. **Pentamidine** binds avidly to transfer RNA, inhibits the ribosomal synthesis of protein, with additional actions on the synthesis of nucleic acids and phospholipids membrane. It interacts with RNA of coronavirus, its synthesis and membrane formation of the virion particles **(as well as DNA)** and stops its reproduction **and spread to** healthy cells. To kill the coronavirus in the respiratory passages and alveoli, use pentamidine through a nebulizer spay. Dissolve 300 mg of pentamidine in 10 ml of sterile water and sprayed every 4-8 hours depending on the symptoms. **It has been used intravenously also.**

Suramin as anti-coronavirus therapy

It is an antiparasitic (trypanosomiasis, onchocerciasis) and has anti-angiogenesis and anti-cancer effects. The mechanism of action for suramin is unclear, however, it is thought that parasites are able to selectively uptake suramin via receptor-mediated endocytosis of the drug that is bound to low-density lipoproteins and to a lesser extent, other serum proteins. Once inside parasites, (here, coronavirus infected cells), suramin combines with proteins, especially trypanosomal and coronavirus infected cell glycolytic enzymes to inhibit energy metabolism that results in inhibition of viral reproduction and death. Its antiviral and other therapeutic effects are fundamentally driven by the metabolic dysfunction it creates in the infected cell, which is detrimental to the coronavirus multiplication in the host cell.

It is thought that it is internalized into the cell where it may affect the activity of various key enzymes involved in the intracellular transduction of mitogenic signals including **protein kinase C** PKC) and reproduction of coronavirus multiplication. It is known to prevent the HIV entry into cells due to its to inhibitory effect of reverse transcriptase and to prevent HIV entry into the cell. In the same way, it prevents the coronavirus attachment and entry of coronavirus into healthy cell to survive, multiply and spread. Now it is touted to treat ASD-autism which we discuss in part II.

Chapter 34

Angiotensin receptor blockers (ARBs) treatment for COVID-19

Researchers have used cryogenic electron microscopy to show that coronaviruses enter human cells through an interaction with angiotensin-converting enzyme 2 (ACE2) receptors located on cell surfaces of the nasal-oral-pharyngeal-laryngeal mucosa, air passages, and then the lung alveolar cells as well as the blood vessels that surround the alveoli. Scientists exploring how coronaviruses like COVID-19 infect human cells have shown that the SARS-CoV-2 spike (S) glycoprotein binds locks to the cell membrane protein angiotensin-converting enzyme 2 (ACE2) receptors to enter human cells (Chapter 21, Figures 2-6).

If the virus doesn't get inside the host cell, it cannot survive to multiply and spread to the rest of the respiratory system. These viruses will be picked up by the immune system and disposed or swallowed with saliva and neutralized by stomach acids. The only way the COVID-19 can create pathology and illness in the lungs and other organs is for them to enter healthy cells and multiply inside using the cells' machinery as a kind of parasite, producing innumerable viruses, which may take a week or two and then shed from the infected cells and reattach to the healthy cells. Finally, the viruses are all over the alveoli and its blood supply micro vasculature, resulting in profound damage to the delicate vascular system that surrounds the alveoli which puts lungs in a state of shock. Indirectly it may affect the left chamber of the heart which receives blood from the infected lungs.

These capillaries start pouring immune system cells, cytokines and chemokines which further damage the blood vessels and the alveoli causing swelling of the alveoli. There is an immense reaction from the host immune system to attack the coronavirus, which also affects the host lungs and boy. This inflammation cells and fluids filling the lungs air sacs will restrict or block the oxygen entry into blood and carbon dioxide elimination exit from venous blood, resulting hypoxia, hypercarbia and low oxygen in the arterial blood resulting in severe respiratory distress syndrome (SARS). Such patients are to be provided with supplemental oxygen, and may need to be intubated and ventilated, may need PEEP ventilation and critical cases consider using extracorporeal membrane oxygenation (ECMO). Such patients are to be provided with supplemental oxygen, and maybe even intubated and ventilated.

The battle rages between the virus and immune system, and the virus becomes a vicious competitor that the immune system has difficulty controlling, the virus can continue to spread and involve more of the lungs. If the immune system cells keep causing damage and producing more pro-inflammatory chemicals such as cytokines, chemokines, and prostaglandins, the fluid builds up in the air sacs of lungs (alveoli). The lungs can't drain normally and fast enough, and the air sacs then fill with fluid, which causes a "drowning feeling" in patients, and the blood's oxygen levels to drop. The need for a ventilator can arise quickly after admission to the hospital, sometimes within less than 24 hours. Here the massive inflammatory response damage its own tissue.

In spite of all the care in the ICU, this can result in high morbidity and mortality especially if the person is aged and has other diseases associated such as obesity, heart, kidney and liver diseases, diabetes, high blood pressure, autoimmune diseases, and such. If the patient has preexisting lung conditions such as COPD, asthma, cystic disease, emphysema, pulmonary hypertension, pulmonary fibrosis and such infections, pulmonary fibrosis and such, the outcome of the COVID-19 disease is even worse.

How to prevent the attachment of the COVID-19 attaching to angiotensin receptors (Chapter 21, Figure 8)

Here is the finding we have based on anecdotal experience, biological viral binding science and treating patients with cold, flu, possible coronavirus and other viral infections. When we use angiotensin converting enzyme receptor blockers (ARBs), there is no place of entry site for the COVID-19 virus to attack the healthy cell. Thus, their entry is delayed or inhibited. The door is completely or partially shut and the virus cannot get inside the cell to multiply and spread to cause the disease.

The effect of taking ACEI inhibitors is the angiotensin receptors on the cells (ATR) are still open and the virus can attack and enter the healthy cell. Now if the COVID-19 virus is inhaled, they attach to the open angiotensin receptors on the nasal-oral-pharyngeal-laryngeal mucosa, air passages, and then the lung alveolar cells and blood vessels that surround alveoli to enter the healthy cells and multiply, shed and spread with respiration to the rest of respiratory and other systems including alveoli to produce the disease. Hence blocking angiotensin receptors can protecting from coronavirus infection, when the respiratory system comes in contact.

I had a number of patients on telmisartan (trade name Micardis®, as well on other ARBs including losartan), first approved in 1998, which is an effective long acting angiotensin II receptor blocker (ARBs) to treat high blood pressure without production of cough as it happens taking angiotensin converting enzyme (ACEi) inhibitors due to their bradykinin production. In 2009, following the results of the ONTARGET trial, ARBs were the first drug in its class that the FDA allowed a claim that it "reduces the risk of heart attack, stroke, or death from cardiovascular disease in patients at high cardiovascular risk who are unable to take ACEi inhibitors." Here again the angiotensin receptor blockers should be used moderately to serve cases of COVID-19 affliction to inhibit or delay or block the entry of the coronavirus into healthy cells.

A little-known side benefit to the class of ARBs antihypertensive drugs known as angiotensin II receptor blockers is that they enhance insulin sensitivity, increase utilization of fat as energy, and improve mitochondrial function and is shown to have anti-aging effects; it restores health and longevity. Of all the drugs in this class, telmisartan stands out as potential longevity enhancement associated with anti-aging effects in addition to acting against elevated blood pressure. As humans age, mitochondrial dysfunction becomes a deadly factor in the development of obesity, insulin resistance, endothelial breakdown, and type II diabetes, and may be even neurodegenerative diseases. Telmisartan helps correct these underlying mechanisms of aging and death. Other benefits of taking angiotensin receptor blockers: a study of 1,000,000 veterans showed reduced risk of lung cancer.

I have been taking angiotensin receptor blockers (ARBs), for years and have prescribed to many of my patients to treat blood pressure. Ever since I started taking this class of anti-blood pressure medication, I rarely suffered from any form of flu. Even during the severe flu season, I may occasionally suffer maybe a few sniffles for a day or two but it is gone afterwards. My children always ask me to take flu shots, and I ignore it. I have taken one flu shot over two decades, but I did not know why I was protected from flu affliction year after year. The same observation was made by my patients who took *angiotensin receptor blocker* hypertensive medications.

When I read that COVID-19 virus enters through this angiotensin receptors on the cell wall, it occurred to me why myself and many of my friends did not develop flu during flu season. It is because angiotensin receptors are occupied by the angiotensin receptor blocker drug, and the virus has no place to enter healthy cells with ease to cause morbidity and mortality. Further we also learned that corona group of viruses need angiotensin receptors to enter the healthy cells. One out of 3 Americans have high blood pressure and ARBs was the drug of choice for all my patients; I would not recommend ACEi or other types of blood pressure medications, my choice is angiotensin receptor blockers. The other studies also show they reduce weight gain, enhance insulin sensitivity, improves mitochondrial functions and such. Now if any one does retrospective study on the patients on angiotensin receptor blockers, we will find that they hardly suffer from flu attacks and may even live longer (anti-aging effect) with less flu, colds, milder form of COVID-19, cardio-vascular diseases, neurodegenerative diseases and even dementia.

Here ARBs act as an anti-hypertensive by blocking the angiotensin receptors (Chapter 21, Figures 7, 8) thus the angiotensin II circulating in the blood cannot bind to these receptors and cause contraction of the blood vessels to increase the blood pressure. Likewise, the coronaviruses are floating in the nasal-oral-pharyngeal-laryngeal mucosa, air passages, and then the lung alveolar cells and blood vessels that surround alveoli during epidemic and pandemic season, but they cannot enter the healthy cells with ease that make these coronavirus to multiply and spread in large number to cause the disease and death.

It is known that that angiotensin receptor blockers block the type I angiotensin receptors. We believe that the type II to which the coronavirus attaches to get inside the cell, however, we are not sure of the role played by type I angiotensin receptors in the entry of coronavirus. It is important to note that the AT2 receptor shares approximately 34% amino acid sequence homology with AT1 receptor [Karnik SS, Angiotensin Receptors: Pharmacol Rev. 2015;67:754–819]. They are two receptors like twins attached to each other on the surface of the cell membrane with some common amino acid sequences, hence the angiotensin receptor blockers attach to both receptors and block the angiotensin receptors II in part which has some effect on binding of viruses and thus plays a role in delaying or blocking the entry of coronavirus inside the cell 34% of the time or delays that much time for the entry.

Now we have two important factors for using angiotensin receptor blockers (ARBs). First, it will prevent the receptors by not allowing the COVID-19 S spikes, (other such viruses belonging to this group also) attached to the cell membrane, second, at the same time it will allow more angiotensin converting enzyme 2 with angiotensin II protein on the surface of all cells including the respiratory mucosa with reduced production of angiotensin II protein which will reduce the possibility of elevated blood pressure in these patients. Elevated blood pressure plays a role inflammation endothelial cell membrane permeability, meaning that the elevated blood pressure makes them leakier, thus increase lung edema, which is prevented by the ARSs.

We recommend telmisartan and losartan to prevent the angiotensin receptor blockers to block the effect of angiotensin II having effect on the alveoli and the blood vessels that surround them to reduce the lung damage by inhibiting BV, and alveolar leakage. Vitamin C, thiamine hydrochloride and hydrocortisone/dexamethasone, auranofin, combined with these therapeutic agents will save the damage to the lungs and save many lives of the aged afflicted with COVID-19.

We describe the methods to prevent the nasal-oral-pharyngeal-laryngeal mucosa attack by coronavirus using angiotensin receptor blockers spray, so that the virus cannot enter the cell. We describe the nasal and mouth spray, mouth wash and oral medication to prevent the coronavirus infection.

Nearly all ARBs contain biphenyl tetrazole moiety except telmisartan and eprosartan that may help to block the angiotensin receptors I and II. Though their chemical structure is similar but the biological half-life is varied. Comparison of pharmacokinetics of angiotensin receptor blockers and their daily dose in adults are as follows:

a) Telmisartan - Micardis, 24 hours: 20-40-80 mg
b) Losartan – Cozaar, 6–9 hours: 50-100 mgs
c) Azilsartan - Edarbi, 11 hours: 40-80 mg
d) Irbesartan - Avapro, 11–15 hours-150-300 mg
e) Olmesartan – Benicar/Olmetec, 13 hours: 10-40 mg
f) Valsartan – Diovan, 6 hours: 80-320 mg
g) Fimasartan – Kararth, 7–11 hours: 30-120 mg
h) Eprosartan – Tevelen, 5 hours: 400-800 mg
i) Candesartan – Atacand, 9 hours: 4-32 mg

Any one of these angiotensin receptor blockers can be used to block the angiotensin receptors on the cells to prevent the entry of coronavirus to cause COVID-19. We had more experience with telmisartan (Micardis) due to its single dose effect lasting 24 hours.

Chapter 35

Preparation of nasal and mouth spray and mouth wash to prevent coronavirus infecting healthy nasal, oropharyngeal mucosa

As the vast majority of SARS-CoV-2, flu and cold viral transmission occurs indoors, most of it from the inhalation of airborne particles that contain the coronavirus. An estimated 40% of cases are asymptomatic and asymptomatic people can still spread the coronavirus to others. Anti-coronavirus sprays we describe could be a very effective prophylaxis to prevent spread of coronavirus, so also the coronavirus ion zapper under development.

Our invention describes the preparation of nasal and oropharyngeal spray which can be used also as a mouthwash with some modifications of the contents to prevent attachment of coronavirus and as anti-microbials to nasal-oral-pharyngeal-laryngeal mucosa, air passages, microbial biofilm, and then the lung alveolar cells and blood vessels that surround alveoli, as the contents of the spray are transported by the breath are described below.

Example

Preparation one, of our formulation:

1. Take 250 ml of sterile water or distilled water, warm it on a stove or microwave
2. Add 500 mg of powered telmisartan and shake it to dissolve it
3. Add 500 mg of dihydrochloroquin or chloroquine phosphate
4. Filter through cheese cloth or sterile coffee filter to remove the fillings in the tablet
5. Add 75 mg of elemental zinc sulfate
6. Add Ebselen to make 2%
7. Interferon beta
8. Then add 10 mg of ethylenediaminetetraacetic acid (EDTA)
9. Add 2 ml of Dimethyl sulfoxide
10. Add bioactive curcumin 2-5%
11. Add 1 gram of xylitol,
12. Auranofin 12 mgs

13. Chlorohexidine to make the final solution 0.25%
14. Pour it into the simple sprayer
15. Store it in the refrigerator until use.

Spray the nose and oropharynx (throat) 3-4 times a day and before going to mix with the public or going to care for the patients in the hospital. It can be used to gargle in the mouth before leaving the house. You can add superoxide dismutase, cimetidine also to enhance the antioxidant activity on the mucus membrane. All sprays and mouth wash contain elemental zinc. We can add ivermectin to any of these sprays and mouth washes. The studies show that the curcumin coats the cell membrane before it is able to infect the healthy cell, by integrating with the viral envelope to 'inactivate' the virus, thus it is viricidal and inhibits its spread to healthy cells.

A study from England (Synairgen pharma) showed that patients with mild COVID-19 who took interferon beta were less likely to develop severe COVID-19, and they got better faster. The doctors gave the patients interferon beta by letting them breathe in a spray, performed on 101 patients, who were 79% less likely to develop a severe case.

Example

Preparation two, of our invention:

1. Use the methods described in Preparation one
2. Substitute Losartan for telmisartan
3. add 500 mgs superoxide dismutase

Example

3-10: Preparations from three to ten using different ARBs of our invention

1. Valsartan – Diovan, 6 hours: 80-320 mg
2. Azilsartan - Edarbi, 11 hours: 40-80 mg
3. Irbesartan - Avapro, 11–15 hours:150-300 mg
4. Olmesartan – Benicar/Olmetec, 13 hours: 10-40 mg
5. Telmisartan - Micardis, 24 hours: 20-40-80 mg
6. Fimasartan – Kararth, 7–11 hours: 30-120 mg
7. Eprosartan – Tevelen, 5 hours: 400-800 mg
8. Candesartan – Atacand, 9 hours: 4-32 mg
9. Losartan / Cozaar, 6-9 hours, 50-100 ms

When using any of the above in the spray, add 3- 5 times their maximum prescribed daily dose. Combination of two angiotensin receptor blocker can be added also such as telmisartan and losartan to have synergic effects. The rest of preparation is same as method one.

Example

Preparation 11 of our formulation

1. Take 250 ml of sterile water or distilled water, warm it on a stove or microwave
2. Add 500 mg of powered telmisartan or losartan, mix and shake it to dissolve it
3. Add 500 mgs of dihydroxy chloroquine or chloroquine
4. Add 25 mg of elemental zinc
5. Add vitamin C 500 mgs
6. Add Ebselen to make it 2%
7. Curcumin 2-5%
8. Chlorohexidine to make it 0.25%
9. Then add 10 mg of EDTA
10. Add xylitol 1 gram
11. Add auranofin 12 mg
12. Filter through cheese cloth or sterile coffee filter to remove the fillings in the tablet
13. Pour it to the simple sprayer shown in the diagram
14. Store it in the refrigerator.
15. Add xylitol 12 grams as a microbial to make it sweet
16. Spray the nose and throat 4 times a day and before going to mix with the public out of the house.

We add chloroquine or dihydrochloroquin to augment the prophylactic effect of angiotensin receptor blocker and antiviral agent remdesivir also which will prevent the multiplication of coronavirus inside the infected or about to be infected cells.

To prevent the attachment of the coronavirus to angiotensin receptors on the healthy cell, we can also use suitable monoclonal antibodies and coronavirus specific antibodies extracted from the recovered patients to block the entry of coronavirus into the cells also. EDTA is an effective adjuvant in the spray to prevent coronavirus and bacteria get into our nasal, oral and respiratory passages and acts synergistically with other therapeutic agents used in our inventive spray and mouth wash.

Example

Preparation 12 of our formulation

1. Take 250 ml of clean water or distilled water, warm it on a stove or microwave

2. Add 500 mg of powered telmisartan or losartan, mix and shake it to dissolve it
3. Add 200 mgs of chloroquine
4. Add remdesivir or another suitable protease inhibitor.
5. Add 25 mg of elemental zinc
6. Add Auranofin12 mg
7. Then add 10 mg of EDTA
8. Cimetidine 800 mgs
9. Xylitol 500 mg
10. Filter through cheese cloth or sterile coffee filter to remove the fillings in the tablet
11. Chlorohexidine to make it 0.25%
12. Pour it into the simple sprayer
13. Store it in the refrigerator.
14. Spray the nose and throat 3-4 times a day and before going to mix with the public.

Any of the above sprays can also be used to wet the face mask also to block the virus entering through the mask. Use a **bioactive mask** as described in the introduction with antiseptics and even lithium titanium nanoparticles. Any one of the ingredients in the spray can be reduce or removed from the spray if there are allergic or other adverse effects user notices without reducing their efficacy. Other antiviral therapeutic agents can be added be sides the one described. This can be sprayed on the face also to prevent the attachment of coronavirus on the face and enter the eyes, nose, and mouth (Chapter 21, Figure 4). There are many types of protease inhibitors in the market widely used in the treatment of HIV. Our choice is remdesivir. As it is a spray, one can increase the 3-5 times the oral dose. Many modifications are possible with addition of other agents and future antiviral drugs that are introduced. These sprays can be used also to spray olfactory mucosa to eliminate any coronavirus attached to it and prevent is entering the CNS. The sprays can be used during the height of the flu season to prevent the influenza virus from attacking our respiratory system.

Example

Aerosolize superoxide dismutase to produce coating of the lungs alveoli.

Take 250 ml of distilled water, add 5 grams of superoxide dismutase, dissolve it. In our invention, we add ivermectin also to make it more effective antiviral and anti-oxidant spray. Fill the pressurized canister, pressurize to desired pressure, and inhale the aerosolized superoxide dismutase when one takes deep breath as you activate the spray (Chapter 21, Figures 9, 10), so that it reaches the alveolar sacs of the lungs and coats the lining cells. This coating will prevent the coronavirus attaching and attacking the lung alveoli, migrating to alveolar microvascular bed, and prevent the production and naturalizing cytokines by removing free radicals from the lung parenchyma, thus preventing the pathological effects of the coronavirus infection. We advocated using this spray by spraying inside the breathing tube if intubated. This method can be used to treat other lung conditions including pulmonary fibrosis, tuberculosis, cancers, cystic diseases, COPD, chronic asthma, and such by using specific therapeutic agents to that particular disease.

Example

Use of our inventive sprays along with corticosteroids:

Studies show that Budesonide cortico-steroid aerosol oral-nasal sprays used in the beginning and during coronavirus infection may lessen the severity of the coronavirus infection. Combined with our inventive sprays and mouth wash described herein, Budesonide corticosteroid could be even more effective as a prophylactic and taming down the virulent coronavirus infection in those suffering from various stages of COVID-19. Japan, Taiwan and other Asian countries have maintained a much lower fatality rate with COVID-19 then we have here in America, in spite of the fact they live in densely populated communities. Many people attribute to and believe that it is due to their preferred method of treatment. They use a cortico-steroid (Budesonide, Pulmicort) spray, available as an inhaler. It is inhaled in a mist through a home use nebulizer. It is already used widely in the long-term management of asthma and chronic obstructive pulmonary disease (COPD) and nasal spray for allergic rhinitis and nasal polyps. There are antidotal reports of dramatic relief of respiratory distress and rapid recovery form COVID-19. As soon the coronavirus tests are positive, administer this aerosol, as well those hospitalized. The results can be dramatic to reduce the severity of the disease. N-acetyl cysteine can be added to any of the sprays to liquefy the thick mucus of air passages to make it easy to expectorate with a cough.

Example

Aerosolize coronavirus antibody nanoparticles, coronavirus antibody containing plasma and monoclonal antibody and ivermectin coating of the respiratory air passages including nose and lungs (Chapter 21, Figure 9). There are going to be many kinds (https://www.israelhayom.com/2020/05/08/israeli-scientists-develop-series-of-coronavirus-antibodies/) of antibodies that will be in the market to attack the coronavirus with passage of time. It is known that infusing antibody containing plasma from the COVID-19 afflicted will tame or abort the coronavirus infection. It acts by antibodies binding to S-spike on the virus (Chapter 21, Figures 2, 3, 5, 6), thus inhibiting or blocking its spread to healthy cells.

We can use the same method delivered directly to the mucus membrane of the respiratory system including the lungs to block the virus attaching to the healthy host cells. We want to coat the mucus membrane of the respiratory system nose and oropharynx with biologic coronavirus antibody nanoparticles or coronavirus antibodies in the plasma or monoclonal antibodies with saline or other suitable media, pour it into a the pressured canister, and then spray aerosolized mist (Chapter 21, Figures 9, 10) to coat the respiratory passages including of treatment for inhibiting and killing coronavirus infection of the host cells mucus membrane of the oral, nasal, laryngeal, pharyngeal; respiratory passages larynx, trachea, bronchi, bronchioles; lung alveoli lining cells and their surrounding microvasculature blood vessels with antibodies. These antibodies will attach on to the S-spike of the coronavirus in the breathed air and prevent them attaching and attacking healthy cells and their entry into the respiratory system mucus membrane.

Fill the pressurized canister, pressurize to desired pressure, and inhale the aerosolized coronavirus antibody nanoparticles (may be even monoclonal antibodies, coronavirus antibody containing plasma), take deep breath so that it reaches the alveolar sacs and coats the lining. Aerosolized sprays can be delivered through the endotracheal tube also if the patient is intubated and cared in ICU. This biologic coating will prevent the coronavirus attaching and attacking the lung alveoli, mucus membrane of the respiratory system, thus preventing or aborting the coronavirus infection and its multiplication. There is a possibility that antibodies will inhibit the production of cytokines by removing free radicals from the lung parenchyma.

The formulation of any of these sprays, mouth and nasal washes can be changed or supplemented by adding other ingredients and /or newly discovered therapeutic agents to make them more effective, thus our inventive method becomes more functional and effective.

Use of corticosteroid spray along with ivermectin in our inventive sprays described herein with the above method can be effective in early cases of coronavirus infection. Japan, Taiwan and other Asian countries have maintained a much lower fatality rate with COVID-19 then we have here in America, in spite of the fact they live in densely populated communities. Many people believe that it is due to their preferred method of treatment. They use a steroid medication that is inhaled in a mist through a home use nebulizer as described below.

Electrically charged photonic ions for neutralizing free-floating coronavirus and other viruses in the droplet, aerosolized air we breathe

In nature, negative ions and ozone are nature's most powerful air-cleansing agents. All of us have noticed how refreshing air is at mountains, water falls, and after a thunderstorm rainfall. This is due to thousands (> 2000 negative electrically charged ions/ cubic centimeter) of negative ions in our breathing air. We already have patented technology to change room air using this nature's modality by using heating-air-conditioning outlets in houses using our device. Now we are developing the method to be used as a prophylactic to wear it around the neck or on the forehead to get rid of infection from droplets and inhaled coronavirus infected air, by sterilization. We are developing the device to be used in class rooms and gathering places. For details, read in Part II of the book. This ionizer and charges make the photons electrically charged ions attach to viral particles and dust to neutralize viruses.

We want to describe the following protocols of our inventions for prophylaxis and treatment for COVID-19 as follows:

So far there are no vaccine and specific therapeutic agents to prevent COVID-19 infection or to treat once infected. Pharmaceutical industry, national institute of health and various university research centers are on a frenetic pace to produce a vaccine and treatment. They will come, but it may be months away. It could by use of beta propiolactone, inactivated viral component, tamed coronavirus vaccine, many types of antibodies from serum weeks after COVID-19 infection. At present we describe prophylactic and compassionate treatment modality for COVID-19 virus affliction. What we describe here can be used with some modification for future epidemic/pandemic viral or non-viral outbreaks.

We know that the high dose intravenous vitamin C has been used for decades as an antiviral, anti-oxidant and anti-cancer therapeutic agent without the FDA stamp of approval, which we describe here. Any licensed physician, physician assistant or nurse practitioner can prescribe and provide the therapy. It has already been used for more than 7 decades to treat viral infections starting from measles to polio to cancers without any adverse effects. We propose using it to treat the COVID-19 infection of the lungs, heart and such using vitamin C. Here we are describing the inventive method of prophylaxis and treatment.

Dihydrochloroquin and chloroquine phosphate: how it blocks the COVID-19 viruses (the names dihydrochloroquin and chloroquine phosphate are used interchangeably)

We know that chloroquine is an effective antimalarial drug used all over the world for decades. It also found its use to treat rheumatoid arthritis, lupus, many autoimmune diseases, even cancers, parkinsonism and such. In my practice, we found that some of the cancer patients we treated with chloroquine hardly developed any flu or colds during the flu season. Then, we started prescribing it as a prophylactic during cold season along with reduced dose of ARBs such as telmisartan to the aged and those with multiple systemic diseases to thwart influenza attacks during flu season. It was an effective and safe drug to use as a prophylaxis.

I believe that it is also safe to use dihydrochloroquin and chloroquine phosphate for this COVID-19 infection to prevent, curtail, stop the progression of and / or cure as well as reduce the symptoms, and cut down the natural course of the disease. Use caution in patients who have heart conduction defects and heart rhythms. The dose to use for coronavirus infection is as follows:

a) As prophylaxis: Use the spray and mouth wash we describe above. Take chloroquine 250-500 mg orally twice a day for 3 days and then maintenance dose of 250 every day till the end of COVID-19 epidemic and other flu season.
b) Once you are infected with COVID-19: Use prophylaxis spray 3-4 times a day to prevent the viral load in the expired air. Take chloroquine 600 mg orally loading dose and then 300 mg every 8-12 hours, with ivermectin one dose be tried with doxycycline.
c) Then taper the dose to 200 mg twice a day for two to four weeks to prevent shedding of the virus from the respiratory system that can infect the healthy. Continue till the symptoms abate and season ends.
d) Lower the dose in people who suffer from seizures, low blood pressure, AV block of the heart, cardiomyopathies and such. Use cautiously in patients with blood disorders or G6-PD deficiency.
e) Administer angiotensin receptor blockers (Chapter 21, Figures 5-13) such as telmisartan to block the entry of coronavirus to healthy cells

as it passes through the inspired air and comes in contact with air passages mucosa and as it buds of to infect the healthy cells.

All the above should take oral vitamin C 1000 mg twice or thrice a day, telmisartan 20 -40 mg a day, bioactive curcumin 1000-2000 mg, zinc 75 mg per day. Vitamin D3, 10,000 IU a day.

***Caution**: Antimalarials are contraindicated in heart block patients.*

How does chloroquine act against COVID-19 and other flu-influenza viruses?

Chloroquine (CQ), N′-(7-chloroquinolin-4-yl)-N, N-diethyl-pentane-1,4-diamine, is widely used as an effective and safe anti-malarial and anti-rheumatoid agent. CQ was discovered 1934 as "Resochin" by Andersag and co-workers at the Bayer laboratories. Ironically, CQ was initially ignored for a decade because it was considered too toxic to use in humans. CQ was "re-discovered" during World War II in the United States in the course of anti-malarial drug development. The US government-sponsored clinical trials during this period showed unequivocally that CQ has a significant therapeutic value as an anti-malarial drug. Consequently, CQ was introduced into clinical practice in 1947 for the prophylaxis treatment of malaria (Plasmodium vivax, ovale and malariae). CQ still remains the drug of choice for malaria chemotherapy because it is highly effective and well tolerated by humans. In addition, CQ is widely used as an anti-inflammatory agent for the treatment of rheumatoid arthritis, lupus erythematosus and amoebic hepatitis. More recently, CQ has been studied for its potential as an enhancing agent in cancer therapies.

Accumulating lines of evidence show that CQ can effectively sensitize cell-killing effects by ionizing radiation and chemotherapeutic agents in a cancer-specific manner. The lysosomotrophic property of CQ seems to be important for the increase in efficacy and specificity. CQ may be one of the most effective and safe sensitizers for cancer therapies. Taken together, it appears that the efficacy of conventional cancer therapies can be dramatically enhanced if used in combination with CQ and its analogs. I have used it to treat cancer of all kinds for years.

Dihydrochloroquin / chloroquine phosphate have decades of safety profile using millions of doses of this drug in treatment of malaria, rheumatoid arthritis, lupus and such. I have used it as anticancer and anti-Parkinson's as an adjuvant therapeutic agent. They are absorbed from the gastro-intestinal track easily. As it circulates in the blood, it binds to DNA, and RNA interfering with protein synthesis. It also inhibits DNA and RNA polymerases, and thus inhibits viral entry and multiplications in the host cell. The drug increases the alkalinity in the lysosomes, thus inhibits coronavirus final assembly in the infected host cells. Besides that, Dihydrochloroquin and chloroquine phosphate sensitize the viral membrane for action by antiviral therapeutic agents and rapid clearance by immune system.

The drug may also antagonize histamine, serotonin and inhibit prostaglandin effects to reduce inflammation, relieve pain and other symptoms of coronavirus. It also may inhibit chemotaxis of polymorphonuclear leukocytes, macrophages, and eosinophils entering lung alveoli from inflammatory reaction from the surrounding leaky blood vessels, thus saving the patient from pneumonia, bronchitis, fluid in the lungs and other lungs related inflammation including SARS or morbidity due to cytokine storm, especially in those afflicted with COVID-19 lung symptoms.

Patients with severe cases of COVID-19 infection with low oxygen saturation need to be admitted to the intensive care unit (ICU) of the hospital, intubated, ventilated and / or put on supplemental oxygen to maintain arterial blood oxygen saturation and may need to use the ECMO method to raise the tissue oxygenation level

Severe symptoms (Chapter 21, Figures 1a, 4) of COVID-19 develop in elderly patients whose immune system is compromised and who also suffer from other systemic diseases of lungs, heart, liver, kidney, diabetes, obesity, those under chemotherapy for cancers, those residing in the nursing homes and extended care due to various health and personal reasons and such. They can develop high fever, headache and moderate to severe respiratory problems such as cough, difficulty in breathing, air hunger (low oxygen saturation measured by oximeter), rapid labored breathing, rapid heart rate, and blood pressure changes with copious sputum productions of COVID-19 infection (Chapter 21, Figures 1a, 4). These patients have a high mortality if they are not treated properly. There is a possibility that there is decreased production of the body's anti-viral interferon (IFN)-α/β thereby increasing the infiltration of inflammatory immune system cells into the lungs with production of cytokine leading to the cytokine storm SARS.

For those who are very sick, Hyperbaric therapy (HBO) should be considered. I have provided HBO therapy for septicemic patients, unhealed leg ulcer, neurodegenerative and cardio-vascular diseases, chronic fatigue syndrome, fibromyalgia, Lyme disease, autism, multiple sclerosis and such.

Chapter 36

Is there a treatment for COVID-19? Yes, there is

Yes, there is.

Our protocol includes administration of vitamin C, thiamine hydrochloride and dexamethasone/hydrocortisone, Ebselen which is a Glutathione peroxidase mimic, protease inhibitor remdesivir, angiotensin receptor blockers telmisartan and losartan, patient antibody plasma, monoclonal antibody. If there is severe cytokine storm, consider rapamycin to calm down the cytokine storm and diminish or settle down the hyperactive immune system. In cases of severe hypoxia in intubated patients, consider using ECMO as long as needed. Usha and Bill Martin and Karim Jabr have used it successfully sometimes for weeks to maintain the tissue oxygenation and saved the life of many patients.

Our protocol also describes treatment by administering one single dose of ivermectin orally, in countries where it is available such as in Bangladesh and Southeast Asia, with a booster dose 7 days later (if needed) as soon as the patient is tested positive or suffering from COVID-19. One can combine it with doxycycline given for 2 weeks or azithromycin to prevent secondary infection.

Although there is no prophylactic vaccine or specific treatment to counter COVID-19 (they will be coming in truck loads within 18 months), there has been a treatment protocol available since 1948 that is backed by medical science and research. Studies were conducted by Klenner, Pauling, Marik, Kim and Shantha for those infected with a viruses and microbes (Klenner, F. R.: Virus pneumonia and its treatment with vitamin C. Southern Med. Surg., Feb. 1948, Massive doses of vitamin C and the virus diseases. Southern Med. Surg., 1951. Pauling, L.: Vitamin C and the Common Cold. W. F. Freeman & Co., 1970. US Patent Pub. No. 2016/0074480 A1, 2016, TR. Shantha, Marik PE et al. Hydrocortisone, Vitamin C and Thiamine for the Treatment of Severe Sepsis and Septic Shock: Chest Journal 2016, 2017;151:1229-38., Journal: Nutrients 2018, Kim WY, et al. J Crit Care. 2018. Faloon, Life Extension, September 2018).

The treatment described in the above research publications includes intravenous administration of high dose vitamin C, with addition of vitamin B1 (Benfotiamine), corticosteroids such as hydrocortisone or dexamethasone, protease inhibitor like remdesivir, and angiotensin receptor blockers telmisartan and losartan on the cell surface, antimalarials, Auranofin and zinc and such described in examples of nasal and oropharynx sprays. This method kills the COVID-19; provides anti-viral, anti-inflammatory and anti-shock therapy to reverse cytokine shock caused by the virus in the lungs, and helps restore functions of afflicted organs such as lungs and heart. In addition, high dose vitamin C, as well as auranofin are antioxidants and also produces hydrogen peroxide with is a viricidal that kills the coronavirus as they are released, budding off from the infected cells surface to spread.

The recent discovery that the coronavirus initiates the sprouting of *filopodia* in infected cells suggests that it has, at some point in its evolution, developed more than one way to get into cells and establish a way to ensure it gets passed quickly from cell to cell though these filopodia as well as cell surface membrane (Chapter 21, Figure 6).

In addition, the above studies and our own studies show that vitamin C is effective in the treatment of numerous viral and bacterial infections and can reduce organ failure and mortality in patients with severe acute respiratory distress syndrome, sepsis and septic shock. Because humans do not produce vitamin C, it must be obtained from food or supplements. With COVID-19, the levels of vitamin C may be even lower as it is observed for vitamin D3, levels. Many elderly and sick people with serious pre-existing medical conditions often do not get good nutrition and are therefore deficient in vitamin C and vitamin D3 also.

Furthermore, vitamin C stimulates immune system cells to produce interferon which is an anti-viral against coronavirus. Vitamin C also produces hydrogen peroxide in the body's cells that kills viruses and inhibits inflammatory cytokines that contribute to lung and cardio-vascular mortality (Kim Y, et al. doi: 10.4110/in.2013.13.2.70). As far back as the 1940s, high doses of intravenous vitamin C with hydrocortisones have been used by Klenner, M.D to treat every kind of viral infection including polio, herpes, flu, measles, mumps, chicken pox, and countless others infections and diseases. In fact, I have used up to 25 grams of intravenous vitamin C effectively in my practice to treat various cancers, bacterial (MRSA), and severe viral infections. Doctors could treat COVID-19 patients with the simple and inexpensive treatment that has been successfully used for many decades, instead of waiting for coronavirus therapeutic agents and vaccines.

Chapter 37

Our method of treatment for severe cases of COVID-19

Those residents of nursing homes, long-term care facilities or in the ICU hospital setting, besides taking oral chloroquine as described above, should take vitamin C as described below. In the hospital setting, or outpatient setting, or in private clinics: start intravenous fluids to hydrate with ringers' lactate, and administer intravenously:

a) Administer vitamin C 1500-2000 mg in normal saline or ringers lactate every 8-12 hours.
b) Combine vitamin C with 100-250 mg of vitamin B1 (Benfotiamine), every 8-12 hours.
c) Administer 25-50 mg hydrocortisone or proper doses of dexamethasone (±10-20 mg, depending on the severity of COVID-19) every 8-12 hours.
d) Add 50-100 mg of zinc sulfate once a day
e) Use 200mgs of chloroquine once a day with protease inhibitor remdesivir
f) Consider using ivermectin single dose as soon as the diagnosis made.
g) Use non-rebreathing oxygen mask to maintain desired blood oxygen saturation
h) If the good oxygen saturation cannot be achieved, intubate the patient (endotracheal tube) and use ventilator to maintain proper tissue oxygen saturation
i) Use positive pressure ventilation if need be to reach the desired oxygen saturation
j) If proper oxygenation cannot be achieved and vital signs are difficult to maintain, consider using extracorporeal membrane oxygenation (ECMO) with therapeutic agents and life support system measures. This vital support method can be used for weeks and months.

Continue to monitor all vital signs, including oxygen saturation of blood, and treat any change in the patient vital signs. Maintain to use various therapeutic, pharmaceutical, biochemical, and biological agents or compounds described above, and use other needed intensive care therapies such as fluid management, antibiotics such as azithromycin and blood pressure support as well as therapies instituted to maintain the vital signs until the patient is out of danger, and fully recovered in lung, heart and kidney functions.

We also add 6 mg of auranofin given orally for one week for ±15 day to ally the cytokine storm, kill the coronavirus by producing ROS and hydrogen peroxide, acts as viricidal conjoined with other therapeutic agents.

We also recommend Ebselen and Remdesivir 200-600 mg a day followed by 100 mg a day for 10 days. Vitamin D3, 5000 IU twice a day is a must, with bioactive curcumin up to 1000 mg twice a day, Benfotiamine 100-250 a day, zinc 50-75 mgs a day, angiotensin receptor blocker telmisartan 20-40-80 mg a day depending on the blood pressure.

Use of intravenous DMSO and EDTA as well as hyperbaric oxygen therapy (HBO), ECMO, can be a game changer to control the uncontrollable coronavirus, microbial infection, and cytokine storm.

Chapter 38

Use of electrical fields to treat coronavirus infection

There are reports of the application of treatment modality using low-intensity, intermediate-frequency, alternating electrical fields to treat infections and cancer (Tumor-Treating Fields-TTF). The method overcomes the multidrug resistance, (MDR) with artificially induced electrical field around the chest wall and boney lesions has been described in Pub. No.: US 2016/0074480 A1 by Shantha. There are reports of the application of this modality of treatment using low-intensity, intermediate-frequency, alternating electric TTF fields for treatment of malignancies in humans (Salzberg M, et al. A pilot study with very low-intensity, intermediate-frequency electric fields in patients with locally advanced and/or metastatic solid tumors. Onkologie. 2008;31: 362–365). This feasibility study to treat coronavirus has the goal to demonstrate the safety and absence of adverse effects of this simple method in patients described in the above patent. Weak electric AC currents generated employing conductive electrodes (Figure 15, below) are known to increase the efficacy of antibiotics against viral, microbial bacterial biofilms, a phenomenon termed "the bioelectric effect." We have used the battery-operated pads, inserted electrodes, ultrasound and magnetic fields for the treatment of breast and lung cancers, and infections, for over two decades. The sound waves and electrical fields are known to kill bacteria, and multiplying cells so the effect of using this method to treat COVID-19.

The proposed main biological basis of this therapy in treating coronavirus infection is an added modality with therapeutic effect of using simple alternating electrical fields on the formation of new viruses in the host cells by inhibiting the formation of new viruses. A secondary effect of this electric therapy is the weakening of host cell membrane and the viral envelope, allowing more antiviral agents disrupting viral multiplications. The energy of electrical fields created by this method (Figure 15, below) acts as an electrical signaling of the cell wall and produces a hormone-like effect by triggering an electrical field induced change to the cell membrane G protein. This change influences the activity of adenylate cyclase, resulting in the formation of cAMP. cAMP induced repair processes are necessary to stabilize the cell membrane and inhibit continued leakage of acids known to trigger pain and inflammatory mediators. Vaso-dilations caused by this method "wash out" of waste products (the products of inflammation such as cytokine) with pH normalization; the reduction in inflammation produces a concomitant reduction in lung parenchymal edema. The application of AC fields around the chest wall as shown in Figure 15, below, will reduce the viral load, and promote healing of the afflicted lung tissue with rapid recovery of the patients.

We advocate the use of the high frequency alternating AC electrical fields around the chest wall (Figure 15, below) as one of the adjuvant methods of anti-COVID-19 treatment. It can reduce the viral load, cytokine storm and promote rapid recovery of the patients. The alternating current created electrical field due to high frequency alternating electrical current applied to the chest wall (Figure 15, below) passing from front to back and side to side applied to the body by means of insulated electrodes acts by blocking the viral multiplication processes (akin to mitotic mechanism) mechanism by changing the pH (changing from acid to neutral and alkaline pH) within the viral infected cells. This change to alkalinity in the viral infected cell will prevent the final assembly of the virus in Golgi-lysosome complex. This method is safe, nontoxic, and has no adverse effects. The treatment is applied by electrodes attached onto the shaved skin of the patient (Figure 15, below) chest, and powered by a mobile generator that is carried by the patient over a time span for one week or more to be used 24 hours a day. This feasibility study to treat COVID-19 has the goal to demonstrate the safety and absence of adverse effects of this simple method in patients.

Weak electric AC currents generated employing conductive electrodes are known to increase the efficacy of antibiotics against viral and bacterial biofilms, a phenomenon termed "the bioelectric effect." We have used the battery-operated pads, inserted electrodes, ultrasound and magnetic fields for the treatment of breast and lung cancers, and infections. For details please read patent # US 2016/0074480 A1 by Shantha available free on the Free Patent Online website.

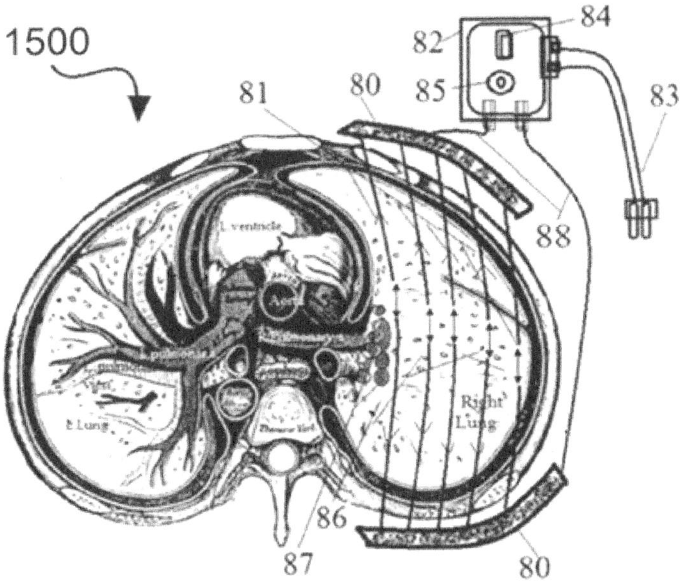

Figure 15

Figure 15 showing the diagram 1500 of the coronal section of the thoracic cavity with lesions of COVID-19 infection of the right lungs 86 with lymph node enlargement 87. It shows the AC electric output generator manipulator 82 with on and off switch 84 and electrical power output adjuster 85 connected to the electrical outlet 83. Insulated electrical conductor 80 are attached to the anterior and posterior chest wall which generated electric field 81 that travels the entire lungs front and back over the lungs 86 with tuberculosis / coronavirus lesions. The alternate current generator 82 is connected by a pair of conductive leads 88. The device combines, and simultaneously delivers, frequency-modulated (FM) and amplitude-modulated (AM) electric cell membrane signaling currents in pulsed electromagnetic fields (EMFs) to inhibit and kill coronavirus. Combined with the above explained therapies, it could save many COVID-19 patients from more severe morbidity and mortality.

Use of anti-cholesterol drug to interfere with production of coronavirus

Over three months, Professor Nahmias and Dr. tenOever conducted research on how the SARS-CoV-2 virus changes patients' lungs to reproduce itself. They found that the virus prevents the routine burning of carbohydrates, resulting in large amounts of fat accumulating inside lung cells. This is a condition the virus needs to reproduce, the Hebrew University explained in a statement. In lab tests, the cholesterol-lowering, FDA-approved drug ***Fenofibrate (Tricor)*** showed "extremely promising results" by interfering with how the SARS-CoV-2 virus – the novel coronavirus that causes COVID-19 – is able to reproduce, according to the research.

The study was led by the Hebrew University of Jerusalem's Professor Yaakov Nahmias, an Israeli biomedical engineer. and Dr. Benjamin tenOever, a professor of medicine and microbiology. They showed that fenofibrate reversed the metabolic changes induced by SARS-CoV-2 blocking viral replication by burning more fat. Thus, this the drug breaks the virus' grip on the cells and prevents SARS CoV-2's ability to reproduce. This needs to be tested in humans, and it may turn COVID-19 into a common cold like symptoms! It is available, has hardly any toxicity, so why not use it in COVID-19 for 2-3 weeks, though not approved?

Chapter 39

Various combination methods to treat COVID-19

A method of treating coronavirus infected person medically, suffering from COVID-19, intended for inhibiting and killing the coronavirus infection of the mucus membrane of the respiratory system by administering 1500 mgs of vitamin C intravenously every 6 to 12 hours to act as an antioxidant to remove the reactive oxygen species created by coronavirus infection of healthy cells that damages cells respiratory system membranes, to act as inhibitor of cytokine production, promoter of antiviral interferons and to act as viricidal and antibacterial by producing hydrogen peroxide; combining the vitamin C with auranofin 6 mg orally every 8 to 12 hours, and one dose ivermectin which is an anti-inflammatory that results in killing of the virus, inhibition of cytokine production and removal of free radicals generated by coronavirus and coronavirus associated bacterial infection and due to immune system response.

A method of treating coronavirus infected person medically, suffering from COVID-19, intended for inhibiting and killing the coronavirus infection of the mucus membrane of the respiratory system by administering 2500 mgs of vitamin C intravenously, every 6 to 12 hours to act as an antioxidant to remove the reactive oxygen species created by coronavirus microbial that damaged cells membrane, to promote of antiviral interferons and to act as viricidal and bactericidal by producing hydrogen peroxide, vitamin C is combined with administration of rapamycin 6 mg loading dose with maintenance dose of 2 mg daily to inhibit the chemotaxis of white blood cell such as polymorphonuclear leukocytes, macrophages, and eosinophils and such entering lung alveoli from inflammatory reaction from the surrounding leaky blood vessels due to coronavirus infection, and prevent or to inhibit cytokine production, and prevent the cytokine storm, thus saving the patient from severe acute respiratory distress syndrome, pneumonia, bronchitis, fluid in the lungs and other lungs related inflammation.

A method of treating coronavirus infected person medically, suffering from COVID-19 of the respiratory system, intended for inhibiting and killing the coronavirus microbial infection of the mucus membrane of the respiratory system by administering intravenous injection 3000 milligrams of vitamin C every 6 to12 hours combining with remdesivir administration to enhance the viricidal effect synergistically inside and outside the host cell and 3 mg of rapamycin to prevent the coronavirus using the cell mTOR mechanism to multiply.

A method of treating coronavirus infected persons suffering from COVID-19 of the respiratory system, intended for inhibiting and killing the coronavirus infection of the mucus membrane of the respiratory system by administering chloroquine combining with 4000 mgs of vitamin C intravenously, combined with angiotensin receptor blockers telmisartan and losartan, rapamycin 2mg, for inhibiting the coronavirus binding on the host cell angiotensin receptors II for its entry and for facilitating the zinc entrance inside the afflicted cells that inhibits coronavirus production within the cells and prevent the chemotaxis of white blood cells.

Tocilizumab (brand name Actemra® and RoAcemtra®), commonly used to treat rheumatoid arthritis, an anti-IL-6 monoclonal antibody, were FDA approved for steroid-refractory CRS based on retrospective case study data. It should be combined with remdesivir with or without auranofin could be very effective in coronavirus infected patients to prevent extreme inflammation in people gravely ill. Lenzilumab, an anti-GM-CSF monoclonal antibody, may also be effective at managing cytokine release by reducing activation of myeloid cells and decreasing the production of IL-1, IL-6, MCP-1, MIP-1, and IP-10.

A method of treating coronavirus infected person medically, suffering from COVID-19 of the respiratory system, intended for inhibiting and killing the coronavirus microbial infection of the mucus membrane of the respiratory system by administering angiotensin receptor II blockers telmisartan and losartan combining with remdesivir to prevent viral docking to the healthy cells in an attempt to infect the healthy cells with the help of the protease and inhibit coronavirus multiplications within the host cell by changing the viral RNA combined with rapamycin 2 mg.

A method of treating coronavirus infected person medically, suffering from COVID-19, intended for inhibiting and killing the coronavirus infection of the mucus membrane of the respiratory system by administering ethylenediaminetetraacetic acid intravenous infusion with angiotensin receptor blockers Azilsartan and zinc sulfate for preventing the virus docking to healthy cells angiotensin receptors II.

A method of treating coronavirus infected person medically, suffering from COVID-19 of the respiratory system, intended for inhibiting and killing the coronavirus infection of the mucus membrane of the respiratory system by administering angiotensin receptor blockers telmisartan and losartan with chloroquine, 3500 mgs dose vitamin C, rapamycin 3 mgs and zinc to block the coronavirus infection of mucus membrane healthy cells of the respiratory system.

A method of treating coronavirus infected person medically, suffering from COVID-19 of the respiratory system, intended for inhibiting and killing the coronavirus microbial infection of the mucus membrane of the respiratory system by administering high dose vitamin C of 5000 mgs intravenously, with rapamycin 4 mgs to act synergistically to reduce activation of nuclear factor kB (NF-kB), for attenuating the releasing of pro-inflammatory mediators (cytokines, chemokines, prostaglandins), reducing the endothelial injury of micro-blood vessels surrounding the alveoli, by combining with administering 50 mgs of hydrocortisone or 4-6 mgs of dexamethasone (brand names Decadron, Dexamethasone, Intensol, Dexasone, ZoDex), mixing with 200 mgs of thiamine hydrochloride, intravenously to stabilize the microvasculature of the alveoli to prevent their leakage so as to prevent white cell infiltration of white blood cells into alveolar sacs and suppress the cytokines production, to prevent severe acute respiratory distress syndrome along with medically effective amount by administering 500 mg azithromycin, intravenously as a coronaviral associated antibacterial.

A method of treating coronavirus infected person medically, suffering from COVID-19, intended for inhibiting and killing the coronavirus infection of the mucus membrane of the respiratory system by orally/parenterally administering antioxidants and free radical scavenger bioactive curcumin 2000 mgs two times a day, Vitamin D3, 5000 IU 3 times a day, zinc sulfate 75 mgs once a day, with intravenous infusion of superoxide dismutase, and ethylenediaminetetraacetic acid.

A method of treating coronavirus infected person medically, suffering from COVID-19, or diagnosed with coronavirus infection in the early stages in the young people, intended for inhibiting and killing the coronavirus infection of the mucus membrane of the respiratory system by administering orally 1000 mgs of superoxide dismutase, vitamin C 1000 mgs three times a day, zinc sulfate 75 mg a day, auranofin 6 mgs a day, auranofin 6 mgs a day, Benfotiamine 250 mg a day.

A method of treating coronavirus infected person medically, suffering from COVID-19, or diagnosed with coronavirus infection using ivermectin.

A method of treating coronavirus infected person medically, suffering from COVID-19, or diagnosed with coronavirus infection using anti-inflammatory agent Colchicine.

A method of treating coronavirus infected persons suffering from COVID-19, intended for inhibiting and killing the coronavirus infection of the mucus membrane of the respiratory system by administrating through the respiratory system, for coating the respiratory system mucus membrane including lung alveoli, using and aerosolized spray or intravenously:

1. administering aerosol and intravenously delivered specific nanoparticle antibodies or,
2. administering aerosol and intravenously delivered diluted plasma from patients recovered from the COVID-19 that contains coronavirus antibodies or,
3. administering aerosol and intravenously delivered viral specific biologic monoclonal antibodies against coronavirus.
4. To kill the coronavirus in the respiratory passages and alveoli, use pentamidine through a nebulizer spray. Dissolve 300 mg of pentamidine in 10 ml of sterile water and spray every 4-8 hours depending on the symptoms.

These biologic agents are delivered by aerosolized spray through the nasal and oral air passages to respiratory passages while a person is breathing while taking a deep breath. They are also be delivered directly to endotracheal tube if intubated, through oral airway devices, or tracheostomy tube if the patient has tracheostomy. These biologic agents are intended for inhibiting and neutralizing the coronaviruses infection on the mucus membrane of the respiratory system especially lungs alveoli and in the systemic circulation which are provided high number of angiotensin receptors II by preventing their binding to host cells and making them biologically infective to attach and enter the host cells.

A method of preventing the coronavirus infection of a person during epidemic, pandemic, or cold-flu season intended for preventing the coronavirus entering the body through the nasal and oral air passages and the rest of the respiratory system by administering and following measures:

1. by wearing an effective mask when you walk out of the house and spraying it with antiviral spray claimed above,
2. adhering to rigid isolation, keeping 6 feet social distancing - away from another person when you walk out of the residence and come across another person,
3. taking vitamin C 1000 mgs, 2-3 times a day,
4. taking bioactive curcumin 1000 mgs 2 times a day, which acts as effective anti-oxidants, anti-inflammatory, anti-septic, anti-coagulant, anti-aging, anti-cancer, anti-cardio- vascular diseases, anti-neurodegenerative diseases, anti-Alzheimer’s and Parkinson’s, anti-scarring, and anti-fat and protein clump accumulate etc.,

5. taking vitamin D3, 5000 IU twice a day, along with zinc 50 -75 mgs a day, selenium, Benfotiamine,
6. Taking melatonin 10-20 mgs before sleep, and add magnesium L-threonate 1000 mg a day,
7. Using the anti-coronaviral antimicrobial nasal spray and mouth wash 2-3 times day described herein,
8. taking angiotensin receptor blocker II telmisartan 20 mg oral, 25% of the antihypertension dose to prevent any virus attaching to the mucus membrane of the respiratory system and other organs as it is transported through the circulation,
9. Taking micro-doses (3-6 mgs) of rapamycin once a week, intermittently after the age of 60-70s,
10. Taking lithium carbonate 300-1000 micrograms a day after the age of 50s.
11. Take mini doses ±250 mgs of metformin whether you are diabetic or not, after lunch and supper to maintain homeostatic glucose and other energy levels at the same time bestow longevity.
12. Take N-acetyl cysteine as mucolytic breakdown the phlegm and antioxidant
13. Nursing homes, extended care facilities, and the aged living independently develop muscle wasting. Whey protein isolate with added creatine and glutamine is a premium food supplement taken with exercise program, to maintain muscle mass, greater strength and daily activities. It is a favorite supplement of body builders.
14. Creatine supplement added. It supports energy production by increasing levels of cells' energy ATP in muscles and helps to maintain healthy muscle mass.

Take some or all these above supplements and other additional vitamin, food, and herbal supplements not listed as needed. Nothing is written in stone; one can make changes according to one's need. The above supplements should be provided in all nursing homes, elderly care centers, and in the hospitalized patients who are advanced in age with co-morbidities.

Chapter 40

Metformin to enhance the therapeutic effectiveness of other drugs

Example of "off label, compassionate use and repurposing old drug" Dihydrochloroquin, chloroquine phosphate, ivermectin, Colchicine, auranofin, metformin and such to enhance the therapeutic effectiveness of other drugs at the same to prevent, curtail, stop the progression of and/or cure diseases

I have also used oral administration of hypoglycemic agent metformin, the fourth largest prescribed medication in US (±250-500-750 mg) 30 to 60 minutes before administering prescribed therapeutic agents/drugs especially chemotherapeutic agents, anti-retroviral therapies, antibiotics, plasma and plasma derived antibodies, monoclonal antibodies, protease inhibitor, health supplements, and such to enhance the pharmacological activity and effectiveness. **I call it *"metformin potentiation therapy*, *MPT*.**" I used this method routinely before administering chemotherapy therapeutic agents, antimicrobial agent, and to treat infections, autoimmune diseases, allergies, headache medication, even before taking supplements, nutraceuticals and such.

By using this method, therapeutic agents prescribed to treat any illness or supplements taken to improve longevity and health span become very effective. The dose of some expensive therapeutic agents can be considerably reduced, decreasing their ill effect, at the same time saving the cost of medication. This method can be easily adopted to treat COVID-19, to treat any and all diseases and to maintain health. I advised most of my patients to take metformin 30-60 minutes before taking most of the prescribed oral or parenteral medications especially antibiotics, hormones, supplements, anti-autoimmune diseases therapeutic agents, health supplements, any and all hormones, and the list is endless.

Using to treat COVID-19 would not be the first-time chloroquine has been **"repurposed"** to treat a condition other than malaria. As noted, chloroquine has anti-inflammatory properties. Consequently, the drug has been prescribed in treating inflammatory diseases, such as rheumatoid arthritis and lupus erythematosus.

I have used Dihydrochloroquin and chloroquine phosphate by local injection, given orally and intravenously also to treat cancers for 5 years along with insulin and specific chemotherapy agent and during radiation therapy. It prevents cell to cell adhesion of cancer cells, which also means it prevents the formation of thrombus inside the blood vessels of the COVID-19 by preventing sticking of platelets, white and red blood cells. It is very effective in reducing the tumor burden. It appears to work by sensitizing the cancers cell for radiation and chemotherapeutic agent through endoplasmic reticulum-lysosome-Golgi complex and may have anti-adhering effects. It also has anti-neurodegenerative diseases effects in Parkinson's disease. I prescribed it in selected cases with good results. Chloroquine probably binds to the endosomes-endoplasmic reticulum -lysosome – Golgi complex (increase pH) and prevents the attack by free radicals that destroy these organelles generated by highly active dopamine producing nerve cells in substantia nigra and ventral tegmental nucleus in the midbrain. I advocate using it in other childhood neurological disorders and chronic intestinal afflictions also. It is inexpensive with no toxicity – why not give it a chance.

Dihydrochloroquin and chloroquine phosphate's anti-viral effects (e.g. Coronavirus) could be due to its misfolding effect of viral genetic material and inhibiting cell to cell adherence virus to cell adherences of coronavirus. Chloroquine can distort the shape of proteins called cytokines, which can reside in the endosomes of immune cells that fight infection. Sometimes, the immune system gets too excited and creates what is called a "cytokine storm," which has been a major complication of COVID-19. By quelling this cytokine storm, chloroquine provides a dual advantage in helping the body combat coronavirus infection

Chloroquine, a well-known antimalarial drug, possesses *pleotropic* effects as well as well: anti-inflammatory, anticoagulant and vascular actions which may prevent the thrombo-embolic phenomenon in COVID-19. It also improves significant increase in urine flow rate, glomerular filtration rate and sodium excretion, as well as stimulation of nitric oxide synthase, thus its use protects from kidney damage or acute renal failure by coronavirus. Rapamycin is another important drug, which I advise in micro doses as an anti-aging agent for the aged and use it also for metastatic cancers. Rapamycin given in mini-doses has great therapeutic value in stage III, IV cancer surgery to prevent metastasis.

Chapter 41

"Compassionate use" "off label use" "repurposing an old drug" as alternate therapies, devices and procedures

Compassionate use of therapeutic agents, devices and alternative therapies

Expanded access, also called "compassionate use," provides a pathway for patients to gain access to investigational drugs, biologics and medical devices for serious diseases or conditions. Investigational drugs and devices have not yet been approved by the FDA and they have not been proven to be safe and effective. Therefore, they may be effective in the treatment of a condition, or they may not. It is important to remember that the drug/biologic/medical device may have unexpected serious side effects and that patients need to consider all the possible risks when seeking access to an investigational medical product. To gain access to an investigational medical product outside of a clinical trial, the sponsors must decide whether to make their experimental medical product available to patients via expanded access. FDA regulations specify two groups of people eligible for expanded access:

a) those with life-threatening diseases or conditions for which "there is a reasonable likelihood that death will occur within a matter of months or in which premature death is likely without early treatment"
b) those with serious diseases or conditions that have a "substantial impact on day-to-day functioning"

In most cases, patients who seek compassionate use must have exhausted all approved therapies for their condition, and be unable to enroll in a clinical trial.

Can a licensed medical practitioner prescribe drugs (that are already FDA approved for a particular illness) for other illnesses off-label

The Answer is an Emphatic Yes. A black box warning or boxed warning is the U.S. Food and Drug Administration's most serious warning for drugs and medical devices—about side effects that may cause serious injury or death. But studies show they do not always impact the patients' decisions as they should and doctors may ignore them when prescribing drugs. Many of therapeutic agents we describe are selected for a particular disease, and may not follow the so-called black box warnings. But, it's not illegal for any qualified doctor to prescribe the medication. "Once the FDA approves a drug, health care providers generally may prescribe it for an unapproved use when they judge that it is medically appropriate for their patient," explains Jeremy Kahn, a spokesman for the FDA. In the year 2016, there were around 4.45 billion medical prescriptions issued all over the United States of which one in five prescriptions today is prescribed for off-label use. Hence the use of dihydrochloroquin and chloroquine phosphate, protease inhibitors remdesivir, angiotensin receptor blockers telmisartan, vitamin C and zinc and such, by a qualified physician that it is medically appropriate for their patient, if there are no other drug and this the only modality available to save the life of the patient.

Drug "repurposing" an old drug for new therapeutic applications

Drug ***'repurposing'*** may be off label use and it is the identification of new therapeutic applications for drugs that have received US FDA approval for another purpose. Repurposing an FDA approved drug is more affordable and achievable than novel drug discovery. The current system of discovery, development, registration and testing of new drugs is estimated to cost ±USD$1.5 billion and often needs 10–17 years to complete [Pessetto ZY, Weir SJ, Sethi G, Broward MA, Godwin AK. Drug repurposing for gastrointestinal stromal tumor. Mol Cancer Ther. 2013;12(7):1299–1309.]. Further, this expensive and time-consuming process of new drug development every so often results in failure, with an estimated 70–90% of drugs failing clinical trials (Woodcock J, Woosley R. The FDA critical path initiative and its influence on new drug development. Annu Rev Med. 2008; 59:1–12.). That is why drug 'repurposing' can play an important role to prevent, curtail, stop the progression of and / or cure human diseases, new and old both. Due to the reduced length and cost of research and trial phases, drug repurposing is more affordable and achievable than novel drug discovery, with patients gaining access to new faster therapies immediately. Such a drug is auranofin, so also vitamin C, anti-malarial dihydrochloroquin, chloroquine phosphate, protease inhibitor and such we discuss for the treatment of COVID-19. Recently, gastro-intestinal track antiparasitic drugs such as ivermectin and Niclosamide as well as Ebselen are emerging as most prominent **repurposing** therapeutic agents to treat the coronavirus infection.

All the above explained therapeutic administration ***"off label, compassionate use and repurposing old drug"*** can be used to treat diseases when there are no medical or surgical therapies available and the physician is left with no alternative that results in prolonged morbidity and ultimate early mortality; the benefits should outweigh the risks. The methods and choice of therapeutic agents administered are weighed between benefits and harm in each case. I have treated hundreds of cases of cancers of all kinds and all stages with local injection of appropriate therapeutic agents with insulin (insulin potentiation therapy, IPT) with excellent results besides administering parenterally which can be easily labelled as "off label, compassionate use and repurposing old drugs" which gave the patient long, pain free, cost effective, quality life. I have injected hundreds of tumor metastasis injection locally with insulin and appropriated chemotherapeutic agent. Insulin enhances the tumor killing effect of anticancer agents hundreds of times (Alabaster O. et al. Metabolic Modification by Insulin Enhances Methotrexate Cytotoxicity in MCF-7 Human Breast Cancer Cells. 1981, Eur J Cancer Clin Oncol, Vol. 17, No. 11, pp. 1223-1228, 1981). Every oncologist should use this method to treat all stages of cancers besides chemotherapy. Now I have further modified the IPT using metformin instead of using insulin, which is easy and practical as described below and can be used at home setting to enhance activity drugs and supplements you take. I believe that ***Ebselen*** may become one of the most important ***repurposed*** drugs ever and is one of the most effective therapeutic agents to treat COVID-19, and many such uses are on the horizon. There are thousands of such compounds considered to be as repurposing.

Epilogue/closure

Numerous modifications and alternative arrangements of steps explained herein regarding our inventive therapeutic agents and their methods of use may be devised by those skilled in the art without departing from the spirit and scope of the present invention and the appended claims are intended to cover such modifications and arrangements. Thus, while the present invention has been described above with particularity and detail in connection with what is presently deemed to be the most practical and preferred embodiments of the invention, it will be apparent to those of ordinary skill in the art that several modifications, including, but not limited to, variations in compounding doses using our embodiments to the invention form function and manner of procedure, assembly and use may be made. While the preferred embodiment of the present invention has been described, it should be understood that various changes, adaptations and modifications may be made thereto. It should be understood, therefore, that the invention is not limited to details of the illustrated invention. Numerous modifications and alternative arrangements of steps explained herein regarding our inventive therapeutic agents and their methods of use may be devised by those skilled in the art without departing from the spirit and scope of the present invention and the appended claims are intended to cover such modifications and arrangements. Thus, while the present invention has been described above with particularity and detail in connection with what is presently deemed to be the most practical and preferred embodiments of the invention, it will be apparent to those of ordinary skill in the art that several modifications, including, but not limited to, variations in compounding doses using our embodiments to the invention form function and manner of procedure, assembly and use may be made. While the preferred embodiment of the present invention has been described, it should be understood that various changes, adaptations and modifications may be made thereto. It should be understood, therefore, that the invention is not limited to details of the illustrated invention.

Chapter 42

What is claimed is

1. A method for the treatment or prophylaxis of coronavirus, said method comprising administering a medically effective amount of an anti-coronavirus agent using a nasopharyngeal spray.
2. A method for the treatment or prophylaxis of coronavirus, said method comprising administering a medically effective amount of an anti-coronavirus agent using a mouth wash.
3. The method of claim 1 further comprising the step of administering a medically effective amount of an anti-coronavirus agent using a oropharyngeal spray.
4. The method of claim 1 further wherein said nasopharyngeal spray is delivered to a patient's larynx, trachea, bronchial tree and lungs through an endotracheal tube and ventilator system.
5. The method of claim 1 further comprising the step of administering said medically effective amount of anti-coronavirus agent using an oropharyngeal spray.
6. The method of claim 3 further comprising the step of administering said medically effective amount of anti-coronavirus agent using a mouth wash.
7. The method of claim 1 wherein said anti-coronavirus agent is a viricidal agent.
8. The method of claim 1 wherein said anti-coronavirus agent is an antibacterial agent.
9. The method of claim 6 wherein said anti-coronavirus agent is an viricidal agent.
10. The method of claim 6 wherein said anti-coronavirus agent is a combination of viricidal and antibacterial agents
11. The method of claim 6 further comprising the steps of:
    a. adding angiotensin II receptor blockers (ARBs) telmisartan and losartan to said anti-coronavirus agent wherein coronavirus is prevented from attaching to angiotensin II receptors on healthy

host cells of the mucus membrane of the respiratory system including lungs alveoli;

b. adding an antimalarial Dihydrochloroquin/chloroquine phosphate inhibitor to said anti-coronavirus agent wherein said coronavirus is prevented from multiplying inside healthy cells;
c. adding a Remdesivir protease inhibitor to said anti-coronavirus agent thereby blocking a host cell membrane protease usage by said coronavirus thus inhibiting said coronavirus and other microbials from attaching, entering and multiplying in respiratory system lining host cells;
d. adding superoxide dismutase to said anti-coronavirus agent, wherein free radicals produced by said coronavirus are destroyed;
e. adding ethylenediaminetetraacetic acid to said anti-coronavirus agent thereby blocking infected coronavirus cells and microbial created free radicals from attacking healthy host cell membranes and blocking the attachment of coronavirus and microbes to said host cells;
f. adding Chlorhexidine to said anti-coronavirus agent, wherein said Chlorhexidine is a pharmacologically active carrier antiseptic, antiviral, antimicrobial and antibiofilm;
g. adding a medically effective amount of N-acetyl cysteine to said anti-coronavirus agent to act as a mucolytic and antioxidant to clear the lungs;
h. adding a medically effective amount of dissolved vitamin C to said anti-coronavirus agent, wherein said dissolved vitamin C acts as an antioxidant and viricidal;
i. adding a medically effective amount of zinc sulfate to said anti-coronavirus agent;
j. adding biologic coronavirus antibody nanoparticles/antibodies from patients recovered from COVID-19 to said anti-coronavirus agent;
k. adding monoclonal antibodies to said anti-coronavirus agent; and
l. adding a selected amount of xylitol to said anti-coronavirus agent to act as an anti-viral, anti-microbial, antibiofilm and as a binder to therapeutic agents to said coronavirus.

12. The method of claim 6 further comprising the steps of:
    a. adding chloroquine phosphate to said anti-coronavirus agent;

b. adding superoxide dismutase to said anti-coronavirus agent;
c. adding xylitol adhesive to said anti-coronavirus agent; and
d. adding a carrier Chlorhexidine to said anti-coronavirus agent.

13. The method of claim 6 further comprising the steps of:
    a. applying said anti-coronavirus agent to a patient's nasal cavities;
    b. rinsing said patient's mouth with said mouth wash after brushing;
    c. retaining said oropharyngeal spray in said patient's mouth for at least 2 minutes and do not allow said patient to eat or drink for at least 10 minutes after applying said nasopharyngeal spray and said oropharyngeal spray;
    d. repeating the steps of applying said nasopharyngeal spray to said nasal cavities and applying said oropharyngeal spray to said mouth every six to eight hours; and
    e. filling an irrigation syringe, and irrigating said nasal cavities every 12 hours.
14. The method of claim 1 further comprising the step of administering intravenously:
    a. a high dose of Vitamin C, (IV); and
    b. a high dose of thiamine hydrochloride.
    c. Zinc sulfate; and
    d. Dexamethasone.
15. A method for the treatment or prophylaxis of coronavirus, said method comprising intravenously administering a medically effective amount of at least one anti-inflammatory /antioxidant drug selected from the group consisting of hydrocortisone, dexamethasone, auranofin, Ethylenediaminetetraacetic acid (EDTA), sodium dismutase (SOD) and rapamycin.
16. The method of claim 15 further comprising the step of intravenously administering a medically effective dose of Remdesivir.
17. The method of claim 15 further comprising the steps of:
    a. administering intravenously a medically effective amount of an antiviral/antibacterial drug;
    b. orally/parenterally administering a medically effective dose of dihydrochloroquin and chloroquine phosphate; and
    c. orally administering medically effective doses of immunosuppressants selected from the group consisting of sirolimus and its analogues rapalogues—everolimus, biolumus, Temsirolimus, ridaforolimus or specific cephalosporin wherein

the production of cytokine is blocked to prevent the development of cytokine storm in COVID-19;

d. Orally administering ivermectin dose based on the weight.

e. Administration of convalescent plasma from COVID-19 recovered.

18. The method of claim 15 further comprising the step of administering intravenously a medically effective amount of zinc sulfate intravenously whereby the cell-mediated immunity is enhanced, including expression of interleukin-2 and interferon-γ which enhances the generation of natural killer and cytolytic T cells for killing free floating coronavirus viruses,

19. The method of claim 14 further comprising the steps of:

    a. providing oxygen through extracorporeal membrane oxygenation (ECMO) and blood purifier CytoSorb, for COVID-19 with severe acute respiratory distress syndrome;

    b. providing hyperbaric therapy (HBO) for reducing hypoxia of the organs and tissue due to severe acute respiratory distress syndrome due to COVID-19 and its associated infections;

    c. administering a medically effective dose of oral auranofin; and

    d. administering through extracorporeal membrane oxygenator and intravenously 2000 mgs of vitamin C, 10 mgs of dexamethasone, 200 mg of thiamine HCL, 25 mg of zinc sulfate every 12 hours.

20. The method of claim 1 further comprising the step of adding a medically effective amount of auranofin, zinc sulfate, xylitol, ethylenediaminetetraacetic acid and superoxide dismutase to said oropharyngeal spray and said mouth wash.

21. The method of claim 17 further comprising the steps of administering one dose of ivermectin orally and a booster dose 7 days later as the diagnosis of coronavirus is made and treatments are instituted in home, hospitals rooms and in ICU settings.

22. The method of claim 14 further comprising the steps of cleaning blood of severely ill septicemic and cytokine storm COVID-19 patients using extracorporeal blood purification (EBP) method by means of a device CytoSorb, and in April, the USFDA granted Emergency Use Authorization (EUA) which is approved in India; besides providing the needed oxygenation using ECMO.

Chapter 43

Protocol to treat COVID-19

Protocol to treat COVID-19 is as follows:

1. Admit the patient to hospital and ICU as needed.
2. Administer main protease inhibitors dubbed ***Mpro*** such as Ebselen given 400-600 mg orally and used in prophylaxis sprays. Protease enzymes are responsible for cutting down proteins into smaller particles and allow it to survive and multiply inside the healthy cells. Blocking Mpro causes the virus to die. It may be more potent than other protease inhibitors.
3. Anti-helminthic: Ivermectin and Niclosamide can be effectively used if you can obtain them. Combine with auranofin and Ebselen antioxidants to prevent coronavirus pathology.
4. Remdesivir is the next protease inhibitor we can use. It is widely used now all over the world. Maybe it can be combined with narlaprevir, an anti-hepacitis C protease inhibitor.
5. Vitamin C, thiamine hydrochloride and hydrocortisone/ dexamethasone are the important to start with all coronavirus infected patients. It will have an effect on the virus on its pathologic effect in producing cytokine storm.
6. Dihydrochloroquin and chloroquine phosphate are effective, contrary to the political motivated reports. It is available and inexpensive; try it as the diagnosis is made.
7. Dexamethasone and/or Hydrocortisone should be used as the disease progresses and starts developing cytokine storm and such.
8. Zinc, administered intravenously, helps the protease inhibitors to work effectively and has an antiviral effect also in its own right.
9. Antibody plasma from COVID-19 recovered patients should be considered when available. Monoclonal antibodies be considered.
10. Azithromycin and other antibiotics should be considered to prevent bacterial infection superimposed on viral onslaught.

11. Fluid and nutrition management, supplemental oxygen, intubation, ECMO should be used as needed to maintain homeostatic levels of tissue oxygenation, maintain cardiac and kidney function.
12. On 8/22/2020, the FDA along with President Trump announced the approval of the use of antibody-containing convalescent plasma from the patients who have recovered from COVID-19 as a method of treatment. According to the president and the FDA's announcement, thousands of COVID-19 patients have received this convalescent plasma treatment with 36% reduction in morbidity and mortality. This is one new tool for the treatment of COVID-19.

Use social isolation. Use a mask all the time whether infected or non-infected, in the house or hospital. In a closed room the small droplets of infection can travel 20 feet as shown in the diagram above. Use a carbon-dioxide monitor in the closed room, sororities, areas of large gatherings, cruise ships, social events, bars, movie theaters, sports events, religious gatherings, and such.

Bars are crowded with people gathering together to drink, talk, laugh and let loose in one another's company, so also let loose the coronavirus which jumps from person to person. When you enter a bar, you can't drink through the mask, so you take off your mask. Social distancing is not practiced in the spaces, and alcohol induces people to change their behavior. Bars are the ideal environment for coronavirus to spread from person to person. The chance of catching the virus through tiny airborne respiratory droplets, known as aerosols, goes up significantly in indoor spaces such as bars.

Even if the bar is open, social distancing and mask wearing should be followed and there should be proper exchange of air in the bars and areas of social gathering as measured by simple inexpensive CO2 monitor. The governments should mandate that all closed gathering places, including bars and schools, should have CO2 monitor, proper ventilation and fresh air exchange between occupants and outside air.

Chapter 44

Other Publications (References)

1. "Anatomy of a Killer: Understanding SARS-CoV-2 and the drugs that might lessen its power". The Economist. 12 March 2020. Archived from the original on 14 March 2020. Retrieved 14 March 2020.
2. Aggarwal BB, Sundaram C, Malani N, Ichikawa H. Curcumin: the Indian solid gold. Adv Exp Med Biol. 2007; 595:1-75
3. Ayres HM, Payne DN, Furr JR, Russell AD. Effect of permeabilizing agents on antibacterial activity against a simple Pseudomonas aeruginosa biofilm. Lett Appl Microbiol 1998; 2:79–82.
4. Baer G, Shantha TR, Bourne GH. The pathogenesis of street rabies virus in rats. Bulletin World Health Org. 1965; 33:783-794.
5. Baez-Santos YM, Barraza SJ, Wilson MW. (2014) X-ray Structural and Biological Evaluation of a Series of Potent and Highly Selective Inhibitors of Human Coronavirus Papain-like Proteases. J. Med. Chem. 57: 2393-2412
6. Behring EA, Kitasato S. 1890. Ueber das zustandekommen der diptherie-immunität und der tetanus-immunität bei thieren. Deutch Med Woch 49:1113–1114.
7. Beniac DR, et al. Architecture of the SARS coronavirus prefusion spike. Nature structural & molecular biology. 2006;13(8):751–752.
8. Bertram S, Glowacka I, Blazejewska P, et al. TMPRSS2 and TMPRSS4 facilitate trypsin-independent spread of influenza virus in Caco-2 cells
9. Bertram S, Heurich A, Lavender H, et al. Influenza and SARS-coronavirus activating proteases TMPRSS2 and HAT are expressed at multiple sites in human respiratory and gastrointestinal tracts
10. Blagosklonny MV. Aging and immortality: quasi-programmed senescence and its pharmacologic inhibition. Cell Cycle. 2006;5(18):2087–2102.

11. Bosch BJ, et al. The coronavirus spike protein is a class I virus fusion protein: structural and functional characterization of the fusion core complex. J Virol. 2003;77(16):8801–8811.
12. Buchholz U, Müller, MA, Nitsche, A. et al. Contact investigation of a case of human novel coronavirus infection treated in a German hospital, October-November 2012
13. Budtz-Jörgensen E, Löe H (1972). "Chlorhexidine as a denture disinfectant in the treatment of denture stomatitis". Scandinavian Journal of Dental Research. 80 (6): 457–64.
14. Bunyavanich S, et al. Nasal gene expression of angiotensin - converting enzyme 2 in children and adults. JAMA. May 20, 2020. Doi:10.10. 1001/jama.2020.8707.
15. Cao Y, Li L, Feng Z, et al. Comparative genetic analysis of the novel coronavirus (2019-nCoV/SARS-CoV-2) receptor ACE2 in different populations. Cell Discovery volume 6, Article number: 11 (2020)
16. Casadevall A. 1996. Antibody-based therapies for emerging infectious diseases. Emerg Infect Dis 2:200–208.
17. Chan JF, Yuan S, Kok KH, et al. A familial cluster of pneumonia associated with the 2019 novel coronavirus indicating person-to-person transmission: a study of a family cluster
18. Chang Y, Gu W, McLandsborough L. Low concentration of ethylene diamine tetraacetic acid (EDTA) affects biofilm formation of Listeria monocytogenes by inhibiting its initial adherence. Food Microbiol 2012; 29:10–17.
19. Chatre C, Roubille F, Vernhet H, Jorgensen C, Pers YM. Cardiac Complications Attributed to Chloroquine and Hydroxychloroquine: A Systematic Review of the Literature. Drug Saf. 2018 Oct;41(10):919-931. doi: 10.1007/s40264-018-0689-4.
20. Chen N, et al. 2020. "Epidemiological and clinical characteristics of 99 cases of 2019 novel coronavirus pneumonia in Wuhan, China: a descriptive study". The Lancet. 395 (10223): 507–513
21. Clarke NE, Clarke CN, Mosher RE. Treatment of angina pectoris with disodium ethylene diamine tetra-acetic acid. Am J Med Sci 1960; 232:654–656.
22. Cole P, Rodu B, Mathisen A. (August 2003). "Alcohol-containing mouthwash and oropharyngeal cancer: a review of the epidemiology". Journal of the American Dental Association. 134 (8): 1079–87.

23. Corman VM, Lienau J, Witzenrath M. Coronaviruses as the cause of respiratory infections.
24. Cui J, Li F, Shi ZL. (2019) Origin and evolution of pathogenic coronaviruses. Nat. Rev. Microbiol. 17, 181-192.
25. de Wit E, van Doremalen N, Falzarano D. SARS and MERS: recent insights into emerging coronaviruses
26. Delmas B, Laude H. Assembly of coronavirus spike protein into trimers and its role in epitope expression. Journal of virology. 1990;64(11):5367–5375.
27. Ding Y, He L, Zhang Q, Huang Z, et al. Organ distribution of severe acute respiratory syndrome (SARS) associated coronavirus (SARS-CoV) in SARS patients: implications for pathogenesis and virus transmission pathways.
28. Doan T, Massarotti E. Rheumatoid arthritis: an overview of new and emerging therapies. J Clin Pharmacol. 2005;45(7):751–762.
29. Dupont et al. (2011) "Autophagy-based unconventional secretory pathway for extracellular delivery of IL-1" EMBO J. 30:4701-11.
30. Ehninger D, et al., and Blagosklonny describes "longevity and aging" Cell Mol Life Sci. 2014; 71(22): 4325–4346.
31. El Sahly HM. "Genomic Characterization of the 2019 Novel Coronavirus". The New England Journal of Medicine. Retrieved 9 February 2020.EMBO J., 24 (2005), pp. 1634-1643. Euro Surveill., 18 (2013), p. 20406
32. Fehr and Perlman Page 9 Methods Mol Biol. PMC 2016. The rodent adapted COVID-19 strains show similar clinical features to the human disease, including an age-dependent increase in disease severity.
33. Fehr AR and Stanley Perlman. Coronaviruses: An Overview of Their Replication and Pathogenesis, Methods Mol Biol. 2015; 1282: 1–23. doi:10.1007/978-1-4939-2438-7_1.
34. Foster M, Samman S (2012). "Zinc and regulation of inflammatory cytokines: implications for cardiometabolic disease". Nutrients. 4 (7): 676–694. doi:10.3390/nu4070676.
35. Fragasso G, et al. Cardiotoxicity after low-dose chloroquine antimalarial therapy. Heart Vessels. 2009 Sep;24(5):385-7.
36. Glowacka S, Bertram MA, Müller P et al. Evidence that TMPRSS2 activates the severe acute respiratory syndrome coronavirus spike protein for membrane fusion and reduces viral control by the humoral immune response J. Virol., 85 (2011), pp. 4122-4134.

37. Goel A, Jhurani S, Aggarwal BB. Multi-targeted therapy by curcumin: how spicy is it? Mol Nutr Food Res. 2008;52(9):1010-30.
38. Gralinski LE, Menachery VD. 2020. "Return of the Coronavirus: 2019-nCoV". Viruses. 12 (2): 135. doi:10.3390/v12020135.
39. Guardabassi L, Houser GA, Frank LA, Papich MG. Guidelines for antimicrobial use in dogs and cats. Guide to antimicrobial use in animals 2008;183–206.
40. Gurtovenko AA, Jamshed Anwar. Modulating the Structure and Properties of Cell Membranes: The Molecular Mechanism of Action of Dimethyl Sulfoxide, J. Phys. Chem.B2007, 111, 10453-1060
41. Hamming W, Timens ML, Bulthuis AT, et al. Tissue distribution of ACE2 protein, the functional receptor for SARS coronavirus. A first step in understanding SARS pathogenesis
42. Harrison DE, Strong R, Sharp ZD, et al., Rapamycin fed late in life extends lifespan in genetically heterogeneous mice. Nature. 2009;460(7253):392–395.
43. Herrera, D (March 2013). "Chlorhexidine mouthwash reduces plaque and gingivitis". Evidence-Based Dentistry. 14 (1): 17–18.
44. Hey A. History and Practice: Antibodies in Infectious Diseases. Microbiol Spectr. 2015 Apr;3(2):AID-0026-2014. doi: 10.1128/microbiolspec.AID-0026-2014.
45. Hilgenfeld, R. (2014) From SARS to MERS: crystallographic studies on coronaviral proteases enable antiviral drug design. FEBS J. 281,4085-4096
46. Hoffmann M, Kleine-Weber H, Schroeder S. SARS-CoV-2 Cell Entry Depends on ACE2 and TMPRSS2 and Is Blocked by a Clinically Proven Protease Inhibitor. Cell. 2020 Mar 4. pii: S0092-8674(20)30229-4. doi: 10.1016/j.cell.2020.02.052.
47. Huang C, Wang Y, Li X, Ren L, Zhao J, et al. Clinical features of patients infected with 2019 novel coronavirus in Wuhan.
48. Imai Y, Kuba K, Rao S, et al. Angiotensin-converting enzyme 2 protects from severe acute lung failure
49. Jiang et al. (2013) "Secretory versus degradative autophagy: unconventional secretion of inflammatory mediators" J Innate Immun. 5(5):471-9).
50. Kean WF, Kean IR. Clinical pharmacology of gold. Inflammopharmacology. 2008;16(3):112–125.
51. Kim WY, et al. Combined vitamin C, hydrocortisone, and thiamine therapy for patients with severe pneumonia who were admitted to the

intensive care unit: Propensity score-based analysis of a before-after cohort study. J Crit Care. 2018;47: 211-218.

52. Kimura et al., "Chloroquine in Cancer Therapy: A Double-Edged Sword of Autophagy", American Association for Cancer Research, 2012, pp. 3-7, vol. 73, No. 1.

53. Kirkland JL. Inflammation and cellular senescence: potential contribution to chronic diseases and disabilities with aging. Public Policy and Aging Report. 2013;23: 12-5.

54. Kleineeber H, Elzayat MT, Wang L. et al. Mutations in the Spike Protein of Middle East Respiratory Syndrome Coronavirus Transmitted in Korea Increase Resistance to Antibody-Mediated Neutralization

55. Klenner FR. Encephalitis as a sequela of the pneumonias. Tri-State Med. J., Feb. 1960.

56. Klenner FR. Virus pneumonia and its treatment with vitamin C. Southern Med. Surg., Feb. 1948Klenner, F. R.: An insidious virus. Tri-State Med. J., June 1957.

57. Kontiokari T, Uhari M, Koskela M. Effect of xylitol on growth of nasopharyngeal bacteria in vitro. Antimicrobe Agents Chemother 1995;39: 1820-3.

58. Kopelovich L, Fay JR, Sigman CC, Crowell JA. The mammalian target of rapamycin pathway as a potential target for cancer chemoprevention. Cancer Epidemiol Biomark Prev. 2007;16(7):1330–1340

59. Kuba K, Imai Y, Rao S, et al. A crucial role of angiotensin converting enzyme 2 (ACE2) in SARS coronavirus-induced lung injury

60. Lambert RJW, Hanlon GW, Denyer SP. The synergistic effect of EDTA/antimicrobial combinations on Pseudomonas aeruginosa. J Appl Microbiol 2004;96: 244–253.

61. Lassi ZS, Moin A, Bhutta ZA (2016). "Zinc supplementation for the prevention of pneumonia in children aged 2 months to 59 months". The Cochrane Database of Systematic Reviews. 12 (12): CD005978. doi:10.1002/14651858. CD005978. pub3.

62. Lau SK, Woo PC, Li KS, et al. Severe acute respiratory syndrome coronavirus-like virus in Chinese horseshoe bats. Science, 309 (2005), pp. 1864-1868

63. Lau YL, Peiris JSM. Pathogenesis of severe acute respiratory syndrome. Curr Op Immunol. 2005;17, 404–410

64. Law HK, et al. Chemokine upregulation in SARS coronavirus infected human monocyte derived dendritic cells. Blood. 2005; 106:2366–2376.
65. Lee DW, Gardner R, Porter DL, Louis CU, Ahmed N, Jensen M, et al. Current concepts in the diagnosis and management of cytokine release syndrome. Blood. 2014; 124:188–195.
66. Letko M, El Sahly HM. et al. 2020. "Genomic Characterization of the 2019 Novel Coronavirus". The New England Journal of Medicine. Retrieved 9 February 2020.
67. Letko M, Munster V. 2020. "Functional assessment of cell entry and receptor usage for lineage B β-coronaviruses, including 2019-nCoV". bioRxiv (preprint). doi:10.1101/2020.01.22.915660.
68. Li F, Li W, M. Farzan M, S.C. Harrison SC. Structure of SARS coronavirus spike receptor-binding domain complexed with receptor.
69. Li Q, et al. 2020. "Early Transmission Dynamics in Wuhan, China, of Novel Coronavirus-Infected Pneumonia". The New England Journal of Medicine. 382 (13): 1199–1207.
70. Li W, Hulswit RJG, Widjaja I, et al. Identification of sialic acid-binding function for the Middle East respiratory syndrome coronavirus spike glycoprotein
71. Li W, Moore MJ, Vasilieva N, Sui J, et al. Angiotensin-converting enzyme 2 is a functional receptor for the SARS coronavirus. November 2003. Nature, volume 426, pages450–454(2003)
72. Li W, Zhang C, Sui J, Kuhn JH, et al. Receptor and viral determinants of SARS-coronavirus adaptation to human ACE2. EMBO J. 2005 Apr 20; 24(8): 1634–1643. Published online 2005
73. Lif, Holgerson P, et al. Effect of xylitol-containing chewing gums on interdental plaque-pH in habitual xylitol consumers. Acta Odontol Scand 2005; 63:233-8.
74. Life Extension: https://www.lifeextension.com/Magazine/2018/SS/Major-Advance-in-Healthy-Longevity/Page-01. Accessed June 11, 2019.
75. Lin JT, Zhang JS, Su N, et al. Safety and immunogenicity from a phase I trial of inactivated severe acute respiratory syndrome coronavirus vaccine
76. Liu W, Fontanet A, Zhang PH, Zhan L, et al. Two-year prospective study of the humoral immune response of patients with severe acute respiratory syndrome. J. Infect. Dis., 193 (2006), pp. 792-795.

77. Lu R, et al. 2020. "Genomic characterization and epidemiology of 2019 novel coronavirus: implications for virus origins and receptor binding". The Lancet. 395 (10224): 565–574.
78. Madeira JM, Renschler CJ, Mueller B, Hashioka S, Gibson DL, Klegeris A. Novel protective properties of auranofin: inhibition of human astrocyte cytotoxic secretions and direct neuroprotection. Life Sci. 2013;92(22):1072–1080. doi: 10.1016/j.lfs.2013.04.005.
79. Marik PE, Khangoora V, Rivera R, Hooper MH, Catravas J. Hydrocortisone, vitamin C, and for the treatment of severe sepsis and septic shock: a retrospective before-after study. Chest. 2017;151: 1229–1238. doi: 10.1016/j.chest.2016.11.036
80. Matsuyama S, N. Nagata N, K. Shirato K, Kawase M, Efficient activation of the severe acute respiratory syndrome coronavirus spike protein by the transmembrane protease TMPRSS2. J. Virol., 84 (2010), pp. 12658-12664
81. Menachery KH, Dinnon III, Yount BL Jr., et al. Trypsin treatment unlocks barrier for zoonotic bat coronaviruses infection J. Virol., 94 (2020)
82. Messori L, Marcon G. Gold complexes in the treatment of rheumatoid arthritis. Met Ions Biol Syst. 2004; 41:279–304.
83. Moskowitz A, Andersen LW, Huang DT, Berg KM, Grossestreuer AV, Marik PE, et al. Ascorbic acid, corticosteroids, and in sepsis: a review of the biologic rationale and the present state of clinical evaluation. Crit Care. 2018;22(1):283. doi: 10.1186/s13054-018-2217-4
84. Munster V J, et al. A Novel Coronavirus Emerging in China - Key Questions for Impact Assessment
85. Murakami Y, Osawa K, Tokunaga T, et al. Modulation of TNF-alpha-converting enzyme by the spike protein of SARS-CoV and ACE2 induces TNF-alpha production and facilitates viral entr. N Engl J Med., 382 (2020), pp. 727-733.
86. Park JE, Li K, Barlan A, Fehr AR, Perlman S, et al. Proteolytic processing of Middle East respiratory syndrome coronavirus spikes expands virus tropism
87. Park YJ, Walls AC, Wang Z. Structures of MERS-CoV spike glycoprotein in complex with sialoside attachment receptors Nat. Struct. Mol. Biol., 26 (2019), pp. 1151-1157
88. Patel A, Verma A, 2020, Nasal ACE2 COVID-19 levels and COVID-19 in children. JAMA June 2020, 323, #23, 2386-2387.

89. Pauling L. Vitamin C and the Common Cold. San Francisco: W. F. Freeman & Co., 1970.
90. Peiris JS, et al. Clinical progression and viral load in a community outbreak of coronavirus-associated SARS pneumonia: a prospective study. Lancet. 2003; 361(9371):1767–1772.
91. Percival SL, Bowler P, Parson D. (2005) Antimicrobial composition. US/WO2007068938A2?cl=endotracheal tube
92. Pessetto ZY, Weir SJ, Sethi G, Broward MA, Godwin AK. Drug repurposing for gastrointestinal stromal tumor. Mol Cancer Ther. 2013;12(7):1299–1309.
93. Pollack, A. (2003) Company says it mapped part of SARS virus. New York Times, July 30, section C, page 2.
94. Prasad, Ananda S. (6 March 2013). "Discovery of Human Zinc Deficiency: Its Impact on Human Health and Disease". Advanced Nutrition. 4 (2): 176–190
95. Riou J, Althaus CL 2020. "Pattern of early human-to-human transmission of Wuhan 2019 novel coronavirus (2019-nCoV), December 2019 to January 2020". Euro-surveillance. 25 (4). doi:10.2807/1560-7917.ES.2020.25.4.2000058
96. Roberts A, et al. Aged BALB/c mice as a model for increased severity of severe acute respiratory syndrome in elderly humans. Journal of virology. 2005; 79(9):5833–5838
97. Roder C, Thomson MJ. Auranofin: Repurposing an Old Drug for a Golden New Age. Drugs R D. 2015 Mar; 15(1): 13–20.
98. Sannella AR, Casini A, Gabbiani C, Messori L, Bilia AR, Vincieri FF, et al. New uses for old drugs. Auranofin, a clinically established antiarthritic metallodrug, exhibits potent antimalarial effects in vitro: mechanistic and pharmacological implications. FEBS Lett. 2008;582(6):844–847. doi: 10.1016/j.febslet.2008.02.028.
99. Shantha TR and Bourne GH: Perineural epithelium: A new concept of its role in the integrity of the peripheral nervous system. Science 154:1464-1467 (1966).
100. Shantha TR Bypassing the BBB: Drug Delivery from the Olfactory Mucosa to the CNS. Drug development and delivery. 2017, Vol 17, # 1, 32-37.
101. Shantha TR, Bourne GH. Perineural epithelium; structure and function of nervous tissues. Academic Press. 1969;1: 379-458.
102. Shantha TR, Bourne. GH. The perineural epithelium: and significance. J Nature. 1963; 4893:577-579

103. Shantha TR, Nakajima Y. Histological and histochemical studies on the rhesus monkey (Macaca Mulatta), olfactory mucosa. Yerkes Regional Primate Research Center, Emory University, Atlanta, Georgia. Z. Zellforsch. 1970; 103:291-319.

104. Shantha TR. Alzheimer's disease treatment with multiple therapeutic agents delivered to the olfactory region through a special delivery catheter and iontophoresis. US20120323214, US20140012182, US20150080785, WO/2015/013252A1, and WO/2009/149317A3.

105. Shantha TR. Peri-vascular (Virchow-Robin) space in the peripheral nerves and its role in spread of local anesthetics. ASRA Congress at Tampa. Regional Anesthesia. March-April;1992.

106. Shantha TR. Presented at Hanoi: Rabies Cure: Nasal and Oral Route of Transmission of Rabies Virus and Possible Treatment to Cure Rabies. Rabies in Asia conference in Hanoi (RIACON). September 10, 2009. US Patent Application Publication Number: 201110020279 Al. Jan. 27, 2011.

107. Shantha, TR. Innovative methods of treating tuberculosis. US Patent # 2016/0074480A1- 78 pages,

108. Shantha, TR, Wieden R. Lower jaw thrusting, mandibular protracting, tongue holding, universal oropharyngeal airway device. WO 2018/200063 A1. 2018, 102 pages.

109. Shieh WJ, Hsiao CH, Paddock CD, et al. Immunohistochemical, in situ hybridization, and ultrastructural localization of SARS-associated coronavirus in lung of a fatal case of severe acute respiratory syndrome in Taiwan Hum. Pathol., 36 (2005), pp. 303-309.

110. Shimabukuro-Vornhagen A, Gödel P, Subklewe M, et al. (15 June 2018). "Cytokine release syndrome". J Immunotherapy Cancer. 6: 56

111. Shirato K, Kanou K, Kawase M, et al. Clinical Isolates of Human Coronavirus 229E Bypass the Endosome for C Entry J. Virol., 91 (2016),

112. Shirato K, Kawase M, Matsuyama S. Wild-type human coronaviruses prefer cell-surface TMPRSS2 to endosomal cathepsins for cell entry. Virology, 517 (2018), pp. 9-15

113. Shulla T, Heald-Sargent G, Subramanya J, Zhao S, et al. A transmembrane serine protease is linked to the severe acute respiratory syndrome coronavirus receptor and activates virus entry. J. Virol., 85 (2011), pp. 873-882

114. Simmons G, Gosalia DN, Rennekamp AJ. Inhibitors of cathepsin L prevent severe acute respiratory syndrome coronavirus entry Proc. Natl. Acad. Sci. USA, 102 (2005), pp. 11876-11881

115. Simon F and Percival, SJ. EDTA: An Antimicrobial and Antibiofilm Agent for Use in Wound Care. Adv Wound Care (New Rochelle). 2015 Jul 1; 4(7): 415–421.

116. Solomon et al. "Chloroquine and its analogs: A new promise of an old drug for effective and safe cancer therapies", 2009, European Journal of Pharmacology, 625(1-3), pp. 220-233 (Year: 2009).

117. Sommerstein R. Preventing a COVID-19 pandemic. BMJ 2020; 368 doi: https://doi.org/10.1136/bmj.m810 (Published 28 February 2020) BMJ 2020;368:m810.

118. Soto-Gamez A, Demaria M. Therapeutic interventions for aging: the case of cellular senescence. Drug Discov Today. 2017 May;22(5):786-95.

119. Spiegel M, et al. Interaction of severe acute respiratory syndrome-associated coronavirus with dendritic cells. J Gen Virol. 2006; 87(Pt 7):1953–1960.

120. St John SE, Tomar S, Stauffer SR, et al. (2015) Targeting zoonotic viruses: Structure-based inhibition of the 3C-like protease from bat coronavirus HKU4-The likely reservoir host to the human coronavirus that causes Middle East Respiratory Syndrome (MERS). Bioorg. Med. Chem. 23: 6036-6048

121. Teachey DT, Lacey SF, Shaw PA, Melenhorst JJ, Maude SL, Frey N, et al. Identification of predictive biomarkers for cytokine release syndrome after chimeric antigen receptor T cell therapy for acute lymphoblastic leukemia. Cancer Discov. 2016

122. Thomsen TR, Hall-Stoodley L, Moser C, Stoodley P. The role of bacterial biofilms in infections of catheters and shunts. In: Bjarnsholt Thomas, Jensen Peter Østrup. Moser Claus, Høiby Niels, eds. Biofilm Infections. New York: Springer, 2011:91–109

123. Tisoncik JR, Korth MJ, Simmons CP, Farrar J, Martin TR, Katze MG. Into the eye of the cytokine storm. Microbiol Mol Biol Rev. 2012; 76:16–32.

124. Wang C, Horby PW, Hayden FG, Gao GF. A novel coronavirus outbreak of global health concern Lancet, 395 (2020), pp. 470-473

125. Wang K, Chen W, Zhou YS, et al. (2020). "SARS-CoV-2 invades host cells via a novel route: CD147-spike protein". Doi:10.1101/2020.03.14.988345

126. Woodcock J, Woosley R. The FDA critical path initiative and its influence on new drug development. Annu Rev Med. 2008; 59:1–12.
127. Wooley RE, Jones MS. Action of EDTA-Tris and antimicrobial agent combinations on selected pathogenic bacteria. Vet Microbiol 1983; 8: 271–280.
128. Wu C, et al. 2020. "Analysis of therapeutic targets for SARS-CoV-2 and discovery of potential drugs by computational methods". Acta Pharmaceutica Sinica B. doi:10.1016/j. apsb.2020.02.008.
129. Wu NH, Yang W, Beinecke A, et al. D-differentiated airway epithelium infected by influenza viruses maintains the barrier function despite a dramatic loss of ciliated cells Sci. Rep. 6 (2016), p. 39668
130. Yang L, Wang XG, Hu B, Zhang L, et al. A pneumonia outbreak associated with a new coronavirus of probable bat origin Nature (2020)
131. Zhou P, et al. (February 2020). "A pneumonia outbreak associated with a new coronavirus of probable bat origin". Nature. 579 (7798): 270–273.
132. Zhou Y, Vedantham P, Lu K, Agudelo J. Protease inhibitors targeting coronavirus and filovirus entry
133. Zhu N, Zhang D, Wang W, et al. A Novel Coronavirus from Patients with Pneumonia in China, 2019.

Part II

Innovative ideas patented, patent pending, and needing experimentation

The following studies are discussed below to prevent, curtail, stop the progression of and/or cure diseases. These are the author's ideas, ready to discuss with any scientist, university, research institution, or charity organization to undertake the project.

Chapter 45

Methods to neutralize coronavirus in the droplets and air

Electrically charged photonic ions for neutralizing free-floating coronavirus and such in the droplet, aerosolized and in the air we breathe.

In nature, negative ions and ozone are nature's most powerful air-cleansing agents. All of us have noticed how refreshing air is at mountains, water falls, and after a thunderstorm rainfall. This is due to thousands (> 2000 negative ions /cubic centimeter) of negative ions in our breathing air. We already have patented technology to change room air using this nature's modality by using heating-air-conditioning vent system in houses using our device. This ionizer charges make the photons, electrically charged ions, attach themselves to contaminants, allergens, and free-floating viruses, which are neutralized. The newly-formed larger particles fall to the ground, and out of the air we breathe.

Based on our patented technology, we are developing an environment-friendly, small-sized, drum-like device which emits electrically charged photons which attack the coronavirus spike protein (as well as pollen and such) and is 99.9% effective in curbing the spread of the coronavirus in an enclosed space in preventing droplet and aerosol spread. The electrically charged electrons ejected from the device actively seek the neutral charge of the virus, which in turn disable the spike protein of the coronavirus.

These electronically charged photons released are not dangerous and have no impact on a COVID-19-infected person, and cannot act as a cure. The devices can be carried individually, or set up in offices, restaurants, classrooms, malls, grocery stores and most importantly, in houses as a viricidal device depending upon the size. The idea is to prevent the spread from an infected person to a non-infected, even if they are next to each other, by neutralizing the virus in droplets and suspended in air. There is a possibility of the use of our technology in chicken and pig farms, to prevent widespread bird flu and swine flu, and it may be of use during flu season also at preschool nurseries and nursing homes, extended care facilities, and the aged living independently.

Chapter 46

Alzheimer's treatment

Direct drug delivery of therapeutic agents delivered to the brain CSF
(A novel approach to great Alzheimer's, Parkinson's, neurodegenerative diseases, rabies, Creutzfeldt-Jakob Diseases / CJD and such)

There are 5.8 million diagnosed with Alzheimer's, besides other neurodegenerative diseases with cognitive impairment to dementia living in the nursing homes, extended care facilities, and the elderly (age of 65 or above) living independently under family supervised care. The elderly above the age of 65, make up 15% of the population and will be 24% by 2060.

Etiology of Alzheimer's is still debated with dozens of theories about the causation of this dead-end disease. The three stages of AD proposed by the guidelines from National Institute on Aging (NIA) and the Alzheimer's Association are:

1. preclinical AD
2. Mild Cognitive Impairment (MCI)
3. dementia.

Even now there are no therapeutic agents to prevent, curtail, stop the progression of and / or cure Alzheimer's disease (so many other dementias) and many other chronic comorbidities which the aged population suffers. Drugs to treat Alzheimer's only alleviate the symptoms of the disease, not the etiology. Recently, a Mayo clinic study by Tyler Bussian, reports zombie cells, aka senescent cells (aged cells), have been found to accumulate in the brain ahead of the toxic protein build-ups that are generally implicated in Alzheimer's disease and dementia. The new research reveals that many pathological signs of this neurodegenerative disease can be eliminated by eliminating these cells from the brain, in which chronic inflammation of the brain leads to neurodegenerative diseases.

It is becoming evident that Alzheimer's is said to be caused by chronic insidious brain inflammation over a period of 2 decades due to any number of aetiologias, before the full-blown disease presents that will continue to the end of their life. Treatment of symptoms and caring for Alzheimer's patients costs billions of dollars, and may reach a trillion dollars by the end of this century.

In my method of treatment, at the early stages, drugs to treat Alzheimer's and other neurodegenerative diseases can be given orally or parentally as a prophylaxis. As the disease advances past stage 2, I recommend an infusion of anti-inflammatory therapeutic agents such as Rapamycin, rapamycin analogues, cephalosporins, IGF-I, lithium, especially prepared curcumin, antioxidant Ebselen, auranofin, Suramin, a neuro-trophic factor such as insulin, monoclonal antibodies such as aducanumab and such therapeutic agents with or without dimethyl sulfoxide (DMSO) as a carrier. Recently (6/24/2020) the FDA approved Leptim being repurposed to be used in Alzheimer's treatment.

These drugs would be delivered through a catheter in the cervical subarachnoid region (cisterna magna or cervical epidural) or through an Ommaya pump in the brain ventricle's chambers (see Figures 1, 2, 3 below) of the brain, and will prevent the progression of the disease and provide partial or maybe complete recovery from memory loss and other neurological tasks to the functional level. This method can be used to treat any and all brain neurodegenerative diseases including brain tumors, CJD and rabies.

Diagrams depicting the direct cerebro-spinal fluid delivery methods

Below are the diagrams depicting how easy it is to deliver therapeutic agents such as Rapamycin in doses of ±5 mg to cerebro-spinal fluid in mini-doses to treat all neurodegenerative diseases, including Alzheimer's and Parkinson's, Creutzfeldt-Jakob disease, Huntington's chorea, fronto-temporal dementias, Lewy bodies disease as well as rabies. Micro- and mini-doses of rapamycin should be taken by all the elderly, as discussed in a chapter in part II in this book, which can also prevent cancers and cardio-vascular diseases, besides dementias of Alzheimer's type.

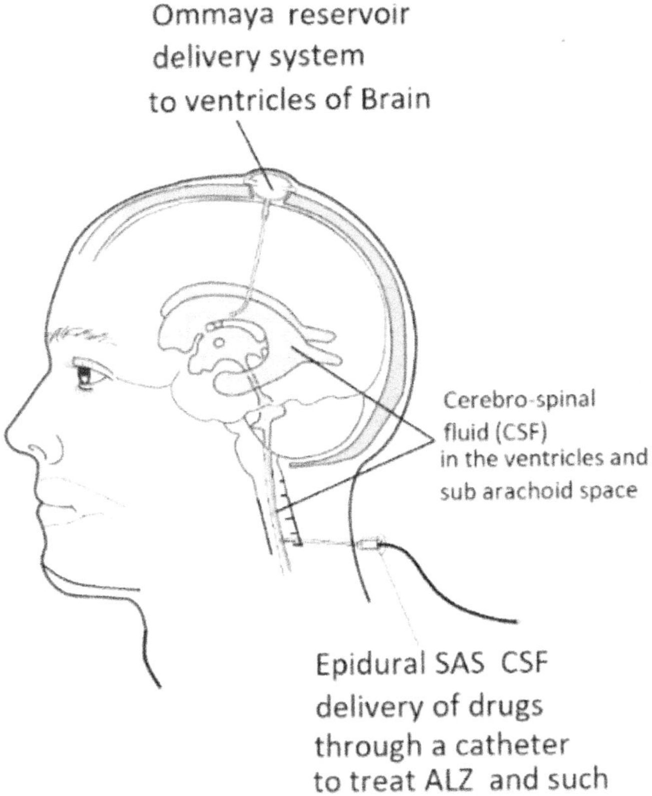
Ommaya reservoir
delivery system
to ventricles of Brain
Cerebro-spinal
fluid (CSF)
in the ventricles and
sub arachoid space
Epidural SAS CSF
delivery of drugs
through a catheter
to treat ALZ and such

Figure 1

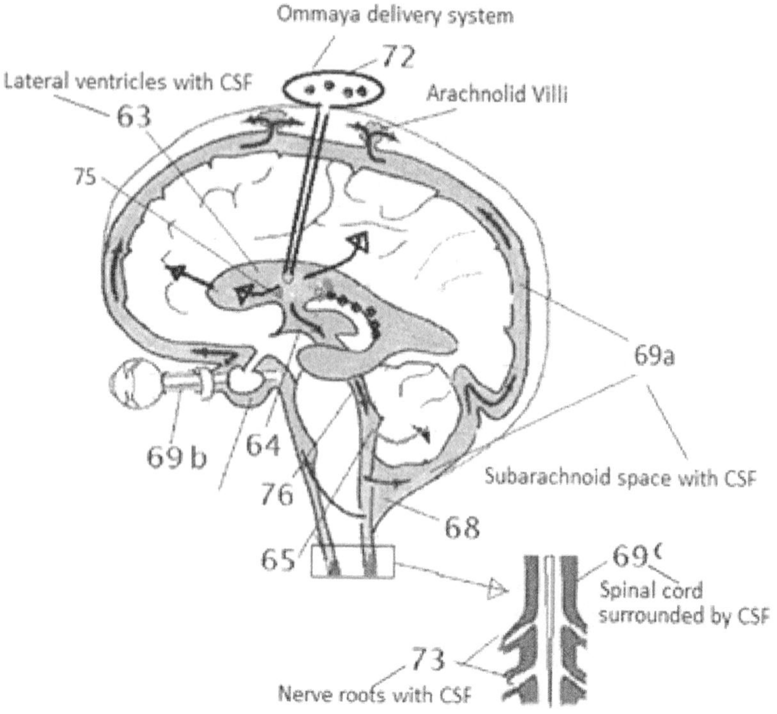

Figure 2

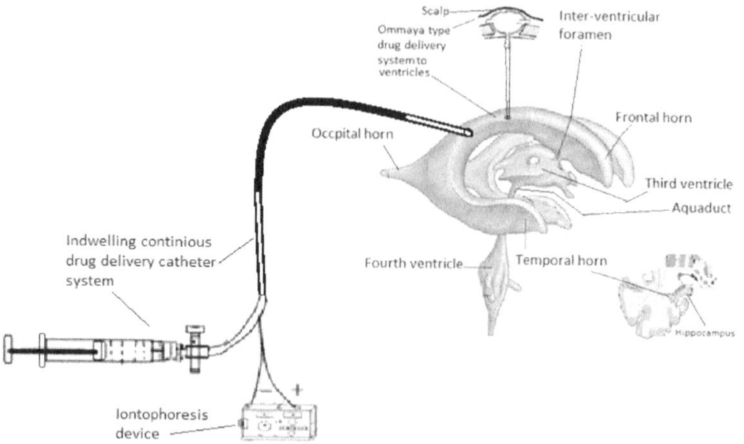

Figure 3

LIST OF PATENTS FILED

US20150073330 Alzheimer's Disease treatment with multiple therapeutic agents delivered to the olfactory region through a special delivery catheter and iontophoresis. This invention describes the administration of multiple therapeutic agents with insulin in conjunction with bexarotene, ketamine, monoclonal antibodies Etanercept, IGF-1, and acetylcholine.

US20150080785 Alzheimer's Disease treatment with multiple therapeutic agents delivered to the olfactory region through a special delivery catheter and iontophoresis. This invention describes the administration of multiple therapeutic agents with insulin in conjunction with bexarotene, ketamine, monoclonal antibodies Etanercept, IGF-1, and acetylcholine.

US20120323214 Alzheimer's Disease treatment with multiple therapeutic agents delivered to the olfactory region through a special delivery catheter and iontophoresis. This invention describes the administration of multiple therapeutic agents with insulin in conjunction with bexarotene, ketamine, monoclonal antibodies Etanercept, IGF-1, and acetylcholine.

Chapter 47

Parkinson's disease

We have developed a new device and methods for treatment of Parkinson's and other neurodegenerative diseases

Incidence

1. Nearly 1.2 million will be living with Parkinson's disease (PD) in the US by 2020, which is more than the combined number of people diagnosed with multiple sclerosis, muscular dystrophy and Lou Gehrig's disease (or Amyotrophic Lateral Sclerosis)
2. Approximately 60,000 Americans are diagnosed with PD each year.
3. Men are 1.5 times more likely to have PD than women.

Signs and symptoms and etiology

Parkinson's disease begins with mild limb stiffness with infrequent tremors, and progresses over a period of ten or more years to frequent tremors, uncontrollable tremors, stiffness, drooling, a mask-like facial expression, difficulty in speaking and dementia as the disease progresses. It is due to continued unabated destruction of dopamine producing cells in the basal ganglia, which comprise the pars compacta of the substantia nigra (see Figures below) and ventral tegmental nucleus. Loss of dopamine producing neurons results in a relative excess of acetylcholine at synaptic relays, resulting in production various movement disorders (Jellinger, K. A., Post Mortem Studies in Parkinson's Disease—ls It Possible to Detect Brain Areas for Specific Symptoms, J Neural Transm 56 (Supp);1-29:1999). Every type of reward studied increased the level of dopamine in the brain. It is in a way an addiction hormone. A variety of addictive drugs, including stimulants such as cocaine, amphetamine, methamphetamine and such act by amplifying the effects of dopamine. ***The method we have is one of the important modalities to treat all kinds of addiction.*** (Figures below)

Dopamine in the CNS is distributed and released primarily into:

1. Basal ganglia (striatum)
2. Frontal lobe

3. Nigrostriatal System, which plays a role in movement
4. The nigrostriatal dopaminergic pathway in the brain accounts for 75% of the dopamine in the brain.
5. Mesolimbic System which plays a role in reinforcement/reward.
6. Mesocortical System which plays a role in short-term memory, planning, and problem solving.

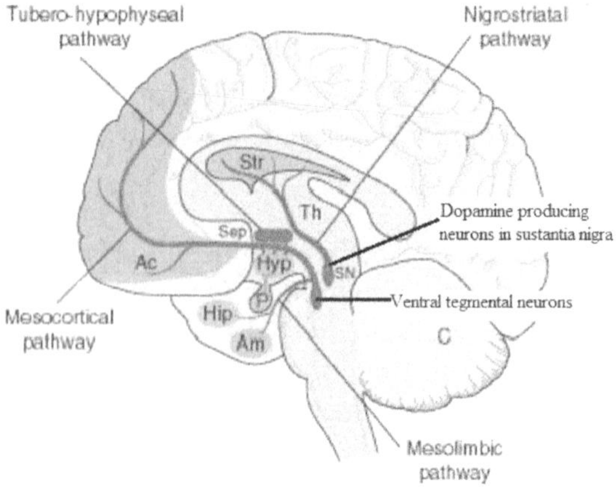

Dopaminergic pathways from the midbrain to other areas in the brain

Figure 1

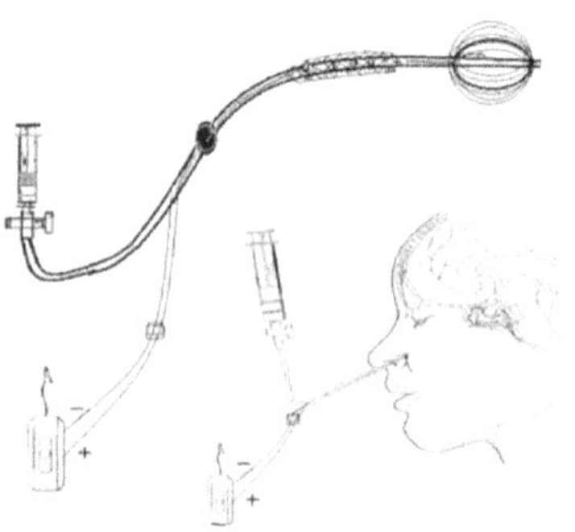

Figure 2 (reproduced from 1998 patent by T. R. Shantha)

We have developed a method to treat PD (Patent pending). We need funding to test this method. Please help us develop this method of treatment which can also be expanded to treat Alzheimer's, Huntington's chorea, restless leg syndrome, ALS, MS, Fronto-temporal dementia, MCI, mental diseases and such besides Parkinson's.

Chapter 48

Autism spectrum disorder (ASD)
New methods of treatment

Autism spectrum disorder (ASD) is a developmental disability that can cause significant *social, communication and behavioral challenges*. There is often nothing about how people with ASD look that sets them apart from other people, but people with ASD may communicate, interact, behave, and learn in ways that are different from most other people. The learning, thinking, and problem-solving abilities of people with ASD can range from gifted to severely challenged. Some people with ASD need a lot of help in their daily lives; others need less.

A diagnosis of ASD now includes several conditions that used to be diagnosed separately: autistic disorder, pervasive developmental disorder not otherwise specified (PDD-NOS), and Asperger syndrome. These conditions are now all ***grouped under "autism spectrum disorder***."

Incidence of ASD

About 1 percent of the world population has autism spectrum disorder. Prevalence in the United States is estimated at 1 in 68 births predominantly males. More than 3.5 million Americans live with an autism spectrum disorder. Prevalence of autism in US children increased by 119.4 percent from 2000 (1 in 150) to 2010 (1 in 68). It is one of the fastest-growing developmental disabilities.

Claims of association of vaccination with ASD are discredited. I do believe that in some rare cases the microbial proteins of vaccines can have an effect on the nerve cells, their processes and the synapses, microglia, astroglia and oligodendroglia cells and their function. The vaccines may initiate a subtle symptomatic inflammatory reaction in the brain in some children.

Medical cost

Autism services cost US citizens $236-262 billion annually. (Järbrink K1. Autism. 2007 Sep;11(5):453-63). The US cost of autism over the lifespan is about $2.4 million for a person and is growing exponentially.

Our new methods of treatment of ASD

We have treated many ASD patients and now have developed and envisioned using the latest therapeutic agents delivered to the ears and olfactory mucosa with a special delivery catheter; the specially compounded therapeutic agents' solution (patent pending) combined with whole body hyperthermia and sulforaphane nutritional supplement.

Hyperthermia treatment for ASD with other therapeutic agents

I have experimented with whole body hyperthermia since 1962 starting for treatment of infections to cancer to HIV-AIDS. I saved the life of a terminal patient with severe peritoneal hyperthermia for 4 hours combined with Diflucan. No one has ever tried this method that we know of. We tried hyperthermia method in Santo Domingo, Dominican Republic to treat cancers and infectious diseases. I have used the whole-body hyperthermia in my practice to treat cancers, infections and such. I have now developed a simple method to induce whole body hyperthermia using a bathtub and far infrared methods as therapeutic modality. I practice whole body hyperthermia, and I can attest that I sleep better, my brain feels clearer, I think better, any small tremors are gone and my hand writing is better.

A fever usually makes you feel foggy, tired, and grumpy in children. But in some children with autism, a fever due to flu, cold, and other viral or microbial infections makes them "better," that is, they become more social, alert, talkative, relaxed, friendly and chatty. As one mother described it, her child's fevers provide "a glimpse into what her child might be like without autism." Researchers have wondered about the so-called "fever effect" since at least 1980, when a viral infection swept through a therapeutic nursery for children with autism at Bellevue Psychiatric Hospital in New York. Hospital staff reported that the children were more social, alert, and talkative while feverish (Sullivan, R. C. (1980). Why do autistic children...? Journal of Autism and Developmental Disorders, 10(2), 231-241. Curran, L. K., et al. (2007). Behaviors associated with fever in children with autism spectrum disorders. Pediatrics, 120(6), e1386-92. doi:120/6/e1386).

Studies do show that 83% of children with autism spectrum disorders were found to improve transiently in association with fever. Improvements were lost with the resolution of the fever. This study showed that, despite the many different causes of ASD, the symptoms were not permanent and could be improved in a substantial fraction of children.

We have a method to induce whole body hyperthermia using various modalities including the method for the thoracic area which we have used for years. It raises the body temperature between 99-100.4 degrees Fahrenheit (38.0 degrees Celsius). We can train the parents how to induce whole body hyperthermia and night time sleeping hyperthermia combined with sulforaphanes (Singh. K, et al. 2014 PNAS October 28, 2014 111 (43) 15550-15555), and other therapeutic agents' administration modalities. The results are that we can effectively treat ASD and reduce symptoms encompassing impaired communication and social interaction, and repetitive stereotypic behavior and language in a majority of children.

We do not know how whole-body hyperthermia works to soothe and subside down the signs and symptoms of autism. We know that fever may cause temporary cellular or metabolic changes that affect the central nervous system that affects behavior. Possible explanations include changes to the action of cells in the brain's hypothalamus, neurons, and glial cells. It may be due to the release of glutamine or taurine, may alter the effect of the neurotransmitter at synapses, and make them more homeostatic. We know that whole body hyperthermia increases "heat-shock proteins" that people produce during fever to protect nerve cells and the heart also. It is also possible that the glymphatic circulation and microvascular circulation of the entire brain and spinal cord increases which removes all the toxic metabolites from the brain, which makes neurons more physically functional. Hyperthermia also dilates the brain blood vessels and delivers more needed oxygen and nutrients to neurons to function normally. Hyperthermia has profound effect on sodium, potassium, and magnesium ions pump. Magnesium enters the axons and dendrites, synapses, and thus inhibits the excessive excitatory action of glutamate involved in many repetitive behaviors of ASD children. Hyperthermia also clears out brain inflammatory cytokines from the brain due to improved glymphatic and blood flow.

The serotonin system has been implicated as a factor in some cases of autism since the finding in 1961 of elevated serotonin (5-hydroxytryptamine) levels in the blood of patients with autism (Schain RJ, Freedman DX: Studies on 5-hydroxyindole metabolism in autistic and other mentally retarded children. J Pediatr 1961, 58:315-320.). This has been clarified as elevation in the platelet content of serotonin. Most individuals with autism who are treated with potent serotonin transporter inhibitors have a reduction in ritualistic behavior and aggression (Current Opinion in Pediatrics 1996, 8:348-354). It is likely that the induced fever therapy acts as a serotonin uptake and transport inhibitor akin to serotonin uptake and transport inhibitor drugs. There is a likelihood that induced fever therapy also acts as angiotensin converting enzyme (ACEi-US 7,276,472 B2, 5,049, 533) and angiotensin receptors which is associated with elevated blood pressure that reduces the self-injury activities of autistic children and adults. Along with fever therapy, we want to administer ACEi such as lisinopril during, before or after hyperthermia. Induced fever therapy, which may act as a natural NMDA receptor antagonist akin to memantine, which is shown to improve autism symptoms. It is also possible that hyperthermia blocks the influx of calcium at synapses into the nerve fiber and may act as 5-HT antagonist. We explored all the possibilities about the mechanisms of fever, in our case whole body hyperthermia and its possible curative effects.

ASD: Purine metabolism disorder and method of treatment

There are reports that ASD and some other psychiatric disorders are due to Purine metabolism and are treatable with anti-purinergic suramin therapy which restores normal social behavior with a single dose. Correction of purine metabolism normalized 17 of 18 metabolic pathways that were disturbed in the experimental models. The effects lasted for two days. As serum, brain and brain stem levels of suramin dropped, symptoms reappeared. Now we have developed an easy method of therapeutic agents' delivery to the brain every day or as often as needed, that can used by the patients and parents. It has been reported that the fever also reduces the signs and symptoms of ASD. We want to combine these methods to treat ASD inducing whole body hyperthermia.

Mode of action of our therapeutic agents

The specially compounded therapeutic agents we propose to administer pass through the blood brain barrier (BBB) and also pass from the olfactory mucosa and internal ears to the sub-perineural epithelial space, then to subarachnoid space to the cerebro-spinal fluid to be distributed to the afflicted brain and brain stem neuronal masses (nuclei) as depicted in the patent. They are transported to the nerve cells, their axons, dendrites, synapses, and glial cells cytoplasm as protein bound complex via endocytosis. Inside the neurons and glial cells cytoplasm, these drugs inhibit the glycolytic pathway through glycosomes, thus suppressing brain inflammation, and the abnormal neuronal firing activity, consequently reducing the signs and symptoms of ASD.

It is advised to induce therapeutic-calming-remedial-restorative hyperthermia for 25 minutes to ± 120 minutes, during sleep and/or day slumber time to improve the brain's function as seen after an attack of fever due to microbial infections. This is a therapeutic fever; we call it an ***"Induced Fever Therapy."*** The advantage of our method is that one need not wait to get an infection to develop fever and observe the relief of symptoms. Our "induced fever therapy" is devoid of other associated symptoms that accompany the febrile illnesses such as sore throat, congestion, headache, coughing, sneezing and such, and any can be imitated and terminated at will. One need not wait for fever to subside.

Secretin, a gastrointestinal peptide hormone, given to children diagnosed with autism resulted in ameliorating the symptoms associated with autism. This finding was published in the article by Horvath et al., Improved Social and Language Skills After Secretin Administration (In Patients with Autistic Spectrum Disorders, Journal of the Association for Academic Minority Physician Vol. 9 No.1, pp. 9-15, January, 1998.).

Administering phytochemical sulforaphane supplements derived from broccoli sprouts, which is non-toxic, upregulates the genes that protect aerobic cells against oxidative stress, inflammation, and DNA damage, all of which are prominent and possibly mechanistic systematic characteristics of ASD. Dietary sulforaphane has a certain capacity to reverse abnormalities that have been associated with ASD, including oxidative stress and lower antioxidant capacity, depressed glutathione synthesis, reduced mitochondrial function and oxidative phosphorylation, increased lipid peroxidation, and decreased neuro-inflammation. Studies show that sulforaphane reduces the symptoms of ASD (Singh et al.) probably due to a correction of these metabolic disruptions.

Eat a diet rich in cruciferous vegetables such as broccoli, and omega 3 fish such as mackerel, lake trout, salmon, herring, sardines, sturgeon eggs, tuna, and blue fish, along with a Mediterranean diet and less wheat, rice, sugar and colas.

Administer bioactive curcumin 250-1000 mg, vitamin C 250-500 mg, and vitamin D3, 2500-5000 IU, magnesium L-threonine, auranofin, Ebselen and other antioxidants discussed under COVID-19 patent.

Give me an opportunity to test and show how effective our patent pending method is against ASD, which torments thousands of young children and tears apart the parents and family.

(19) **United States**

(12) **Patent Application Publication** (10) **Pub. No.: US 2012/0128683 A1**

Shantha (43) **Pub. Date: May 24, 2012**

(54) **AUTISM TREATMENT**

(76) Inventor: **Totada R. Shantha**, Stone Mountain, GA (US)

(21) Appl. No.: **13/302,648**

(22) Filed: **Nov. 22, 2011**

A61P 25/00 (2006.01)
A61K 38/30 (2006.01)

(52) **U.S. Cl.** **424/141.1**; 514/5.9; 514/6.5

(57) **ABSTRACT**

A safe and effective treatment to curtail and cure autism

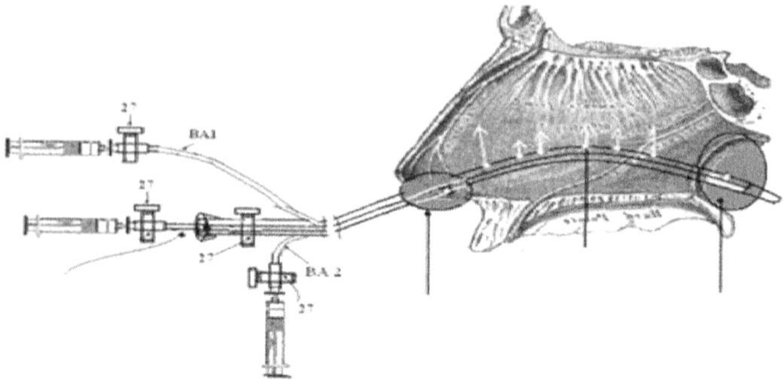

Figure 1

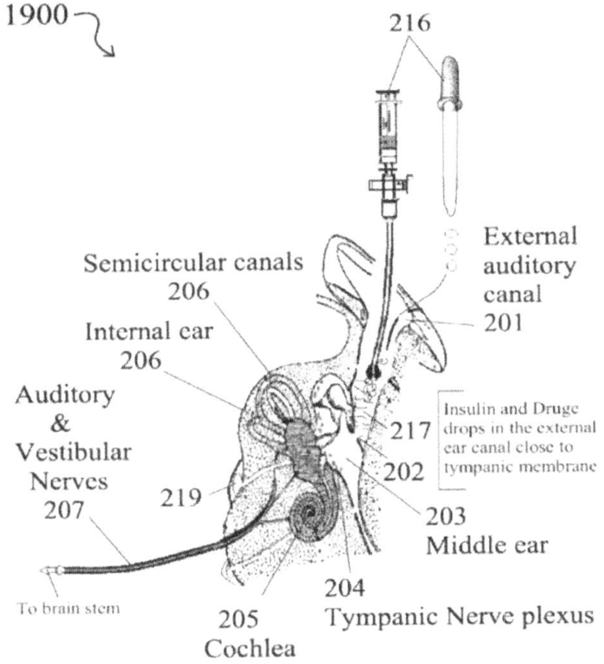

Figure 2

Figures 1 and 2 show how easy it is to deliver therapeutic agents such as insulin, folic acid, oxytocin, IGF-1, Suramin, Ebselen, and such to ear and on olfactory mucosa to be transported to the brain as described in the patent and shown in the above diagrams.

Chapter 49

Benign prostate hypertrophy and prostate cancer

Incidence

It has been reported that the majority of men will develop benign prostatic hyperplasia-hypertrophy (BPH) by the time they reach their 60s. Prostatic diseases have a significant morbidity and mortality affecting over a billion men throughout the world. The lifetime risk of a man developing histologically confirmed benign prostate hyperplasia (BPH) has been reported to be 50% in those in the 51-60 years age group and increasing to 70% in the 61-70 years age group.

With the exclusion of non-melanoma skin malignancy, prostate cancer (PCa) is the second most prevalent cancer in men globally. The International Agency for Research on Cancer reported that PCa accounted for 14% of cancers diagnosed in men and over 900, 000 cases were diagnosed throughout the world in 2008 alone. In only 20% of cases do BPH and PCa co-exist in the same prostatic zone. With men living longer, the PCa incidence will increase.

Historical autopsy data has shown that 83.3% of PCa arises with the concomitant presence of BPH. Chronic inflammation is thought to be one of the numerous driving factors for the development of PCa. Incidence of prostate cancer was 20% in men harboring inflammation at baseline compared to 6% in those with no inflammation at baseline biopsy. Surprisingly, NSAIDs in the form of aspirin and ibuprofen in nearly 5000 men had no protective effect on the development of BPH and lower urinary tract symptoms (LUTS) as seen in colon cancer.

Etiology

A common and well-established driving factor co-linking the advancement and progression of both BPH and PCa is the action androgens (steroid testosterone) have on prostatic glandular tissue. Not only are androgens explicitly required for pathological growth of the prostate, they are vital for normal growth and development of this gland. The iso-enzymes ***5-α reductase type 1 and 2 in the prostate*** convert testosterones into its active form ***dihydrotestosterone*** (DHT), which in turn facilitates cellular differentiation and proliferation of prostatic tissue. Surprisingly, a similar hormonal mechanism is involved in male baldness.

Abnormal prostatic growth occurs when there is a disturbance on the androgen profile of the prostate, which encourages a proliferative state of the gland augmented by chronic inflammation. This disrupted androgen cycle overcomes apoptosis within the prostate, therefore, disrupting the homeostatic regulation of prostate cell proliferation and cell death. The suppression of androgen activity on the prostate forms the basis of pharmacological treatment of both BPH and PCa and it is lifelong therapy.

We have developed a patented (pending) method to prevent the development of BPH and PCa which inhibits prostate gland multiplication and inhibition of inflammation resulting to prevent, curtail, stop the progression of and / or cure BPH, PCa, and prostatic inflammation. It can be used as a prophylactic and as an adjuvant treatment for all these conditions. For patients with advanced PCa or any cancer with metastasis, this method will stop and/or slow down further progression, thus giving a good quality pain-free extended life. For example, I had a patient on my protocol, with reduction of Alkaline phosphatase, PSA, and suppression of further growth. Any patient, urologist, scientist, and pharmaceutical company experimenting on this subject can contact me to use my simple method. We also have treated many cases of interstitial cystitis with reduced pain and related problems in many female patients using my method to treat bladder cancer with BCG.

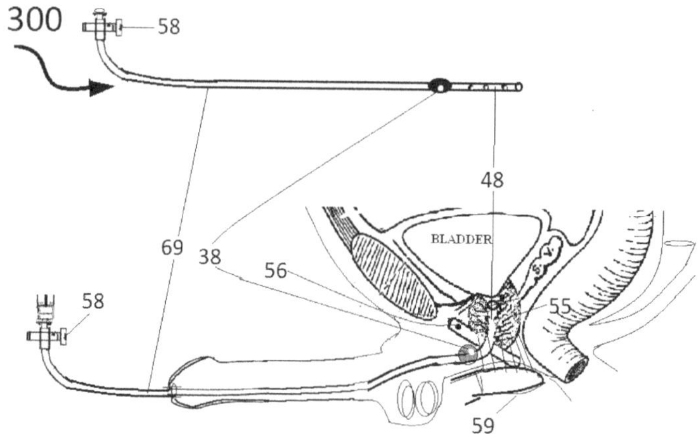

Figure 1

Chapter 50

Breast cancer—prophylaxis and treatment

Incidence

One in 8 US women (about 12%) will develop invasive breast cancer over the course of her lifetime. In 2020, an estimated 276,480 new cases of invasive breast cancer are expected to be diagnosed in women in the US, along with 48,530 new cases of non-invasive (in situ) breast cancer. About 2,620 new cases of invasive breast cancer are expected to be diagnosed in men in 2020 with a lifetime risk of breast cancer is about 1 in 883.

About 42,170 women in the US are expected to die in 2020 from breast cancer. Death rates have been steady in women under 50 since 2007, but have continued to drop in women over 50. The overall death rate from breast cancer decreased by 1.3% per year from 2013 to 2017. These decreases are thought to be the result of treatment advances and earlier detection through screening.

For women in the US, breast cancer death rates are higher than those for any other cancer, besides lung cancer. As of January 2020, there are more than 3.5 million women living with a history of breast cancer in the US This includes women currently being treated and women who have finished treatment. Besides skin cancer, breast cancer is the most commonly diagnosed cancer among American women. In 2020, it's estimated that about 30% of newly diagnosed cancers in women will be breast cancers.

In women under 45, breast cancer is more common in Black women than white women, and Black women are more likely to die of breast cancer earlier also. For Asian, Hispanic, and Native-American women, the risk of developing and dying from breast cancer is lower. Ashkenazi Jewish women have a higher risk of breast cancer because of a higher rate of BRCA gene mutations. A large study called the Women's Health Initiative published in 2002 shows a connection between hormone replacement therapy (HRT) and increased breast cancer risk.

A woman's risk of breast cancer nearly doubles if she has a first- degree relative (mother, sister, daughter) who has been diagnosed with breast cancer. Less than 15% of women who get breast cancer have a family member diagnosed with it. About 85% of breast cancers occur in women who have no family history of breast cancer. These occur due to genetic mutations that happen as a result of the aging process and life in general, rather than inherited mutations (from the book Your Breast and its Care by Dr. T. R. Shantha, published by Fedrick Fell).

Approximately 5-10% of breast cancers can be linked to known gene mutations inherited from one's mother or father. Mutations in the BRCA1 and BRCA2 genes are the most common. On average, women with a BRCA1 mutation have up to a 72% lifetime risk of developing breast cancer. For women with a BRCA2 mutation, the risk is 69%. Breast cancer that is positive for the BRCA1 or BRCA2 mutations tends to develop more often in younger women. An increased ovarian cancer risk is also associated with these genetic mutations.

Every patient with breast cancer with or without metastasis should take the immunosuppressant which inhibits subtle but chronic breast inflammation and the growth of cancer cells. Taking aspirin and such related drugs can prevent the spread of breast cancer during development and after surgery, and prolong the life according to Dr. Vikas Sukhatme, dean of the Emory University School of Medicine. Dr. Jessica Shantha attended a luncheon meeting with Dr. Shantha on Aug 1, 2018, where he presented the topic "A Simple, One-Time, Inexpensive and Non-Toxic Intervention to Improve Cancer Survival." Dr. Vikas Sukhatme with his wife are founders of Global cures; for details contact: Kamal at global-cures.org (781) 269-2602.

We add to this list micro and mini-dose rapamycin – now it is the most important part of our protocol, besides Ebselen, cimetidine, vitamin C, vitamin D3, melatonin, thalidomide, aspirin, Celebrex and other anti-inflammatories. ***I have prescribed micro and mini dose rapamycin before and after breast cancer surgery and in between, taken intermittently (pulsed administration) to inhibit and prevent its spread of cancer far and wide.*** I have prescribed aspirin and cimetidine for colon, breast and other cancers for decades, before and after surgery. I have developed a breast cancer prevention patch as shown in the diagram and would love to develop it in collaboration with an interested company, foundations, pharmaceutical company and/or party. Its use will reduce the incidence of cancer (it acts as a prophylactic), fibrocystic diseases, mastodynia and such.

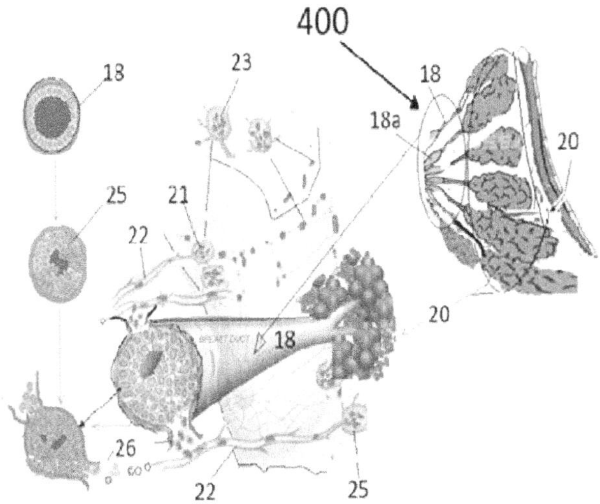

Figure 1 shows breast cancer and its progression

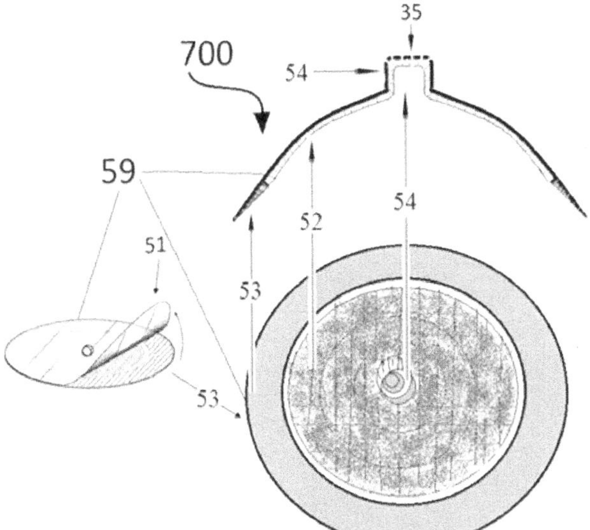

Figure 2 shows our treatment method of delivering therapeutic agents to prevent many afflictions of the breast including breast cancer and its prophylaxis

Chapter 51

Tuberculosis—innovative method to cure rapidly

Latest patent on the treatment of tuberculosis:

(19) **United States**
(12) **Patent Application Publication** (10) **Pub. No.: US 2016/0074480 A1**
SHANTHA (43) **Pub. Date: Mar. 17, 2016**

(54) INNOVATIVE METHODS OF TREATMENTING TUBERCULOSIS

(71) Applicant: **Totada R. SHANTHA**, Stone Mountain, GA (US)

(72) Inventor: **Totada R. SHANTHA**, Stone Mountain, GA (US)

(52) U.S. Cl.
CPC ***A61K 38/28*** (2013.01); ***A61K 9/0019*** (2013.01); ***A61K 31/375*** (2013.01); ***A61K 31/496*** (2013.01); ***A61K 31/366*** (2013.01); ***A61K 31/44*** (2013.01); ***A61K 31/7048*** (2013.01); ***A61K 45/06*** (2013.01)

(57) ABSTRACT

A Safe and effective treatment methods to cure and curtail

In spite of all the advances in medicine, tuberculosis (TB) remains one of the top 10 causes of death worldwide. Millions of people continue to fall sick from TB each year even now. Globally, 7 or more million people were diagnosed and treated for TB in 2018, up from 6.4 million in 2017. WHO's latest Global TB Report says that 2018 also saw a reduction in the number of TB deaths: 1.5 million people died from TB in 2018, down from 1.6 million in 2017. Around 10 million people developed TB in 2018. The report released highlights that the world must accelerate progress if it is to reach the Sustainable Development Goal of ending TB by 2030. The report also notes that an estimated 3 million of those with TB still are not getting the care they need.

Tuberculosis incidence in the United States has steadily declined since 1993, but the pace of decline has slowed in recent years. **8,920** is the number of diagnosed or reported TB cases in the United States in 2019. **Up to 13 million** are the number of people in the United States living with latent TB infection.

The fight against TB remains chronically underfunded. The WHO estimates the shortfall for TB prevention and care in 2019 at US$3.3 billion. **The World Health Assembly approved Global TB Strategy** aims for a 90 per cent reduction in TB deaths and an 80 per cent reduction in the TB incidence rate by 2030. The UN Political Declaration on TB in 2018 included 4 new global targets:

- Treat 40 million people for TB disease in the 5-year period 2018-22 (7 million in 2018)
- Reach at least 30 million people with TB preventive treatment for a latent TB infection in the 5-year period 2018-22
- Mobilize at least US$13 billion annually for universal access to TB diagnosis, treatment and care by 2022
- Mobilize at least US$2 billion annually for TB research
- I urge *funding be made available for research for my patent published method of treatment.* This treatment would cost less, and cure or curtail in a shorter period of time compared to the billions now spent in prolonged treatment.

Tuberculosis facts

1. Tuberculosis (TB) is caused by bacteria (Mycobacterium tuberculosis) that most often affect the lungs. Tuberculosis is preventable and curable. Robert Koch discovered the cause of TB in 1882 and received the Nobel Prize in 1905, and the Waksman in 1952 for the discovery of streptomycin to treat TB.
2. About one-quarter of the world's population (2 billion people) have latent TB, which means people have been infected by TB bacteria but are not (yet) ill with the disease and cannot transmit.
3. The highest incidence of TB in 2018 is in 8 countries: Bangladesh, China, India, Indonesia, Nigeria, Pakistan, the Philippines, and South Africa. Brazil, China, the Russian Federation and Zimbabwe, which all have high TB incidences, achieved treatment coverage levels of more than 80 per cent.

The method I describe in my patent can prevent, curtail, stop progression of and / or cure TB within 1-3 months. Even now millions are affected and millions die from this disease all over the world. I encourage anyone to try my method, and am happy to participate with anyone who wants to take on the project. ***If the WHO contacts me, I would love to work with the organization to eliminate this scourge which affects so many.***

Chapter 52

Leprosy
Our new approach to eliminate this chronic affliction rapidly and possibly banish it from the earth

Leprosy has affected humans for thousands of years. There are 37 verses about leprosy in 7 books of the Bible, starting with Leviticus. The main reason that leprosy is talked about so much in the Bible is that it is a graphic illustration of sin's destructive power. Jesus heals 10 lepers by touching them (Luke 17:12) and untold numbers in the gathered.

In 1200 AD, an estimated 19,000 leprosy hospitals existed all over Europe. At that time, no one knew what caused leprosy. In 1873 G. H. Armauer Hansen in Norway discovered the causative agent of leprosy, *Mycobacterium leprae.* ***This was the first bacterium to be identified as causing disease in humans***. Today, about 180,000 people worldwide are infected with leprosy, according to the World Health Organization, most of them in Africa and Asia. About 100 people are diagnosed with leprosy in the US every year, mostly in the south, California, Hawaii, and some US territories. However, leprosy is actually not that contagious. You can catch it only if you come into ***close and repeated contact*** with nose and mouth droplets from someone with untreated leprosy. Children are more likely to get leprosy than adults. Leprosy's long incubation period, 5-20 years, makes it very difficult for doctors to determine when and where a person with leprosy got infected. Fears of ostracism, loss of employment, or expulsion from family and society may contribute to a delayed diagnosis and seeking treatment.

My method of treatment combined with whole body hyperthermia, high dose vitamin C and anti-leprae therapeutic agents including rifampicin or rifamycin-Dapsone combination (applied for provisional patent) will eliminate the leprae bacteria rapidly within weeks instead of years, and make patients sterile, meaning no organism to cause the disease. ***Give our method a chance and the WHO goal of eliminating leprosy from the world is at hand. I wrote an email to WHO about my treatment and am still waiting for a reply.***

Chapter 53

Rabies and potentially CJD cure

Ever since the distant past, rabies has been one of the most feared diseases. Human rabies remains an important public health problem in many developing countries where dog rabies is endemic.

Rabies is a lethal disease caused by neurotropic RNA viruses that are endemic in nature. The rabies virus belongs to the Lyssavirus genus, which include other similar viruses found in bats. Lyssaviruses have helical symmetry, with a length of about 180 nm and a cross-sectional diameter of about 75 nm. These viruses are enveloped and have a single stranded RNA genome with negative-sense, akin to coronavirus.

The advances in scientific medicine make rabies control possible by prophylactic pre or post exposure prevention vaccine (PEP). Rabies vaccine was discovered by Louis Pasteur. The first person to receive vaccine was Joseph Meister, in July 1885, who was bitten in 14 places by a huge mad dog. Pasteur injected the vaccine prepared from the spinal cord of a rabbit that had died of rabies. Meister did not develop rabies and lived till 1940. The rest is history.

Prompt wound care and the administration of rabies immune globulin (RIG) are highly effective in preventing human rabies following exposure. That is the protocol that is posted on the CDC website. ***But it has no effect in those who develop the disease due to not getting a timely vaccination.*** Even the well-publicized successful treatment in one teenager (that was used in Wisconsin to treat bat rabies) failed in subsequent multiple cases (Willoughby RE Jr, Tieves KS, Hoffman GM, et al. Survival after treatment of rabies with induction of coma. N Engl J Med 2005; 352: 2508--14). This indicates that there is no cure for this fatal disease, which even now kills more than 35,000 around the world.

My patented method of treatment proposes the use of insulin with therapeutic, pharmaceutical, biochemical and biological agents or compounds delivered to the SAS CSF (intrathecal) through a catheter inserted to the SAS or to cisterna magna described in the following patent.

Now we have modified the treatment further since the patent publication. The most important therapeutic agents we administered are high dose vitamin C with thiamine, along with Colchicine, vinblastine, and immunosuppressant rapamycin and their analogue delivered to the brain as discussed in the Alzheimer's chapter, with high dose vitamin C along with Ketamine, doxycycline and such. Other suitable immune suppressants can be included with rapamycin. The therapeutic agents are delivered intravenously, by olfactory mucosa and sub arachnoid space routes to reduce the cytokine induced by rabies virus, which induce fatal viral encephalopathy, and to disorganize the neuro-skeleton network including neurotubules to prevent the spread of the virus in the brain jumping from one neuron to another. I believe that *rapamycin and their analogues can be easily used for treating many neurodegenerative diseases including CJD and Krabbe disease in children.*

Cool the brain by using the intranasal cooling (INC) device developed by Bene Chill of Santiago called the RhinoChill™ which has greatly facilitated the therapeutic application of cooling of the body and organs to save the life of many, including those with rabies. It is expensive and is not available in countries where rabies is prevalent. In such cases use other methods of cooling the brain. The patient needs to be treated in isolation in ICU with all the supportive care, but whether you cool the brain or not, use the method we have developed.

(19) **United States**
(12) **Patent Application Publication** (10) **Pub. No.: US 2011/0020279 A1**
Shantha (43) **Pub. Date: Jan. 27, 2011**

(54) **RABIES CURE**

(76) Inventor: **Totada R. Shantha**, McDonough, GA (US)

A61P 23/00	(2006.01)
A61P 25/04	(2006.01)
A61P 3/00	(2006.01)
A61P 3/02	(2006.01)

Chapter 54

HIV-AIDS
Human Immunodeficiency Viruses (HIV)

Incidence: 2016 global HIV statistics

HIV continues to be a main global public health issue. The following statistics illustrate this.

- In 2016, an estimated 36.7 million people were living with HIV (including 1.8 million children) and >5000 individuals worldwide are newly infected each day.
- Since the start of the epidemic, an **estimated 78 million people have become infected with HIV and 35 million people have died of AIDS-related illnesses.**
- A major milestone was achieved in 2016 where, for the first time, it was found that more than half of all people living with HIV (53%) now have access to life-saving treatment. In 2016, 19.5 million people living with HIV were receiving antiretroviral treatment (ART) - up from 17 million in June 2016 and 7.5 million in 2010.
- More than 1.1 million people in the US are living with HIV today, and 30% of them don't know it. Women are the fastest growing segment of the population becoming HIV-positive.

So far there is no cure for HIV, but I believe ***within a decade we will have a cure***. Treatment with multiple antiretroviral therapeutic agents has changed the outlook on AIDS and made it a chronic disease rather than a death sentence. The following are combination drugs which amalgamate anti-viral medications from different groups into one oral pill form taken once a day.

- efavirenz/emtricitabine/tenofovir disoproxil fumarate (Atripla)
- emtricitabine/rilpivirine/tenofovirdisoproxil/fumarate (Complera)
- elvitegravir/cobicistat/emtricitabine/tenofovir/disoproxil fumarate (Stribild)
- abacavir/dolutegravir/lamivudine (Triumeq) Nucleoside/nucleotide reverse transcriptase inhibitors (NRTIs)

Now the dual therapy consists of two active ingredients instead of the usual three. One possible advantage of dual therapy is that it contains fewer active ingredients than triple therapy, and will cause fewer side effects and be less expensive than triple therapy. The first approved combination is Dolutegravir and Rilpivirine (Juluco®), and I sure more will follow.

Anti-HIV drugs inhibit the viral replication at many different phases of the HIV replicative cycle (see Figure 1, below). Drugs have been developed that inhibit the fusion of the HIV viral envelope with the plasma membrane of the host T4 cells (fusion inhibitors), the conversion of its RNA genome into double-stranded DNA (reverse transcriptase inhibitors), the integration of the viral DNA into the host genome (integrase inhibitors), and the processing of viral proteins within the infected cells (protease inhibitors).

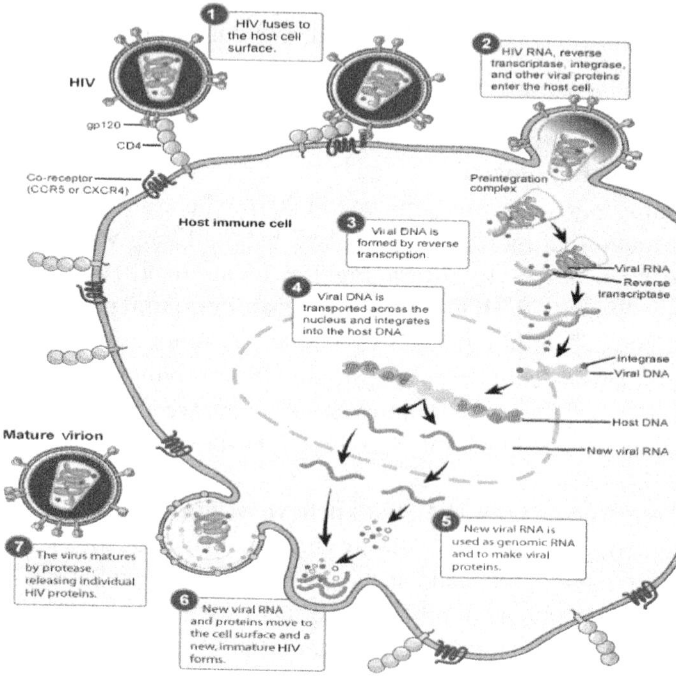

Figure 1

Figure 1 is the diagram of the HIV reproduction (Source: Wikipedia-internet) shows various stages of HIV integration in the host cell, cellular mechanisms involved and therapeutic agents developed to block it.

HIV is not a living organism and is 60 times smaller than red blood cell in our body. It needs living host cells, in this case T4 helper cells of the immune system cells. There are approximately 1 million T4 helper cells per single milliliter of blood. That is why it takes years for HIV to reduce the number of the cells and develop AIDS in the HIV infected patients. HIV attaches to the T4 helper cell host receptors through GP120 spike on the virus and the CCR5 receptor on the cell surface helps to draw the virus closer to the T4 cells surface and get inside the cell. Upon entry into the target cell, the viral RNA genome is released, then it is converted (reverse transcribed) into double-stranded DNA by a virally encoded enzyme, reverse transcriptase, that is transported along with the viral genome in the virus particle.

The resulting viral DNA is then imported into the cell nucleus and integrated into the cellular DNA by a virally encoded enzyme, integrase, and host co-factors. Once integrated, the virus may become latent (hiding), allowing the virus and its host cell to avoid detection by the immune system and spew out viral particles. The HIV virus can remain dormant in the human body for up to ten years after primary infection; during this period the virus does not cause symptoms. Alternatively, the integrated viral DNA may be transcribed, producing new RNA genomes and viral proteins, using host cell resources, that are packaged and released from the cell as new virus particles that will begin the replication cycle anew.

Anti-HIV drugs inhibit viral replication at many different phases of the HIV replicative cycle (see Figure 1). Drugs have been developed that inhibit the fusion of the HIV viral envelope with the plasma membrane of the host cell (fusion inhibitors), the conversion of its RNA genome into double-stranded DNA (reverse transcriptase inhibitors), the integration of the viral DNA into the host genome (integrase inhibitors), and the processing of viral proteins (protease inhibitors).

Why antiretroviral therapy (ART) does not eliminate the HIV virus from the body

The problem with the present ART therapy is that it fails to eliminate integrated copies of HIV-1 pro-viral DNA from the host genome as shown in the diagram above (Lorenzo-Redondo, R. et al. Persistent HIV-1 replication maintains the tissue reservoir during therapy. Nature 530, 51–56 (2016)). As such, the virus persists in a latent, dormant, or hidden state within infectious reservoirs all over the body, mostly in the lymphoid and immune system, and in other cells also. To stop taking ART readily and rapidly leads to viral reactivation and disease progression to acquired immunodeficiency syndrome

(AIDS - Deeks, S. G. et al. International AIDS Society global scientific strategy: towards an HIV cure 2016. Nat. Med. 22, 839–850 (2016). ***Thus, a major issue for any HIV-1 curative strategy is the means to eliminate either integrated pro-viral DNA or the cells that harbor the virus without collateral cytotoxic reactions to healthy cells.*** So far in the entire history of HIV-AIDS infection, elimination of HIV-1 infection in its infected human host is documented only in two individuals.

1. One was a Berlin patient (Hutter, G. et al. Long-term control of HIV by CCR5 Delta32/Delta32 stem-cell transplantation. New Engl. J. Med.360, 692–698 (2009)).
2. The second one was called the London patient (Gupta, R. K. et al. HIV-1 remission following CCR5Delta32/Delta32 hematopoietic stem-cell transplantation. Nature 568, 244–248 (2019)).

New gene editing method as a cure for HIV

Do you know that an estimated 1% of the people in the world are naturally immune to HIV? The reason is a genetic mutation on the gene that encodes CCR5, a protein on the surface of immune cells that the HIV virus uses to enter and infect them. People with this mutation are missing part of the CCR5 protein, making it impossible for HIV to bind to it. Using gene editing, it would, in theory, be possible to edit our DNA and introduce this mutation to stop HIV. US-based Sangamo Therapeutics is one of the most advanced developers of this approach. The company extracts the patient's immune cells and uses zinc finger nucleases to edit their DNA to make them resistant to HIV. Sangamo reported in 2016 that four out of nine patients treated with this gene therapy in one of the arms of a phase II trial were able to remain off antiretroviral therapy with undetectable levels of HIV, and full results of the trial are expected soon.

In the future, this could be done using CRISPR-Cas9 (clustered regular interspaced short palindromic repeats), a gene editing tool that is much easier and faster to make than previously possible. But there is a lot of controversy around CRISPR gene editing after it was used to create the world's first gene-edited babies. These 'CRISPR babies' carry a mutation that protects them against HIV infection. Ethics of altering human DNA without fully understanding the possible consequences is a concern because, there is now proof that people carrying these mutations might be at risk of catching certain infections and dying younger.

Possible breakthrough to eradicate HIV by CRISPR gene editing method from reservoir pool and rest of the body

Even with ART, the virus still lingers in the body in cells in the lymph nodes, spleen, liver, lungs, intestines, bone marrow, and brain (all over the body) by replicating and copying itself into the DNA of infected cells. This year, researchers described a new form of ART which, in combination with CRISPR gene editing, eradicated the virus during testing on mice - a big step toward new therapeutic avenues for AIDS patients to eradicate HIV from the hiding places. This could be a method to get rid of this scourge.

To mimic a human infection in mice, the researchers used "humanized mice," which make human immune cells, and use of retroviral nanoparticles that take more time to dissolve and persist longer in the body, to prevent the virus from replicating. Researchers found that LASER ART, administered every few weeks rather than daily, stopped 99 percent of the virus' replication, but viral HIV is still in the genetic material.

When will we have an HIV cure? We have hope that it is coming within a decade.

Our approach to treatment of HIV-AIDS:

While a cure will come within the next decade, keep HIV in check by taking ART and maintain good health with exercise and life style changes described in another chapter so that one can use them when curative therapy is developed. I believe using whole body hyperthermia with dimethyl sulfoxide (DMSO) and insulin combined with high doses of intravenous or oral ART with or without CRISPR or other such agents can be curative. With repeated attempts, the virus becomes antigenic akin to polio virus and other vaccines which stimulates the immune system to produce anti-HIV antibodies that can eliminate new virus that emerges and infect the cells and those hiding in the nucleus of lymphoid and other tissue. It may also induce epigenetic changes in the viral DNA in the genome of the infected cell to prevent the incorporation of the reverse transcribed viral DNA into cellular chromosomes DNA and make it ineffective to produce viral copies. Using autoimmune therapy as described in the book "HIV-AIDS" Prescription for Survival by Dr. T. R. Shantha, (International Publishing House 1992) needs to be revisited/considered along with this therapy we recommend to activate the immune system to get rid of HIV from the hiding places.

Chapter 55

New method of treatment of migraine

The word migraine derives from the Greek ἡμικρανία (hemikrania), "pain on one side of the head," from ἡμι (hemi-), "half," and κρανίον (kranion), "skull." Many people fail to realize that migraine is a neurological disease, like epilepsy. Migraine ranks in the top 20 of the world's ***most disabling*** medical illnesses. Nearly ***one in four US households*** includes someone with who has migraine headaches on a regular basis. Most sufferers experience attacks once or twice a month, 14 million people or about 4% have chronic daily headache, when attacks occur at least 15 days per month. In addition to the attack-related disability, many sufferers live in fear knowing that at any time an attack could disrupt their ability to work or go to school, care for their families, or enjoy social activities. More than 90% of sufferers are unable to work or function normally during their migraine attacks. The highest rates of migraine are found in women, between puberty and menopause. It will affect 12-28% of people during their lifetimes. (Eur J Neurol 2006; 13:333-45). This translates to over 37 million people in the United States alone! Studies in twins have shown a genetic component of migraine susceptibility, and it is twice as prevalent in the families of epilepsy sufferers.

Characteristically the headache affects one-half of the head, is pulsating in nature, and lasts from 2 to 72 hours. The pain is made worse by physical activity. Infrequently a visual or auditory aura can occur with little or no headache following it. Interestingly, 82% had previously been given a diagnosis of tension headache (Goadsby, PJ, Lipton RB, Ferrari, MD. Migraine. Current Understanding and Treatment. N Engl J Med 2002; 346:257-270. Lipton RB et al. *Headache*. 2001; 41:638-64. Kaniecki R. Neurology, 2002, Suppl 6, 15-20. Hadjikhani N, Sanchez Del Rio M, Wu O, et al. Mechanisms of migraine aura revealed by functional MRI in human visual cortex. Proc Natl Acad Sci USA. 2001;98(8):4687-4692. Silberstein SD, Lipton RB, Goadsby PJ. Headache in Clinical Practice. 2nd ed. London, England: Martin Dunitz; 2002. Olesen J, Tfelt-Hansen P, Welch KMA. The Headaches. 2nd ed. Philadelphia, PA: Lippincott Williams & Wilkins; 2000).

The migraine headache is similar in severity as in trigeminal autonomic cephalalgias, which includes cluster headache, paroxysmal hemicranias, and short-lasting unilateral neuralgiform headache attacks with conjunctival injection, tearing (SUNCT) and its close relative short-lasting unilateral neuralgiform headache attacks with cranial autonomic symptoms (SUNA).

Migraine is a complex primary brain disorder that involves a cascade of events that lead to recurrent inappropriate activations of the trigeminocervical pain system. The events involved in activation and processing of the nociceptive depolarization within the trigeminocervical pain system may provide therapeutic targets for acute and preventative treatments.

The trigeminal nerve, which innervates the meninges, is intricately involved in migraine. How the migraine is triggered and the cascade of events following the original activation of migraine are not completely understood. However, there is increasing evidence that events intrinsic to the cerebral cortex are capable of affecting the pain sensitive dural vascular structures. If this is the case, then this might explain one way in which the headache is activated in individuals experiencing the aura.

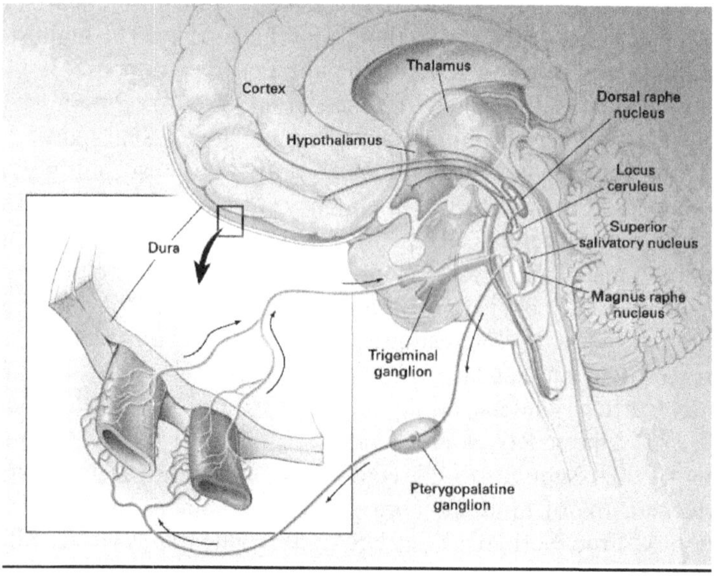

Figure 1

Figure 1 from Goadsby, PJ, Lipton RB, Ferrari, MD publication explains the possible mechanism of migraine related to neurometabolic events (Figures courtesy of authors and the editors of NEJM. Migraine. Current Understanding and Treatment. N Engl J Med.2002; 346:257-270). The primary mechanism is said to be due to Somatosensory input to the head, which involves pseudo unipolar trigeminal and upper cervical branches. Activation of the trigeminal nerve evokes a series of meningeal and brainstem events that appear to be consistent with what is seen during a migraine attack, to a long-lasting blood flow increase within the middle meningeal artery. This increase in blood flow is dependent upon trigeminal and parasympathetic activation. In addition, plasma protein leakage occurrence in the dura. This demonstrates that vasodilation during headache is possibly linked to a series of neurometabolic brain events, including transmission of pain via the trigeminal nerve.

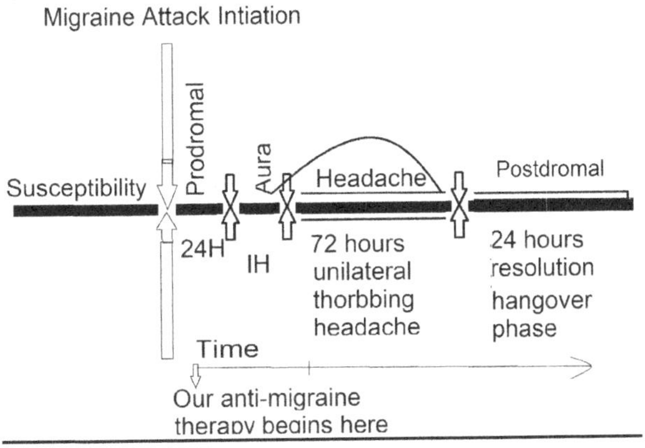

Figure 2

Figure 2 shows the time line of migraine development from prodromal to post-prodromal phase. Our patent pending method begins as soon as the prodromal migraine symptoms begin. I believe that it will revolutionize the treatment of migraine and deliver a pain-free functional patient. I need any migraine organization or research institutes that are interested to please contact me. Let us find a new method to mitigate migraine affliction.

Chapter 56

New oral airway, anti-snoring, and anti-obstructive sleep apnea devices

We have received 6 patents on obstructive sleep apnea, snoring and oral airway recently. The oral airway is used *351 million times every year by anesthesiologists as well as by EMTs and such.* We want to collaborate distribution all over the world which would provide better ventilation during and after surgery without obstruction of the oral air passage during the sleepy state after anesthesia.

We also want to develop devices for obstructive sleep apnea and snoring based on the patent we have received. They affect more than 50 to 300 million around the world. The demand for them worldwide is more than a 100 million. Our inventive device will enable people to stop using a cumbersome non-complaint obstructive sleep apnea device (CPAP), which is disliked by all sleep apnea patients. It will save these sufferers from cardio-vascular diseases, neurodegenerative diseases, stroke, accidents, high blood pressure and untold misery during working hours. Please contact us or Wedge Therapeutics if you are interested in our innovative oral airway, anti-snoring and anti- sleep apnea devices (www.wedgetherapeutics.com or www.indian-gold.com)

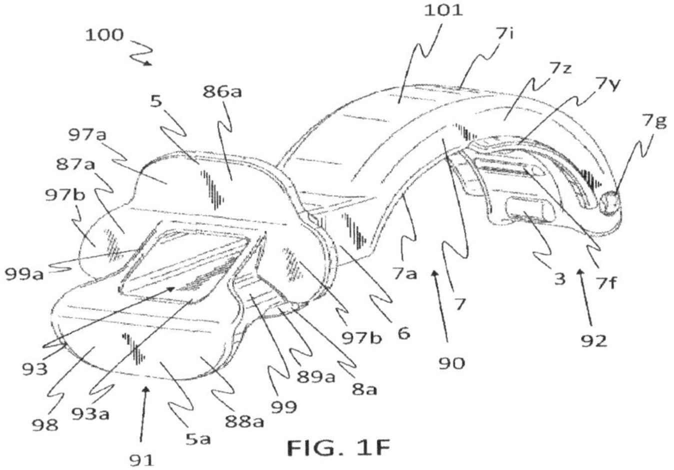

FIG. 1F

Figure 1F: New Oral Airway Device that holds the tongue and lower jaw protracted without using another hand or assistant to prevent obstruction to air-flow as described in the below patents.

International Application Published under the Patent Cooperation Treaty (PCT) International Publication Number WO 2018/200063 A1 November 1, 2018

United States Design Patent No. US D885,558 S May 26, 2020 Oral Airway Device

United States Design Patent No. US D849,233 S May 21, 2019 Oral Airway Device

United States Design Patent No. US D849,234 S May 21, 2019 Oral Airway Device

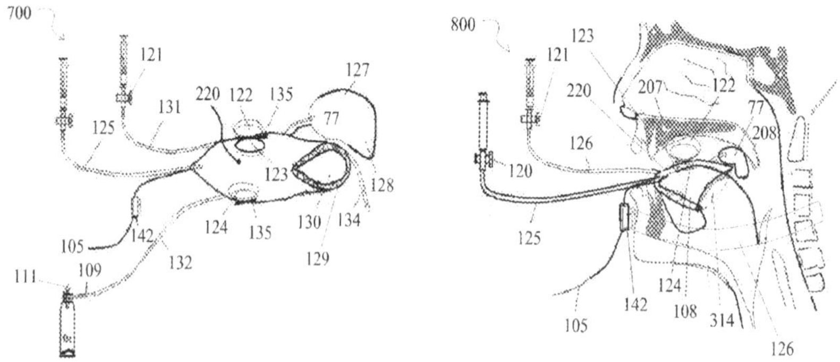

Figure 2

United States Patent No. US 9,072,613 B2 July 7, 2015 Device for Snoring and Obstructive Sleep Apnea Treatment

United States Patent No. US 9,254,219 B2 Feb. 9, 2016 Snoring and Obstructive Sleep Apnea Prevention and Treatment Device

Chapter 57

The power of sleeping left side down (SLD): to be healthy and live longer

Sleeping on the left side of the body (SLD) means touching the bed left side down on the bed with right side up. It can be used as a simple preventative health promoting and health preserving measure against many unwanted diseases, and at the same time promote longevity and health span. It prevents sudden infant death syndrome (SIDS), stomach burns (GERD), and Barret's esophagus—a precancerous condition at the junction of the stomach and esophagus (food pipe) and such.

SLD improves brain lymphatic (glymphatic) flow to clear the brain metabolic debris while sleeping to prevent neurodegenerative diseases such as Alzheimer's and Parkinson's. Most importantly, it provides an easy lymph flow from the thoracic duct (main lymph draining vessel), from the rest of the body and brain to systemic circulation. Lymph fluid is generally similar to blood plasma, which is the fluid component of blood. Lymph returns proteins and excess fluid between trillions of cells to the bloodstream and filters the bacteria that enter the lymphatic system. Lymph also transports fats from the digestive system (beginning in the lacteals) to the blood via chylomicrons.

The flow of lymph in the thoracic duct (main transportation vessel of the lymph in our body) in an average resting person approximates 100 ml (3.5 ounces) per hour. This is accompanied by another ~25 ml per hour in other smaller lymph vessels. The total lymph flow in our body is about 4 to 5 liters (170 ounces) per day. This flow can be increased several folds while exercising and during whole body hyperthermia. It is estimated that without lymphatic flow, the average resting person would die within 24 hours. This shows how important it is to maintain proper lymphatic flow from brain and body by SLD. Unlike blood, it has no pump to circulate it and is an open system, hence SLD will help in its collection, transport, and finally emptying it to systemic blood circulation.

SLD is a must for all pregnant woman to prevent blood flow impediment from the aorta to the developing fetus in the uterus. If you sleep on right side down, pregnant uterus will fall on the aorta and obstruct the blood flow to the fetus, affecting its normal growth. SLD also reduces the postprandial blood sugar (glucose) elevation, the insulin outpouring to liver after a meal and its related effects on the body and brain. It promotes easy and slow digestion in the gastro-intestinal track, providing unobstructed movement of digested food to the rectum for morning evacuation. Sleeping left side down reduces the incidence of appendicitis, improves oxygenation by right lungs, provides for easy flow of venous blood from all over the body to the right side of the heart, and from the lung to the left side of the heart. It promotes and improves coronary circulation without being impeded by the left pulmonary artery, and puts less strain on the arch of the aorta and such (Journal: Inspirator, Nov 1997). This is a form of ***"sleeping yoga"*** for health and longevity. Sleeping on one side of the body (SLD) I call simple effortless SLEEPING-YOGA PRACTICE at night during sleeping cycle.

The comprehensive book on this subject which I am self-experimenting and exploring during my decades of medical research and practice will be published in 2021. I am looking for a publisher.

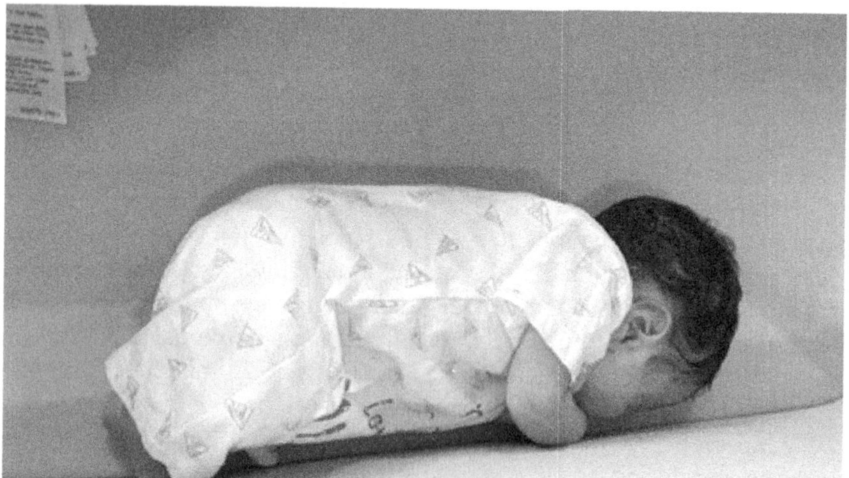

GIFT AND THE POWER OF
SLEEP
ING
LEFT SIDE
DOWN
Dr. T. R. Shantha
MD, Ph.D, FACA

Book under preparation to be published in 2021
By Dr. T. R. Shantha, MD, Ph.D., FACA
in collaboration with:
Usha S. Martin, BSc, PA, CPT, ECMO expert,
Dr. Jessica G.S. Makwana, MD, Dr. Erica M. S. Tarabadkar, MD,
Lauren A. S. Woodling, BSc, BcRA.

Chapter 58

New organ: in the peripheral nerves

Fluid containing sub space around the cranial and spinal peripheral nerves and their endings as route of transport to and from CNS and peripheral nerves—can you call it a new organ?

There was headline news about the discovery of Interstitium, a newfound "organ." With all that's known about human anatomy and histology, you wouldn't expect doctors to discover a new body part in this day and age. But now, researchers say they've done just that: They've found a network of fluid-filled spaces in tissue that hadn't been described before.

A similar tissue fluid or seeping CSF fluid filled space found all around the peripheral and cranial nerves and nerve roots emerging from the brain and spinal cord that extends all the way to the sensory and motor end organs is named sub-perineural epithelial space. This is tissue space and is called it sub-perineural epithelial space—is it a new found "organ?" (see below Figure)

According to these scientists, the new finding reveals that, rather than a "wall," this tissue is more like an "open, fluid-filled highway," said co-senior study author Dr. Neil Theise, a professor of pathology at New York University Langone School of Medicine. The tissue contains interconnected, fluid-filled spaces that are supported by a lattice of thick collagen "bundles," Dr. Theise said, so also the sub-perineural epithelial space we have been describing over >5 decades.

Our sub-perineural epithelial space is exactly what they describe as a new found organ in tissue spaces; then you can call sub perineural epithelial space an " organ." When scientists prepare tissue samples containing the peripheral nerve, they cut them into thin slices and stain them to highlight key features. But this fixing process drains away fluid and causes the newfound fluid-filled spaces to collapse. The same thing happens in peripheral nerve fasciculi in small nerve fibers, but it is open in large nerve fasciculi as shown in the figures in our publications under perineural epithelium in my bio in part IV.

These fluid-filled spaces were discovered in connective tissues all over the body, including below the skin's surface; lining the digestive tract, lungs and urinary systems; and surrounding muscles, so also they are found around the brain, around the blood vessels entering the brain (Virchow-Robin space), inside the brain as aquaporins, and neuropile—can they be called new "organs?" Dr. Theise and his group call this network of fluid filled spaces an organ—the interstitium, so also sub-perineural epithelial space and its fluid. Further, sub-perineural epithelial space directs the new sprouting axon after of Wallerian degermation to proper destination. It also prevents the infection and inflammatory products entering the nerve fasciculi and disrupts the axon function.

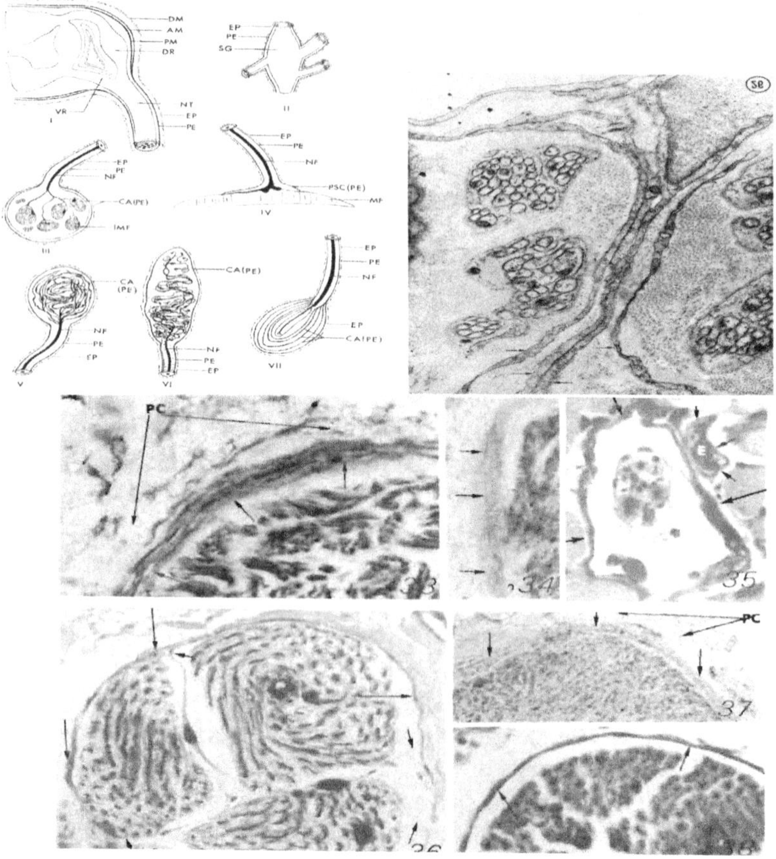

These figures show sub-perineural epithelial space of the peripheral nerves. For details read J Nature 199, 4893:577-579 (1963). Science 154:1464-1467 (1966). Anesthesiology. 37:543-557, 1972 Histochemie 16: 1-8, (1968). 22.PE—92 pages. Str.Fu. Ner. Tissue, Academic Press, GHB edi, 1970.

Part III

Methods of treatment and prophylaxis for other diseases using traditional and complementary adjuvant therapies

Chapter 59

Diseases and death (morbidity/comorbidity and mortality)

How to protect from infirmities and live longer (anti-aging)

Coronavirus death toll in the nursing home, extended care facilities and in the aged has brought the old age and age-related disease labelled as co-morbidities to the forefront, but many of these can be delayed or reduced by following the protocol introduced in this and other chapters that follow.

Gompertz prediction of risk of dying

British self-educated mathematician Benjamin Gompertz is now best known for his "Gompertz law of mortality" —a demographic probability of death model published in 1825. **He found that the risk of dying rises exponentially with age; in humans, for instance, it doubles roughly every 8 years after the age of 30** (Gompertz, Benjamin 1825). He also expounded that a person's **resistance to death decreases as his years increase, as we see in COVID-19**. This model is a refinement of a demographic model of Robert Malthus and used by insurance companies to calculate the cost of life insurance.

Death totals in the United States and longevity revolution

Based on the Gompertz formula, we know the probability of death after the age of 30. Here we have the rate of death as we age. In 2014, a total of 2,626,418 resident deaths were registered in the United States. The age-adjusted death rate, which accounts for the aging of the population, was 724.6 deaths per 100,000 US standard population. Life expectancy at birth for the total population was 78.8 years in 2014, unchanged since 2012. Life expectancy for females was 4.8 years higher than for males since 2010.

The Longevity Revolution (LR) is a term used to characterize the recent and rapid increase in life expectancy (**or lifespan**) overall which has not been paralleled by the same increase in **health span** (period of life spent in good health, free from the chronic diseases and disabilities of ageing).

Important predictor of diseases and death

There are no tests for predicting impending illness and possible early death. A recent research study might have solved this problem: "Our study showed that participants with lymphopenia were at high risk of dying from any cause, regardless of any other risk factor for all-cause mortality including age," said Stig Bojesen, from Copenhagen University Hospital in Denmark. The study, published in the Canadian Medical Association Journal, January 13, 2020, included 108,135 people of Danish descent aged 20-100 years. The study found that low lymphocyte count was associated with a 1.6-fold increase in the risk of death from any cause and a 1.5- to 2.8-fold increased risk of death from cancer, cardiovascular disease, respiratory disease, infections and other causes, the researchers said. This is due to reduced immune capacity to survive potentiality lethal diseases, such as the coronavirus infection. *Lymphopenia* also increases the frailty and those in nursing homes, extended care facilities, and the aged living independently who can benefit from additional surveillance, the researchers noted.

Both men and women die an average of almost three years earlier in the US than they do in Canada, UK, Japan and Australia. Here are the biggest preventable risk factors (for death), according to the University of Washington: smoking, obesity, high blood pressure, high low-density lipoprotein (LDL) cholesterol, high dietary trans fatty acids, high salt intake, low dietary omega-3 fatty acids, high blood glucose, low intake of fruits and vegetables, alcohol abuse, and lack of physical inactivity. I can add to that list, eating grilled high meat diet instead of Mediterranean diet with millet and fish, a diet rich in omega 3 instead of wheat, rice and sugar.

Leading causes of death

Below are the leading causes of death out of 2,626,414 total deaths in US, in the year 2014 (Data from National Vital Statistics Reports: National Vital Statistics Reports Volume 65, 2016).

1. Diseases of heart (heart disease) Arteriosclerotic vascular diseases: 614,348
2. Malignant neoplasms (cancer) (95% of cancers are preventable): 591,700
3. Chronic lower respiratory diseases: 147,101
4. Accidents (unintentional injuries): 135,928
5. Cerebrovascular diseases (stroke): 133,103
6. Alzheimer’s disease: 93,541
7. Diabetes mellitus (diabetes): 76,488

8. Influenza and pneumonia: 55,227
9. Nephritis, nephrotic syndrome and nephrosis (kidney disease): 48,146
10. Intentional self-harm (suicide and such): 42,826
11. Septicemia—blood and body infection: 38,940
12. Chronic liver disease and cirrhosis: 38,170
13. Essential hypertension and hypertensive renal disease (hypertension) kill more than this: 30,221
14. Parkinson's disease: 26,150
15. Pneumonitis due to solids and liquids: 18,792
16. All other causes: 535,737

3.3 million people die due to use of alcohol around the globe. 5% accidents, and illnesses are blamed on alcohol. Drug addiction kills 50,000-70,000 annually in the United States (2017-18-19). Aging can be called a disease in progress. Cancer has been linked with aging since genetic mutations accumulate with age with suppressed immune system and delayed aging has been associated with lesser incidence of cancers.

Additional causes of death due to illnesses, malnutrition, accidents, addiction, and death are as follows:

- **AIDS**—1.1 million AIDS related deaths in the year 2016 around the world.
- **Firearms**—In 2014, 33,594 persons died
- **Poisoning**—In 2014, 51,966 deaths occurred as the result of poisonings including food poisoning. Don't eat stale food; avoid old canned foods.
- **Injury-related**—a total of 199,756 deaths were classified as injury-related at work, driving, and at home etc.
- **Motor-vehicle traffic**—In 2014, motor-vehicle traffic-related injuries resulted in 33,736 deaths.
- **Accidental fall**—In 2014, 33,018 persons died as the result of falls. Watch when you climb stairs, ladders or are in the bathtub.
- **Drug addiction**—In 2014, 49,714 persons died of drug-induced causes in the United States. Estimated 128,000 die every year from prescription drugs, that is one American every 4 minutes.
- **Global poverty** kills millions. COVID-19 killed more poor and mal-educated.
- **In the US, the bacteria resistance** to the existing antibiotics kills 35,000 people annually according to a CDC study.

- **Autistic people** are twice as likely as those in the general population to die prematurely. (Smith DeWalt L. et al. *Autism* Epub ahead of print. 2019).

How do you prevent diseases and death, save health care costs to individuals and health care providers, and at the same time keep functioning longer (health span)?

It is often quoted "you have no control over your birth or death." But I would say "you do have control over what you do in-between to maintain longevity and health span."

- Do not smoke. The food you eat, water you drink and air you breathe has an immense effect on your health and longevity.
- Do not eat saturated fat, avoid grilled or fried foods to prevent AGEs deposits, eat Mediterranean diet, eat more fish rich in Omega 3 (mackerel, lake trout, salmon, herring, sardines, sturgeon eggs, tuna, blue fish) instead of red meat, restrict cheese, avoid margarines and such. Avoid a heavy meal loaded with calories, carbohydrates and sugar. Calorie restriction (CR) is the rule. Use curcumin curry powder with or without garam masala in your food preparations.
- Reduce or cut off wheat, rice and sugar from the diet; substitute with a millet-based diet. Try a different variety every day; choose from 5-6 varieties (see the Chitra food book web site).
- Induce whole body hyperthermia >3 times a month.
- Take bioactive curcumin 1000 mg after the age of 40-50-60-70-80s, and after a heavy meal immaterial of your age.
- Take metformin 250-500 mg after lunch and dinner after you are past your 50-60-70s. It is immaterial whether you have diabetes or not. It has an anti-aging effect on the body, and increases longevity.
- Take multivitamins as needed, vitamin D3 5000 IU and vitamin C 1000-2000 mg every day. May include Ebselen as an antioxidant and inti-inflammatory.
- Exercise daily and practice calorie restriction to prevent diseases, obesity, other comorbidities and live longer.
- Sleep 8 hours day turned left side down most of the time, avoid stress, and avoid jobs that harm your brain and body.
- Take mini doses of intermittent rapamycin as explained in another chapter 7-10 days at once with some interruptions.
- Take felodipine up to 5 mg past the age of 50-60s to clean the nerve cells by autophagy. If you are taking blood pressure medication, ask

your doctor to prescribe this calcium channel blocker with telmisartan and losartan. If you are on an anti-cholesterol drug, take simvastatin instead of other statins because it crosses the blood brain barrier to enter brain and prevent neurodegenerative diseases.

- Take beta-glucan, a special type of nutritional sugars found in the cell wall of bacteria, fungi (mushroom), yeasts, algae, lichens, and plants, such as oats and barley. They have an effect on high cholesterol, diabetes, cancer, HIV/AIDS, HPV infection, high blood pressure, and canker sores to boost the immune system, reduce colds, influenza (flu), swine flu, respiratory tract infections, allergies, hepatitis, Lyme disease, asthma, aging, gastro-intestinal track afflictions, multiple sclerosis and such.
- Incorporate 5 important life style changes as anti-aging methods described in another chapter.

Other measures include: taking micro-dose of lithium, inhibitors of angiotensin II signaling (by taking Losartan, Lisinopril) for BP, statins (simvastatin,), propranolol, aspirin and such (Blagosklonny MV, From Rapalogs To Anti-Aging Formula, Oncotarget. 2017; 8:35492-35507).

Heartburn drug and early death

About ±50 million people in the US are on PPIs blocker. Unfortunately, heartburn drugs are associated with increased risk of death, and the longer a person uses the drugs, the greater the risk, new research suggests. They are called proton pump inhibitors or PPIs, and have been tied to a wide range of side effects including fractures, dementia, heart disease, pneumonia and kidney disease, the study's senior author Dr. Ziyad Al-Aly of Washington University School of Medicine in St. Louis told Reuters Health in a telephone interview. (Thomson Reuters Jul 6.2017 report in BMJ Open.).

BCG vaccination to prevent diseases

I advise those in nursing homes, extended care facilities, and the aged living independently, including those with Alzheimer's and Parkinson's, to take 2 doses of BCG vaccination administered intradermally, 4 weeks apart to enhance the regulatory trigger T cells and eliminate auto-reactive T cells that contribute to more than 152 auto immune diseases and other ill effects on the body. It may even cure or reduce the dose of insulin in type I diabetes or have an effect on diabetes (Faustman DL edit. The value of BCG and TNF in Autoimmunity. 2018. Academic Press). Studies on multiple sclerosis show BCG vaccine reduces signs and symptoms by 57% post BCG vaccine, and at the same time it favors repair mechanisms, and has protective action against neurodegenerative diseases such as Alzheimer's and Parkinson's, diabetes. I have used it by locally instilling it inside the bladder to treat bladder cancer with good results. The BCG vaccination might trigger mechanisms that mediate immunoregulation and immune training at the same time, along with tissue repair, in diverse pathological conditions that share chronic inflammation and tissue damage as seen in aging.

Measles, mumps and rubella (MMR) vaccines to resist other infections

The recent report on measles, mumps and rubella (MMR) vaccine also augments the resistance against viral infections including COVID-19 by taming down the affliction. Maybe the nursing homes and extended care facilities and aged should get booster doses of MMR vaccination, even though they had it in childhood, to boost up the immune system nonspecifically and to ward off coronavirus and other viral-microbial infections as they age due to depressed immune system and comorbidities.

Chapter 60

Aging

The words aging, the aged, elderly, nursing homes, extended care facilities, comorbidities, hot spots, incubation period, ICU and such are in the news due to the coronavirus pandemic which is deadly to the aged segment of the population.

Why does one should think about aging and adopt anti-aging modalities we describe in here? Because by slowing down aging, one can also delay the development and progression of aging-associated diseases and disorders prevented and/or delayed. There are two aging clocks running in your body: 1. First one is the gerontological clock, we can't change it. It starts ticking the minute you are conceived in your mother's womb. 2. The second clock is the biological clock which can be accelerated or slowed down depending on life style changes and other measures described here. We discuss here the life style changes, supplements and drugs you take to prolong life as discussed in other chapters also. How long you live is in your hand, you decide how long you live, barring genetic diseases, pandemics, accidents and such.

As we advance in years, we think of the vocabulary of aging and diseases, with more worrisome words such as: cancer, heart disease, dementia, diabetes, obesity, high blood pressure, besides noisy creaky joints, and aching weak muscles. There is acute inflammation in response to injury or infection and chronic inflammation, a condition caused by a misfiring of the immune system that keeps your body in a constant, long-term state of high alert. You may have it, but don't even know you are suffering from chronic inflammation. ***It is the root cause of almost 95% of the diseases we suffer as we age, commonly labelled co-morbidities.*** Add to it the coronavirus and such infections in the aged, and you are adding fuel to the fire—it can become deadly—killing thousands and maiming millions. As Ashley Montagu (1905-1999) said, "The idea is to die young as late as possible."

Definition of aging, longevity and definition of health span

What do we actually mean by aging? The simplest definition is the loss of ***homeostatic*** ability with the passage of time. Homeostatic ability is the ability to maintain internal stability. That is, the ability of an organism to maintain a stable internal environment in the face of environmental challenges such as changes in temperature, humidity, air quality, and so on. At the most basic level, the loss of this ability is the primary deficit of aging.

The study of the ***biology of aging, or biogerontology,*** has as its principal objective understanding the basic processes that underlie aging and age-related diseases. The drugs that increase the life span are called ***geroprotectors*** and the most important such agents are rapamycin, metformin, and lithium, besides calorie restriction and whole-body hyperthermia.

Calorie restriction (CR) is the only known intervention in mammals, including humans, that has been shown consistently to increase life span, reduce incidence and retard the onset of age-related diseases, including cancer and diabetes. CR has also been shown to increase the resistance to stress and toxicity. Mahatma Gandhiji practiced it for decades for a total of 14 times, had a healthy body, did not have an ounce of fat under his skin and retained a sharp mind, and even participated in negotiated division and independence of India in 1947.

The crucial event of the action of CR is the reduction in the levels of insulin. Insulin-like growth factor-1 (IGF-1) increases insulin sensitivity in our cells that make our body and brain, and alleviates age related metabolic immunosuppression, besides reducing the production of growth hormone from our pituitary gland. For some this means increasing human life span, for others it means increasing human health span.

Biogerontology defines aging as: "The progressive failing ability of the body's own intrinsic and genetic powers to defend, maintain and repair itself in order to keep working efficiently." Aging has been understood to be an inherent, universal, progressive natural phenomenon. Aging as a process, or set of processes, affects virtually all of our bodily organs and systems. At the most basic level, some portions of our aging patterns are set in our genes. The reliable differences in the life spans of various species are clear evidence of genetic "programs" that set the general boundaries of species life span. The environment and our life style modify these boundaries. Genes are important in the differences in longevity between individual members of a single species (e.g., long- and short-lived humans), but environmental factors and our life style play a major role how these differences are expressed - we can modify these by our life style changes. Living in a toxic environment or making deleterious lifestyle choices as discussed under life style chapter can have a significant effect on individual longevity.

The World Health Organization (WHO) defines health as ***"a state of complete physical, mental and social well-being and not merely the absence of disease or infirmity"*** i.e. health span. They do not define the aging healthy or aging with comorbidities. Studies in lower organisms such as yeasts and fungi show that there are genes, called longevity assurance genes, that ensure that cells function long enough for the organism to live out its normal life span. Such a system also operates in humans.

What do some famous people think about aging and death?

There are hundreds of quotes on the internet about living and dying. The following quotes touch the philosophy of aging, life and death: It is not death that a man should fear, but he should fear never beginning to live (Marcus Aurelius), If a man has not discovered something that he will die for, he isn't fit to live (Martin Luther King, Jr.), The fear of death is the most unjustified of all fears, for there's no risk of accident for someone who's dead (Albert Einstein), Live as if you were to die tomorrow. Learn as if you were to live forever (Mahatma Gandhi), Many people die at 25 and aren't buried until they are 75 (Benjamin Franklin), No one wants to age, suffer from disease and die. Even people who want to go to heaven don't want to die to get there (Steve Jobs). That being the case, one needs to know and practice methods how we prevent diseases, prevent early death and prolong life as the gerontological clock is ticking. The following will describe some of the methods to age with grace, slow down the gerontological clock or make it run faster and to be free from comorbidities as much as we can.

Now let us examine how to prolong our life, and ward off secondary afflictions that torment our body as we age

Everything around us including animals, plants, earth, planets, and heavenly bodies in the vast universe are all aging. The day we started life, from ovum and sperm union, we started aging and it continues until the end of our life. There are many theories put forward to explain why we age, but none of them are perfect. They incorporate many known aspects and complex processes involved in aging. As we know now, there is no single theory that can explain aging.

There are many medical, quasi medical, non-medical associations, organizations, and web sites promoting and promising longevity, restoring memory loss with anti-aging herbs, anti-wrinkle creams, supplements, exercises, diet, including hyperbaric oxygen therapy and ozone. Caloric restriction, with adequate nutrition, is the only self-experimental life style changing methodology currently known to extend life span without costing a dime. Understanding how caloric restriction produces extended longevity could provide valuable clues to basic aging processes as well as suggest new therapies for age-related diseases.

The following therapy besides other supplements for the aged has shown to prolong life experimentally. Most important anti-aging methods whether you are in the nursing home or at home are as follows:

a) Take bioactive curcumin 1000 mg a day. Curcumin is a very effective anti-oxidant, anti-inflammatory, anti-septic, anti-cancer, anti-cardio-vascular diseases, anti-neurodegenerative diseases, anti-scarring, anti-coagulant (makes blood thin and prevents clotting inside blood vessels) and has anti-fat protein clump accumulation (anti-Alzheimer's) effects.

b) Take mini-dose lithium carbonate 300-1000 micrograms: As we age, dysfunctional GSK-3 enzyme activity raises the risk of many chronic diseases of older age including Alzheimer's, type II diabetes, mood disorders, cancer, cardio-vascular diseases, eye changes and others. Lithium has been shown to ***inhibit overactivity of GSK-3***. It a must supplement in nursing homes, extended care facilities, and the aged living independently. It is one of the most important therapeutic agents that should be used as preventive medicine. It improves memory and protects the telomers at the end of chromosomes, a key to our health.

c) Take micro-dose intermittent (pulsed) rapamycin, 3-4-6 mg once a week as described in another chapter. It ***inhibits mTOR*** and has multiple health benefits akin to calorie restriction extending longevity with reduced comorbidities.

d) Take metformin 250-500 mgs after lunch and dinner whether you are diabetic or not. It ***activates AMK,*** making cells more sensitive to insulin and changing the metabolism of the cell.

e) Take 25 mg dehydroepiandrosterone (DHEA). it is a hormone, derived from cholesterol and pregnenolone, produced by the adrenal glands, and some in the testes of males and ovaries in females. It is a precursor—the starting material—for production of the sex hormones testosterone and estrogen. NEJM study found that 100 ug/dLin DHEA-S reduced the risk of death from any cause by 36%. (LE July

2020). Premenopausal and menopausal women, and men above the age of 60s to 80s should take this.

f) Vitamin D3, 5000 IU as an anti-inflammatory
g) Deprenyl (selegiline): Deprenyl is a monoamine oxidase B (MAO-B) inhibitor. The MAO-B is an enzyme that degrades the neurotransmitters dopamine, epinephrine, nor-epinephrine and serotonin in the brain. MAO-B levels rise with age and it was believed that this caused a decrease in these neuro-transmitters, which resulted in depression, dementia, neurodegenerative diseases such as Alzheimer's, Parkinsonism. By selectively inhibiting MAO-B, Deprenyl maintains these neuro-transmitters at more youthful levels. All the animals treated with deprenyl lived much longer.
h) Take simvastatin as an anti-cholesterol and as an anti-inflammatory of the brain.
i) Take Ebselen as you age past sixties. It is anti-viral, anti-inflammatory, anti-oxidant, and anti-aging and a protease inhibitor.
j) To soften your arteries for longevity: regular exercise, stop smoking, control your cholesterol and make better decisions around your diet by eating Mediterranean diet with rich in omega 3 such as mackerel, lake trout, salmon, herring, sardines, sturgeon eggs, tuna, blue fish, no saturated fats or margarine, calorie restriction a must and no grilled food.
k) Eat blueberries, a fist full every day. Grind them and drink the ground berries.
l) Whole body hyperthermia or sauna > 3 times a month to keep the blood vessels of the brain body free of cholesterol, and clots. Every time I have done whole body hyperthermia, the next day my brain felt clear, clean and my hand writing improved. Raise the body temperature between 99.5-100° F.
m) Maintain oral hygiene without any gum disease—a must. Every nursing home and extended care facility should have a dental hygienist or dentist visit all residents 2 times a year.
n) Oral 1000 mg of vitamin C with fruit juice,
o) Take oral Ebselen, Ponstel / ibuprofen / Celebrex / baby aspirin and such for reducing inflammatory cytokine such as interleukin-1, tau, beta of beta-amyloid deposits, alpha synuclein, and reduce chronic body inflammation. They are taken only for a limited time off and on.
p) Take melatonin 3-5 mg before sleep which is also a neuroprotective, immune system enhancer, and in addition helps you sleep.

q) Eat a Mediterranean diet with olive oil rich in millets, take Phosphatidylcholine (Lecithin), antioxidants, alpha lipolic-acid, Coenzyme Q10, N-Acetyl-Cysteine, DHEA, vitamin B12, vitamin C. vitamins B1 (Benfotiamine) and B6. Calorie restriction and fasting are important components of life extension.

r) Add miso in your soup and diet as part of an anti-aging diet. It is a fermented soy and comes in dozens of varieties. It comes in ready to use container; it is a natural source of probiotics (good bacteria). It is said to increase longevity and reduce comorbidities as you age. At its most basic, miso is a fermented paste that's made by inoculating a mixture of soybeans with a mold called koji (for you science folks, that's the common name for Aspergillus oryzae, Japan's national mold) that's been cultivated from rice, barley, or soybeans. Over weeks (or even years!), the enzymes in the koji work together with the microorganisms in the environment to break down the structure of the beans and grains into amino acids, fatty acids, and simple sugars; that is miso.

s) Take choline proteins, the precursor molecule to acetylcholine, a memory neurotransmitter, as well as pantothenic acid (vitamin B5), an important cofactor for choline. Thus, it is possible to cover all the bases in providing the means to enhance the levels and effectiveness of acetylcholine in the brain of Alzheimer's disease patients. Take lactoferrin and whey protein supplements and reduce wheat, rice and sugar in the diet.

t) Exercise at least 3-5 times a week, such as walking 30 minutes, and combine with whole-body hyperthermia. Continue with other therapeutic agents prescribed for Alzheimer's disease including Deprenyl, acetylcholinesterase inhibitors and NMDA inhibitors;

u) Sleep 8 hours a day to clear the brain and body of metabolites produced during daily activities, incorporating 5 important life style changes for life span and health span as described in another chapter.

Triple therapy for anti-aging in humans

Castillo-Quan, et al. study shows that the following act additively to increase longevity in Drosophila by 48%. They are implicated in body and brain inflammation, with reduction of cancer, bipolar disorder and such:

1) the mitogen-activated protein kinase (MEK) inhibitor trametinib (drug that inhibits the mitogen-activated protein kinase enzymes MEK1 and/or MEK2.- active in some cancers)

2) the mTOR complex 1 (mTORC1) inhibitor such as rapamycin which inhibits all the metabolic processes in the cells
3) the glycogen synthase kinase-3 (GSK-3) inhibitor of glycogen synthesis by lithium.

Trametinib is expensive. We replace it with ***metformin, which is the 4th most commonly prescribed drug in the world.*** It is cheap and available, used by millions of type 2 diabetics. ***Add Ebselen, a powerful anti-oxidant and anti-inflammatory also.***

Chapter 61

Inflammation

Inflammation, aging and development of comorbidities

Research and clinical data have linked chronic inflammation to nearly every disease that afflicts us as we age, resulting in many co-morbidities. Add to this list, low lymphocyte count (lymphopenia) in the blood as we age is associated with high risk of dying from any cause, regardless of any other risk factor, according to Stig Bojesen, from Copenhagen University Hospital in Denmark.

The term comorbidities is especially in the forefront of the news due to the coronavirus pandemic. COVID-19 involves acute inflammation of the body as opposed to chronic inflammation which all of us suffer from, but don't even notice we have it, and which is responsible for co-morbidities we suffer as we age. During this coronavirus pandemic almost 95,000 of these nursing home patients were affected with death of more than 25,923 with 449 deaths among facility staff according to Seema Verma of Medicare – Medicaid service administrator, showing the full toll among the most vulnerable Americans (May 2020).

In the vocabulary of aging, there are far more bothersome words: cancer, heart disease, dementia, diabetes, neurodegenerative diseases, any one of the 152 autoimmune diseases, obesity, kidney and lung diseases, depression, and such. Researchers have suspected all of these health problems have one common trigger: low-to high grade inflammation attacks on the aged immune system of our body. Without hesitation, we can state that there are proofs that anti-inflammatory medicine can lower the countless co-morbidities we suffer as we age. The most effective is taking micro-doses of ***rapamycin***.

What are the inflammations of the body that contribute to various diseases as comorbidities?

From the day of our birth, till the end of our life, we are attacked by microbes, physical exposure, and endless accidents that result in inflammation all over our bodies besides the stress of growing up, working at a job, and aging. They are named acute when the inflammation is only temporary and is gone after a certain time like flu, sore mouth, physical cut, or injury. They are called subacute where the afflictions stay for a while and then disappear within weeks, such as bronchitis, sore joints and muscles, back pain and such. There are other inflammations which are chronic, lingering year after year within the body doing its damage and ultimately expressing its effect on the heart, lungs, vascular system, and brain and any other organ exposed succumbs to it.

Here we want to discuss only the chronic inflammation in our body that we normally are not aware of, and that we can prevent, delay, curtail and/or cure to promote health and longevity and at the same time ward off comorbidities.

Some of the inflammatory mediators inside our body

The scientific literature reveals inflammations are mediated with inflammatory mediators which are produced in our body and can be blocked or reduced by anti-inflammatories such as:

- lipoxygenase
- cyclooxygenase-2 (COX-2)
- leukotrienes
- thromboxane
- prostaglandins
- nitric oxide
- collagenase, elastase, hyaluronidase
- monocyte chemoattractant protein-1 (MCP-1)
- interferon-inducible protein, and interleukin

Some of the significant inflammation triggers within our body are due to:

- high levels of homocysteine
- high levels of C-reactive protein
- high levels of insulin, iron, fibrinogen, high levels of low-density lipoprotein (LDL), low levels of high-density lipoprotein (HDL) uric acid crystals, and triglycerides, high levels of testosterone
- Increased levels of tumor necrosis factor-α
- vascular endothelial growth factor up regulating adhesion molecule
- high blood pressure
- high blood sugar in diabetes
- obesity producing inflammatory cytokines, a lack of exercise resulting in obesity, and a host of other ill effects

- toxins from cigarette smoking

Body inflammation and free radical scavengers

Our body protects us by producing antioxidants that thwart the effect of inflammation on body and brain, that are produced within our body. They are:

- superoxide dismutase
- catalase
- glutathione
- glutathione peroxidase

All these above anti-inflammation, free radical scavengers within the body are enhanced by use of curcumin, vitamin D3, whole-body hyperthermia, micro-doses of intermittent rapamycin, lithium, metformin and such.

How chronic inflammation affects our body

There are more than 152 autoimmune diseases affecting our body and they are expressed as inflammations lingering year after year. The inflammations contribute to endothelial dysfunction of 60,000 miles of blood vessels lining and 400 miles of blood vessels in the brain with the subsequent development of atherosclerosis, heart attack, strokes, cancer, neurodegenerative diseases and such. Chronic inflammation due to diabetes (glycation) results in diabetic retinopathy, nephropathy, and neuropathy. Chronic inflammation of the brain results in neurodegenerative diseases of every kind from Alzheimer's disease to Parkinson's disease.

As we age, we suffer from chronic inflammation which results in an increase of inflammatory cytokines (destructive cell-signaling chemicals involved in autocrine, paracrine, and endocrine signaling) which include chemokines, interferons, interleukins, lymphokines, monokines, colony stimulating factors and such, that contribute to degenerative diseases including autoimmune diseases such as rheumatoid arthritis, type 2 diabetes, gut diseases, arteriosclerotic vascular diseases, neurodegenerative diseases, cancers, kidney and liver diseases and such. In these disease cytokines such as tumor necrosis factor-alpha (TNF-a), interleukin-6 (IL-6), interleukin 1b [IL-1(b)], and/or interleukin-8 (IL-8) are known to cause or contribute to the inflammatory reaction resulting in these diseases.

Obesity also produces chronic inflammation, self-induced comorbidities resulting in type 2 diabetes associated with an untold number of diseases which it brings with it. If you want to live longer without comorbidities and with body and brain intact, reduce the body weight by practicing calorie restriction and exercise.

Our method to reduce silent chronic inflammation of the body to lead a disease-free long life without comorbidities:

- Take bioactive curcumin 1000-2000 mg every day without fail after the age of forties and fifties all your life. It will counteract the inflammatory free radicals, inhibit inflammatory related cytokines such as TNF-α, VEGF, expression of NF-kB and COX-2 levels and reduce blood sugar, also.
- Keep your teeth clean to prevent chronic inflammation of the gums and oral cavity which has been blamed for body inflammation and related diseases.
- Induce whole-body hyperthermia once a week or more to flush out free radicals and toxic metabolites, and inflammation causing agents. It acts as a drain cleaner of the blood vessels.
    - It dislodges and reduces all the inflammatory molecules discussed above from the body.
    - Hyperthermia increases the heat shock protein that protects the heart and brain.
    - Hyperthermia clears the toxic metabolites from the brain.
    - Hyperthermia removes the tumor necrosis factor-alpha (TNF-a), and reduces the C-reactive protein, fibrinogen, homocysteine, and IL-6 that can cause and predict inflammation.
- Metformin taken orally 250-500 mg after lunch and supper reduces the insulin resistance and prevents glycation; it also has anti-aging effects.
- Take vitamin D3, 2500-10000 IU a day, as an anti-inflammatory.
- Stop cooking foods at high temperatures to prevent formation of **gliotoxins** in the food that produce chronic inflammation (AGEs).
- Food should include fish rich in omega3.
- You can also add DHEA, DHA, omega-3, zinc, magnesium, selenium and such supplements to reduce the effect of inflammatory agents.
- It is important to control BP to prevent damage to the blood vessels wall including the endothelium.
- Check ***blood pressure every night before going to bed,*** and don't sleep till you bring it down to below 120mm Hg. It does not take much time by using Nadolol beta-blocker and telmisartan and/or losartan angiotensin receptor blockers.

- Keep blood sugar level low, between 85-100 mg%, control weight. Exercise regularly, and practice yoga and meditation to reduce stress induced cortisol level and block its inflammatory effect.
- Add to this our micro-doses of intermittent rapamycin and lithium as described.
- Add vitamin C, vitamin D3, and Benfotiamine (vitamin B1) to strengthen the immune system to ward off the effects of comorbidities.

Chapter 62

Comorbidity

Explanation of the term "comorbidity"

In medicine, comorbidity denotes the presence of one or more additional health conditions co-occurring with (that is, concomitant or concurrent with) a primary condition. Additional comorbidities within a patient may also be behavioral or mental disorders. Comorbidity describes the effect of all other conditions an individual patient might have other than the primary condition of interest. I would say that aging (associated with senescence cell accumulation in our body and immune system dysfunction) is a disease, and any other affliction of the body and brain is a comorbidity. The number of comorbid diseases increases with age.

Comorbidity is a term you might hear frequently in psychology as well as in medicine more broadly. It's important to understand comorbidity and how it works so that you can evaluate and help **comorbid patients**, or patients that are suffering from multiple, related or unrelated diseases or disorders at the same time.

Epidemiology of comorbidly

Comorbidity is widespread among the patients admitted at multidiscipline hospitals especially past the middle age and those living in nursing homes and extended elderly care facilities. During this coronavirus pandemic almost 95,000 of these patients were affected, with death of more than 25,923 or more due to comorbidities besides aging. During the phase of initial medical testing, the patients having multiple diseases simultaneously are a norm rather than an exception, especially those in nursing homes. Prevention and treatment of chronic diseases, as a priority project for the second decade of the 20th century, are meant to better the quality of the global population. Treating the comorbidities in the nursing homes and extended care facilities or the aged living at home may prevent many viral afflictions such as coronavirus and flu, thus reducing the morbidity and mortality. It is hard not to notice the absence of comorbidity in the taxonomy (systematics) of disease, presented in ICD-10.

The term "comorbid" has three definitions:

1. to indicate a medical condition existing simultaneously but independently with another condition in a patient (this is the older and more "correct" definition)
2. to indicate a medical condition in a patient that causes, is caused by, or is otherwise related to another condition in the same patient (this is a newer, nonstandard definition and less well-accepted) [Valderas, Jose M. et al. 2009. "Defining Comorbidity: Implications for Understanding Health and Health Services". Annals of Family Medicine. 7: 357–63. 2].
3. to indicate two or more medical conditions existing simultaneously regardless of their causal relationship. [Jakovljević M, Ostojić L 2013. "Comorbidity and multimorbidity in medicine today: challenges and opportunities for bringing separated branches of medicine closer to each other". Psychiatr Danub. 25 Suppl 1 (25 Suppl 1): 18–28].

Different types of comorbidity:

1. Polymorbidity
2. Multimorbidity
3. Multifactorial diseases
4. Polypathy
5. Dual diagnosis, used for mental health issues
6. Plural pathology.

Comorbidity in psychology

In psychology, **comorbidity** refers to more than one disorder or disease that exists alongside a primary diagnosis, which is the reason a patient gets referred and/or treated. You can remember the word because the prefix, co-, indicates that things go together, and when we think of serious illnesses, it can sometimes make us feel morbid. There are some diseases and disorders that are more likely than others to be comorbid with one another such as depressive disorders, anxiety disorders, schizophrenia, PTSD, concussion, traumatic brain injury, chronic traumatic encephalopathy, substance abuse and addiction including alcohol and such.

Chapter 63

Metabolic syndrome

What is metabolic syndrome?

In medicine, comorbidity denotes the presence of one or more additional health problems. We discuss metabolic syndrome here because it is a comorbidity if a patient becomes infected with coronavirus. Metabolic syndrome has become endemic now; it increases the risk of cardiovascular diseases, kidney disease, cancer, diabetes, neurodegenerative diseases, decline in cognition (dementia, Alzheimer's), muscular/skeletal afflictions, as well as increased morbidity and mortality. Metabolic syndrome increases the incidence of type II diabetes all over the world—425 million at present, and estimated to swell to 629 by the year 2045. We all know the answer to the metabolic syndrome: calorie restriction—eat less and exercise more.

A person must have **three of the five following conditions** to be diagnosed with metabolic syndrome:

1. Fasting glucose over **100 mg/dL**
2. Blood pressure **≥130/85 mm Hg**
3. Serum triglycerides **≥150 mg/dL**
4. HDL cholesterol ("good cholesterol") **<40 mg/dL** in men or **<50 mg/dL** in women (or drug treatment for any of the above)
5. Waist circumference **≥40 inches** in men or **35 inches** in women - belly fat (slightly lower measurements apply for Asian individuals)

We consider ***metabolic syndrome to be a self-made pandemic.*** One-third of all adults in the US had/have metabolic syndrome (Moore JX, Chaudhary N, Akinyemiju T. Metabolic Syndrome -Survey, 1988-2012. Prev Chronic Dis. 2017). If metabolic syndrome persists as we age, it becomes a comorbidity resulting in increased morbidity and mortality.

Outlines of how to control and/or not develop metabolic syndrome

One has to make sensible life style changes as described in another chapter. Studies defined 5 low-risk lifestyle longevity factors such as: never smoke, maintain body mass index, exercise ≥30 min/daily, moderate or no alcohol intake, a healthy diet.

Calorie restriction is a must to control syndrome X. Skip a meal, have a fasting day once a week and eat only 40% of the meal, don't go for seconds. Gandhiji was not only a sage of nonviolence, he also practiced nonviolence against eating by practicing calorie restriction and repeated fasting. I believe that eating excess food is a form of violence again nature and the needy. Adopt the above five life style changes and follow the following simple method:

- Adopt Mediterranean diet rich in Omega 3 fish such as mackerel, lake trout, salmon, herring, sardines, sturgeon eggs, tuna, blue fish, less or no wheat, rice and sugar. Avoid cola; two colas a day will increase morbidity and mortality.
- Calorie restriction is a must if you are overweight or have belly fat. Practice calorie restriction diet (CR - described in another chapter in another book) besides above measures. Try ketogenic diet also with low proteins.
- Intermittent rapamycin intake 3-4-5 mg once 7-10 days a week will be one of the most important therapeutic measures one can take to live longer and healthier besides CR and avoid most of the afflictions associated with metabolic syndrome.
- Take 2 BCG vaccinations intradermally two weeks apart. It will improve health by preventing or taming many inflammatory diseases including autoimmune diseases.
- Take 250-500 mg metformin after lunch and dinner even if you are not diabetic.
- Take Ebselen orally as an anti-viral, anti-inflammatory, anti-oxidative, bactericidal, anti=aging agent, with cell protective properties as described.
- Take 1000 mgs of bioactive curcumin every day and vitamin D3, 5000 IU. I get it from Life Extension Foundation of Florida.
- Take micro doses (300 – 100 micrograms) of lithium carbonate a day.
- Do whole body hyperthermia >3-4 times a month besides daily exercise 5 days a week. Get a mobile infrared sauna tent and use it every day if need be and if possible.
- Exercise > 30 minutes 5 days a week. Sleep 8 hours a day.
- Do not smoke, do not indulge in excessive alcohol drinking, maintain proper body weight, eat a Mediterranean diet with less saturated fats, sleep 7-8 hours a day, lead a stress-free life, and work in a clean environment.

Chapter 64

Dementia
Miscellaneous causes of memory loss

Many patients affected by the **coronavirus infection** resulting high morbidity and mortality are aged, in nursing homes and extended care facilities, who have some degree of cognition difficulties such as MCI, some stages of Alzheimer's and other co-morbidities. Hence, we want to discuss causes of dementia (mild to severe) and possible methods to prevent or delay or stop the progression in the home or nursing home setting, to improve the quality of life and live longer with brain and body intact.

What is dementia?

The dementia term is derived from the Latin for "away" and "mind" that describes a diversity of symptoms which stem from as many as dozens of brain disorders or diseases, from Alzheimer's to Parkinson's, from stroke to senility and many unrelated afflictions of the body and brain. All involve nerve cell degeneration or death in the brain. Dementia is a generic term that describes a progressive and irreversible loss of higher mental function, particularly memory. Cognitive decline is associated with other significant alterations in mood and behavior that lead to complete disintegration of the personality, which affects independent living and leads to the patient being more and more dependent on care givers.

The Journal of the American Medical Association defined dementia as ***"significant loss of intellectual abilities, such as memory capacity, that is severe enough to interfere with social or occupational functioning."*** According to ***alz.org,*** the word "dementia" describes as "a set of symptoms that include memory loss, difficulties with thinking, problem-solving or language." Patients may experience changes in their mood or behavior.

Dementia is a loss of mental ability severe enough to interfere with normal activities of daily living, lasting more than six months, not present since birth, and not associated with a loss or alteration of consciousness. Dementia is diagnosed only when both memory and other cognitive functions are each affected severely enough to interfere with a person's ability to carry out routine daily activities.

Our recommended preventive (prophylaxis) therapeutic measures to improve longevity and reduce dementia

There is no a treatment for dementia itself. Treating the symptoms depends on the etiology and the dominant symptoms themselves. Some forms of cognitive impairment are potentially treatable and should be included in the investigation of a patient with suspected dementia. These include nutritional deficiencies (e.g. vitamin B12, folate), endocrine disturbances (such as hypothyroidism), alcohol-related dementia (reversible with abstinence), syphilis (now rare in the developed world), depressive pseudodementia (treatable with antidepressant drugs) and normal pressure hydrocephalus treated with shunt.

Mild cognitive impairment (MCI) is said to lead to Alzheimer's, in most of the cases with passage of time, and hence can be labeled as a harbinger of Alzheimer's or other form of dementia. A number of drugs have been approved by the Food and Drug Administration to *treat the symptoms of Alzheimer's and other dementias, but not the etiology* of Alzheimer's. There are no such drugs on the market yet. The following are simple methods to prevent, curtail, or stop the progression of the diseases and/or to alleviate the dementia symptoms:

1. Take oral bioactive curcumin 1000 mg. Eat curcuminoids laced curry food, use curcumin mixed in mayonnaise, ketchup and mustard to hamburger meat, hot dog, roasting meat, jambalaya, or any and all vegetarian and non-vegetarian food preparation.
2. Eat every day a fist full of blue berries – a must, as a neuroprotective
3. Take vitamin D3, 5000 IU a day. It is a sunshine vitamin, regulated by the parathyroid gland in the neck, embedded in the thyroid gland. The U.S. Preventive Services Task Force recently issued new recommendations that adults at risk for winter blues should take vitamin D.
4. In addition, take Omega 3 supplement, almost 30% of the brain is made of omega 3 fatty acid DHA.
5. Take oral glutamine and progesterone. These are needed for aging and/or traumatic brain injuries (concussion, traumatic brain injury, chronic traumatic encephalopathy, PTSD) They reduce the neural inflammation of the brain due to myriads of etiologies which increases synaptic plasticity.

6. Take micro-doses of lithium (300-1000 micrograms) as soon as one notices cognitive impairment, for neurogenesis and for telomeres protection of chromosomes, which contain genetic code. I advise to take it past the age of 60.
7. Take metformin 250-500 mgs after each meal even if you are not diabetic and add Ebselen to it as described in the patent.
8. Take benfotiamine (thiamine, B1) up to 250 mg a day. It is a fat-soluble vitamin and absorbed efficiently from the gastro-intestinal track. It is needed for proper neuronal function and their membranes.
9. Take micro-doses of Memantine (Namenda, by prescription) to prevent excitotoxicity due to excess calcium entering nerve cells, that results in nerve cell degeneration
10. Take intermittent rapamycin micro-doses 3-4-6 mg once every 7-10 days once you are diagnosed with dementia and/or getting old with mental and physical disabilities or past the age 60-70-80s (read the chapter under rapamycin). It is the most effective anti-aging therapeutic agents there is in the market besides calorie restriction, lithium and metformin. It should be prescribed as the diagnosis of Alzheimer's is made and for those who are developing cognitive deficit (Majumder S, Caccamo A, et al., Lifelong rapamycin administration ameliorates age-dependent cognitive deficits by reducing IL-1beta and enhancing NMDA signaling. Aging Cell. 2012;11(2):326–335.).
11. Take angiotensin receptor blockers—ARBs (telmisartan and losartan) and Felodipine (a calcium channel blocker) if you have elevated BP and are developing memory loss. Felodipine will help to clean the nerve cell toxic deposits by swallowing and self-digesting (autophagy) such as Aβ plaques, alpha synuclein, tau protein, lipofuscin, Lewy bodies and such to reduce nerve cell degeneration and death, thus reducing chances of developing dementia along with above measures.
12. Practice whole-body hyperthermia or sauna 2-3-4 times a month or as often as you can. It removes the clogging of the brain blood vessels and augments the glymphatic to remove toxic neuronal metabolites, improves heart function, clears clogging in 60,000 miles of blood vessels that carry nutrients and oxygen.
13. Try to get 2 BCG vaccination intradermal 4 weeks apart if you can.
14. Control blood pressure, blood sugar, and obesity, quit smoking, drink no more than 2-3 alcoholic drinks a week if need be, adopt a

Mediterranean diet with millets replacing wheat, rice and sugar, eat omega 3 rich food (mackerel, herring, tuna, halibut, sturgeon eggs, salmon, cod liver) and control elevated cholesterol by simvastatin, daily exercise, good 8 hours of sleep, lower blood pressure with telmisartan and Niludipine (a calcium channel blocker that causes brain blood vessels dilatation) and reduce blood cholesterol and blood sugar between 85 -125 mg%.

Alzheimer's disease is responsible for 60-80% of all dementias.

Alzheimer's and some other forms of dementias with other comorbidities are responsible for filling nursing homes and extended care facilities. This type of dementia is caused by degeneration that also contains hippocampus memory center in the cerebral cortex that contains billions of nerve cells, the main part of the brain responsible for thoughts, memories, actions, senses, and personality. The death of brain nerve cell begins and continues in the memory centers (hippocampus) and rest of the brain, which leads to cognitive impairment ultimately to dementia. The cost of caring for those with Alzheimer's runs into billions with no treatment of the cause in sight, only symptomatic therapies. ***Now we have rapamycin to treat the etiology of Alzheimer's, not just the symptoms – think about it and prescribe it.***

Criteria for the diagnosis of dementia

Criteria for diagnosis of dementia include: impairment of attention, orientation, memory, judgment, language, motor and spatial (three dimensional) skills, and other routine function. These changes are often small to start with but become severe enough to affect daily life with lapse of time. By definition, *dementia is not due to major depression, schizophrenia or PTSD,* though these patients can all develop dementia as these disease advances.

Classification of types of dementia

1. **Primary dementia:** Alzheimer's (AD), Lewy bodies disease, frontotemporal dementias, Pick's disease falls into this category. They are not the result of any other diseases.
2. **Secondary dementia**: This is the result of brain injury such as concussions, Posttraumatic stress disorder (PTSD), traumatic brain injury (TBI), Chronic traumatic encephalopathy (CTE), multiple sclerosis, brain infection, tumors, post stroke, alcoholism and such.

3. **Progressive dementia:** Dementia progresses and becomes unmanageable such as senile dementia, vascular dementia, Alzheimer's disease, Lewy body dementia, frontotemporal dementia, and chronic alcoholic dementia
4. **Subcortical dementia:** This is due to problems in the part of the brain beneath the cortex (brain stem). The ability to start activities and speed of thinking are affected. Parkinson's disease, Huntington's disease and HIV can cause these types of dementia
5. **Cortical dementia:** Dementia is due to brain damage by disease to the cortex. The patient develops severe memory loss and has problems understanding language. Creutzfeldt-Jakob (CJD) and Alzheimer's disease (AD) are two types of cortical dementia.
6. **Combination dementia:** Dementia could be due to a combination of any of the above diseases.

Cortical dementias—etiology due to diseases affecting brain—cerebral hemispheres and their nerve cells and their connection within

There are 7 known major causes of dementia; most of the cases are from Alzheimer's disease. The list is as follows:

1. Alzheimer's disease (AD) is the number one cause of dementia and 6th leading cause of death in the US affecting 5.8 million (2020)
2. Vascular dementia is due to pathology in brain blood vessels, causing brain damage from large or multiple mini strokes, atherosclerosis, endocarditis, or amyloidosis, and such makes up 15-20% of dementia cases.
3. Lewy bodies dementia constitutes 10 to 25 percent of dementia cases. Lewy bodies are found in Alzheimer's and Parkinson's dementia also.
4. Leukoaraiosis—a white brain mater vascular related disease - multiple mini strokes due to atrial fibrillation and such afflictions.
5. Frontotemporal dementia also includes Pick's disease and many types are described.
6. Later stage of Parkinson's disease patients experience dementia.

A new study from the University of Kentucky (JAMA Nerology:6-28-20) has described a novel form of dementia characterized by the toxic accumulation of four different proteins amyloid-beta, tau and TDP-43, but also alpha synuclein, a protein known to aggregate into toxic structures called Lewy bodies. in the brain, named ***quadruple misfolded proteins dementia, or QMP*** says Abner of the University of Kentucky. The research suggests many patients diagnosed with Alzheimer's disease may be suffering from this different, more complex, and difficult to diagnose neurodegenerative condition.

Preventable dementia is due to traumatic cortical and subcortical brain injuries (TBI) that include Chronic traumatic encephalopathy (CTE) as seen in football players and boxers. Recent studies indicate that concussion, TBI, and CTE can lead to Alzheimer's disease, so also repeated head trauma in boxers.

Less common brain cortex (cortical) dementias are:

1. Alcohol-Induced Persisting Dementia
   a. Korsakoff's syndrome
   b. Wernicke's encephalopathy
2. Frontotemporal dementia - Frontotemporal lobar degenerations (FTLD - including Pick's disease)
   a. Frontotemporal dementia (or frontal variant FTLD—Pick's disease, it is the most common form of dementia below the age of 60)
   b. Semantic dementia (or temporal variant FTLD)
   c. Progressive non-fluent aphasia
3. Creutzfeldt-Jakob disease (CJD) - mad cow disease
4. Dementia pugilistica - otherwise known as "punch-drunk syndrome" or "boxer's dementia," originates with repeated concussions or other traumatic blows to the head.
5. Moyamoya disease - blood vessel (vascular) disorder in which the carotid arteries in the skull becomes blocked or narrowed, reducing blood flow to your brain resulting in lack of oxygen and nutrition to nerve cells
6. AIDS dementia complex due to damage of inflammatory cells (microglia) and neurons.
7. Uncontrolled chronic diabetes.
8. Dementia due to Cortico-basal degeneration owing to loss of brain cells in the cerebral cortex, and basal ganglion due to genetic mutation and aging

Subcortical dementias due to systemic and other etiology affecting CNS

1. Dementia due to Huntington's disease passed from parent to child, where nerve cells breaks down
2. Dementia due to Hypothyroidism

3. Dementia due to Parkinson's disease – late or end stage
4. Dementia due to vitamin B1, B6 deficiency; dementia due to vitamin B12 deficiency; dementia due to folate deficiency
5. Dementia due to vasculitis, a form of meningitis, syphilis, Lyme disease, or systemic lupus erythematosus (SLE) and such
6. Dementia due to Subdural / subarachnoid / intracerebral bleeding, emboli and hematoma—stroke,
7. Dementia due to Hyper-calcaemia—High blood calcium levels
8. Dementia due to low blood sugar—Hyperglycemia, diabetes related
9. Pseudo-dementia (associated with clinical depression and bipolar disorder)
10. Dementia due to multiple etiologies such as stroke, repeated concussions, Posttraumatic stress disorder (PTSD), traumatic brain injury (TBI), Chronic traumatic encephalopathy (CTE), hydrocephalus, multiple sclerosis, Amyotrophic lateral sclerosis (ALS), also known as motor neuron disease (MND), or Lou Gehrig's disease, brain tumors, post brain surgical procedures, and such.
11. Dementia due to other general medical conditions, i.e. end stage renal failure, cardiovascular disease, etc.
12. Early onset dementia has been associated with neurovisceral porphyria's, a genetic metabolic defect.
13. Post-Operative Cognitive dysfunction (POCD) after prolonged anesthesia for surgery in the aged with co-morbidities.
14. Dementia in children due to Niemann-Pick disease, Batten disease, Lafora body disease, and certain types of poisonings and brain damage during birth.
15. Substance-induced persisting dementia (related to psychoactive use and formerly known as Absinthism).
16. A form of dementia called MELAS for mitochondrial - encephalopathy lactic acidosis and stroke-like episodes.

Numerous other conditions can cause dementia, including:

1. Low oxygen in the blood (anoxia/hypoxia), either from a specific incident (heart attack, stroke, surgical complications, high altitude anoxia), or chronic disease (heart disease, asthma, COPD/emphysema, chronic obstructive sleep apnea) can cause damage to brain tissue. Hypoxia can result in elevated carbon dioxide built up leading to hypertension resulting in stroke, and stroke related death while sleeping.

2. Side effects from medications taken for other medical conditions such as chemotherapy, severe clinical depression.
3. Electrolyte abnormalities—severe dehydration or over hydration (excess of fluid intake).
4. Poisoning such as exposure to lead, heavy metals, with or without alcohol, recreational drugs, or other poisonous substances use.
5. A study published in the journal JAMA Internal Medicine (June 2019) links 50% odds of developing dementia for certain classes of anticholinergic drugs, particularly antidepressants, bladder antimuscarinics, antipsychotics and antiepileptic drugs.
6. Mental fog after Lyme disease transmitted by ticks.

Aged patients who recover from severe COVID-19 disease may not have the same mental ability as they had before. They may develop mild to serious memory problems.

Diagnostic tests for dementias

Mild cognitive impairment (MCI) may be a harbinger of future dementias development such as Alzheimer's and senile dementia due to multiple causative factors. There is no diagnostic test to diagnose dementia. There has been a major understanding of the cellular, molecular, genetic, and biochemical mechanisms of many chronic diseases over the last 6 decades including dementia due to neurodegenerative diseases such as Alzheimer's and Parkinson's disease. Chronic Inflammation of the brain and body plays a major role in neurodegenerative diseases including Parkinson's and Alzheimer's, cancer, cardiovascular diseases (CVDs), metabolic diseases such as diabetes, obesity, autoimmune diseases, pulmonary diseases, and it is even blamed for psychological illnesses including deep depression.

There is no definite test for diagnosis of dementia. It is diagnosed by exclusion, which means doctors rule out other conditions that can cause the symptoms that resemble dementia. Tests to diagnose dementia include:

1. Complete physical examination with exhaustive neurological testing
2. A useful tool in the assessment of a person with suspected cognitive decline (dementia) is the mini mental state examination (MMSE). It is a basic test of orientation, memory, attention, language and visuospatial ability. There are many other tests including computerized testing are available to test dementia.
3. Brain scans (CT scans, MRI, functional MRI (measures brain activity by detecting changes associated with blood flow) scans, special PET scans) —a must

4. Blood tests, urinalysis, toxicology screen, thyroid tests, inflammatory C reactive protein, homocysteine, fibrinogen, lymphocyte count and such should also be utilized
5. Cerebro-spinal fluid (CSF) analysis with microbial analysis with culture of CSF and blood
6. Psychiatric evaluation and more formal testing can be carried out by clinical psychologists and neurologists.
7. Genetic testing: It is now known that the activity and products of nearly ±25,000 different genes regulate the human brain and body function. It is becoming obvious that most diseases are caused by dysfunction (dysregulation) of multiple genes leading to chronic inflammation by their activity and the end results they create. It has been estimated that as many as 300 to 500 different genes may play a role or control leading to any given chronic illness including neurodegenerative diseases such as Parkinson's and Alzheimer's disease. It has opened new methods to diagnose and treat these chronic diseases targeting these dysfunctional genes and their end result of inflammation.
8. Biomarker: Neurofilament light chain (NFL) is a protein that is released as a result of brain cell damage. It is one of the most promising early-stage biomarkers for a variety of neurodegenerative diseases, including Parkinson's disease, ALS and multiple sclerosis. Elevated CSF NfL levels, compared to a control group, turned out to be the most prominent early sign of the disease. revealed increasing NfL levels could be detected at the early 22 years before symptom appear according to a study published in Lancet Neurology May 2020.
9. Even now, tests are being conducted using sensor embedded in the diamond core attached to a computer which analyses about 550 samples of blood to diagnose neurodegenerative diseases including Alzheimer's by passing through infrared light. We hope it is successful and becomes available to all.

Follow the therapies we describe under the chapter Aging and Inflammation (Chapters 60, 61) to prevent, curtain, stop the progression of and/or cure the afflictions.

(This chapter is from the forthcoming book on Alzheimer's and other dementias)

Chapter 65

Our gut to brain connection and Parkinson's

Route of transport of Parkinson's disease causative agents and such from the gastro-intestinal track to the brain

What is Parkinson's disease?

Parkinson's disease (PD) is a progressive neurodegenerative disease of the central nervous system that principally affects the motor system. The symptoms develop slowly over time. Early in the disease, the most obvious are shaking, rigidity, slowness of movement, difficulty with walking, an effect on thinking and behavioral problems, depression and anxiety with sleeping problems. Dementia sets with advancing stage of the disease. It is caused by degeneration of brain nerve cells in the midbrain susbstantia nigra that control movement due to host of possible etiological factors.

How many are affected with this incurable neurodegenerative disease?

Approximately 10 million people worldwide suffer from Parkinson's disease and as many as one million people in the United States according to Parkinson's disease foundation.

In 2003, Parkinson's disease was added to the top 15 causes of death; it entered the list as the 14th leading cause of death (Hoyert DL, National Center for Health Statistics; 2006).

What causes Parkinson's disease?

- Etiology of the condition is not known. But one change we see in the brain of the Parkinson's patient is the malfunction and death of nerve cells in the mid brain susbstantia nigra nerve cells.
- In Parkinson's disease (PD), gastrointestinal features are common and often precede the motor signs such as constipation, gut inflammation, increased gut membrane permeability, dysbiosis of microbiomes and such. In the early 2000s, German scientist Heiko Braak and colleagues proposed that PD may start in the gut, triggered by a pathogen, and spread to the brain. They hypothesized that non-

inherited forms of PD are caused by a pathogen that can pass through the mucosal barrier of the gastrointestinal tract and spread to the brain through the enteric nervous system which we explain based on our research.

- A new study has strengthened Braak's proposition by identifying a type of overabundant pathogen in the guts of Parkinson's disease patients, a novel finding that opens up new lines of enquiry in understanding the root causes of the condition (Zachary D. et al.2020). The research looked at 520 cases of Parkinson's and more than 300 controls, making up what the authors describe as the largest microbiome-wide association study of the disease to date and its relation to Parkinson's.
- In 2005, researchers reported that people with Parkinson's disease who had these proteins called alpha synuclein clumps in their brains also had the similar clumps in their guts.

Whether it is a microbiome or other components or alpha synuclein from the gut, how do they reach the brain to cause this disabling disease? Our decades of studies might have solved part of the conundrum.

How do the causative factors of Parkinson's travel from the gut to the brain?

The alpha synuclein type proteins found in Parkinson's-causing nerve cells had been found in the gut lumen, or the space inside the gastrointestinal tract, but "nerves are not open to the lumen," said gastroenterologist Dr. Rodger Liddle, senior author, appearing June 15 in the journal JCI Insight, and professor of medicine at Duke University in North Carolina. Their study shows that alpha-synuclein is able to travel quite far through the body, passing from one neuron to another and using long nerve fibers as conduits.

Our histological findings (see References below) show that diseases associated with alpha-synuclein and microbiome and other micro molecules components can originate in the gut and travel along the nerve sub perineural epithelial space described by Drs. T. R. Shantha, and G. H. Bourne 5 decades ago from Emory University (Figure 1 and references below and in biodata). Our studies show that there is direct causeway connection between the endings of the vagus nerve (as well as sympathetic chain) on the gut wall and brain through sub perineural epithelial space (diagram and References below) centripetally that acts as a route that is taken by these etiological factors (alpha synuclein, microbiomes particulates and other biological micro molecules), of the Parkinson's disease spread (Ulusoy A, et al.).

A large Danish epidemiologic study supports the theory that Parkinson's disease (PD) may begin in the gastrointestinal tract and spread through the vagus nerve to the brain. Researchers found that patients who have had the entire vagus nerve cut (severed) were less apt to develop PD. "Their risk was halved after 20 years," Elisabeth Svensson, PhD, from the Department of Clinical Epidemiology, Aarhus University in Denmark, said in a statement. "However, patients who had only had a small part of the vagus nerve severed were not protected which supports our finding that the causative agents of Parkinson's disease are transported and transmitted under the sub perineural epithelial and inter-axonal space, not directly through axoplasm. This also fits the hypothesis that the disease process is strongly dependent on a fully or partially intact vagus nerve to be able to reach and affect the brain," Dr. Svensson noted.

These findings, say the researcher, suggest that *having an intact vagus nerve increases the risk of developing Parkinson's disease,* and they need the intact sub perineural epithelial space to be free to conduct it to the brain as shown in our research. The finding is in accord with a primary pathological process being initiated in the gastrointestinal mucosa, which then uses the vagus as a major entry point into the brain. "This study is an "important piece of the puzzle" in terms of the causes of Parkinson's," she told Medscape Medical News. The study was published online July 17 in Annals of Neurology.

In the brain, the misfolded protein molecules stick to each other and clump up, damaging neurons, and some of them may form Lewy bodies. The study also reveals a preferential route of alpha-synuclein transmission via the vagus nerve. Our studies also show the therapeutic agents, drugs, rabies virus, coronavirus, meningococcus, and brain eating amoeba are transported through the sub perineural epithelial space around the 20 olfactory mucosal nerves directly connected to the olfactory bulb, brain, and cerebro-spinal fluid, thus by-passing the blood brain barrier (T.R. Shantha, 2017).

Vagus nerve acts as a pipe line to conduct alpha synuclein and such from gut to brain based on our study of nerve membranes

The vagus nerve is composed of two kinds of fibers: 20% of fibers are "efferent" whereas the remaining 80-90% are "afferents." Efferent fibers are projections of motor neurons that control the gut's motility. In contrast, afferent fibers relay sensory input from the gut to the brain.

As the gut is constantly moving (gut motility), the alpha synuclein molecules or their precursor microbiomes particulate matter find their way to nerve endings, Virchow-Robin space and to sub perineural epithelial space found in in all peripheral cranial and spinal nerves including sympathetic nervous system and reach the brain to cause Parkinson's and other neurodegenerative diseases.

Studies by Holmqvist et al. (2014) demonstrate the first experimental evidence that different α-synuclein forms can propagate from the gut to the brain, and that microtubule-associated transport is involved in the translocation of aggregated α-synuclein in neurons. I believe that sub perineural epithelial and inter axonal space plays an important role compared to microtubules.

Braak and coworkers (2006) hypothesized that Lewy pathology primes in the enteric nervous system and spreads to the brain, suggesting an active retrograde transport of α-synuclein (the key protein component in Lewy bodies), via the vagal nerve. Our sub perineural epithelial space plays a major role in its transport.

Appendix and Parkinson's disease connection

Latest studies show that those who have their appendixes removed in young adulthood run a nearly 20 percent lower risk of developing Parkinson's decades later, according to a study published in October 2018 in Science Translational Medicine. They also found that those who had undergone appendectomies developed the disease 3.6 years later on average than those who still had an appendix. They also found the protein alpha-synuclein, which is misshapen in Parkinson's patients, found in the appendixes of 48 out of 50 healthy people—suggesting the protein may play a useful (but still unknown) function there. According to Labrie, who reported the findings, the deformed clumps of alpha-synuclein might travel up the vagus nerve, which connects the digestive system to the brain, and then seed the brain with this destructive protein and block the production of the chemical dopamine, causing the tremors and stiffness that define Parkinson's. Our published studies (Shantha and Bourne 1962-2016) over decades show that there is space around the vagus nerve that can act as a transportation route connecting the gut to the brain. Note that the appendix is densely innervated compared to the rest of the intestine making its association. The appendix has high levels of alpha synuclein in the richly supplied nerve fibers, and develops inflammation often from the adjacent gut, so we conclude that the appendix may have some role in Parkinson's disease etiology.

Conduction of hunger control hormones, enzymes, and molecules to the brain by sub perineural epithelial space

There are many hunger-inducing and hunger control biologics besides various neurotransmitters. Without going into detail, some of these protein enzymes, hormones and neurotransmitters from the gut wall are transported to the brain through sub perineural epithelial space described below. Centrally acting hunger-weight control biologics transported from the gut wall are: Lorcaserin (5-HT2C agonist), Bupropion/Naltrexone (Contrave) opioid antagonist and amine uptake inhibitor, Phentermine/Topiramate an Adrenergic and amine, Bupropion/Zonisamide and Tesofensine of uncertain action, Y5 antagonist: Inhibition of NPY Y5 receptor, Y4 antagonist-Y4 antagonist, CB1- antagonist; Gut hormone-based therapies such as Intranasal PYY- a PYY agonist, Metreleptin/Pramlintide - a Leptin and amylin agonist OAP- Oxyntomodulin antagonist, and Glucagon/GLP-1 agonist - a Glucagon/GLP-1 agonist.

Most medications in late-stage clinical trials are centrally acting drugs that interfere with the neurocircuitry involved in appetite and reward pathways. You can see from the above list how the gut influences the central nervous system by transporting them through ***sub perineural epithelial space*** of the vagus nerve and sympathetic chain (see Figure 1 below).

Flow route from gut to brain and from brain to gut

Research is supporting how our body functions, organs connected to our body functions, and nutrients to keep them working all the time are all interconnected. The GI (gastro-intestinal) tract is one example where researchers have documented communication between the GI tract and the brain as shown in the diagram below in our studies due to retrograde and antegrade flow of nutrients, enzymes, hormones, neurotransmitters, protein molecules, trophic factors, microbiomes components, alpha synuclein, seeping cerebro- spinal fluid, and such from brain to gut and from gut to brain. We now know that the brain and gut inter communicate constantly through coming and going traffic along the nerve tracts and sub perineural epithelial spaces around the nerve fibers (see Figure 1 below). Gut neurons and sympathetic chain may also play a similar role.

In more than 5 decades of work, Dr. T. R. Shantha and Dr. G. H. Bourne showed that the entire vagus nerve and sympathetic chain and all other cranial and spinal nerves from their exit from the brain stem and spinal cord to their end on the gut wall are covered by perineural epithelium continuously with the pia-arachnoid membranes of the brain and spinal cord.

We also found that there is sub perineural epithelial space (which can be easily called a third organ as described recently) below this covering, which is a potential space with micro-tissue fluid and micro-molecules to travel to reach the brain and spinal cord through cerebrospinal fluid. This ***sub perineural epithelial space also communicates with inter-axonal spaces around axons.***

Further, the authors also showed the Virchow-Robin space in the peripheral nerve as seen in the CNS. Hence any macromolecules whether they are alpha synuclein or immune products or brain derived nerve growth or neurotransmitter from the periphery are transported to the subarachnoid space. From sub perineural epithelial space, and cerebrospinal fluid in the subarachnoid space, the substances are distributed to substantia nigra and tegmental nuclei that produce dopamine and other cranial and spinal nuclei.

Once in CSF, alpha synuclein and microbiomes particulate as well as other neurotransmitters, enzymes, hormones, even the digested broken-down nutrients, and such travel to Virchow-Robin space of brain, then they are transported to glymphatic and finally reach the neurons. The alpha synuclein produce Lewy bodies as seen in Parkinson's disease and may even contribute to the formation of Aβ plaque and tau proteins of Alzheimer's. That is why when the vagus nerve is cut, the incidence of Parkinson's disease decreases drastically because the ***sub perineural epithelial space of Shantha and Bourne*** is connected to subarachnoid space and brain stem nuclei is interrupted. The causative factors from the gut wall cannot travel to the brain. No alpha synuclein and other micromolecules that enter the nerve endings are conducted centrifugally from brain to gut wall, if any it is minimal.

Only neurotrophic factors, neurotransmitter and electrical impulses flow from nerve cells through the axon to reach the end of the axon as it splits and supply the gut wall through their endings.

Figure 1 below shows show how the alpha synuclein and micro molecules of microbiomes and such from the gut wall are conducted from the nerve ending opening on the gut wall through sub perineural epithelial and inter-axonal spaces on nerve fasciculi to vagus nerves and massive sympathetic nerves, then reach the cerebro-spinal fluid of subarachnoid space. From cerebro-spinal fluid, they travel to enter the neurons passing through the Virchow-Robin space and glymphatic to the brain stem, susbstantia nigra in the midbrain, cerebral cortex and spinal cord, based on our published studies as described in the below references.

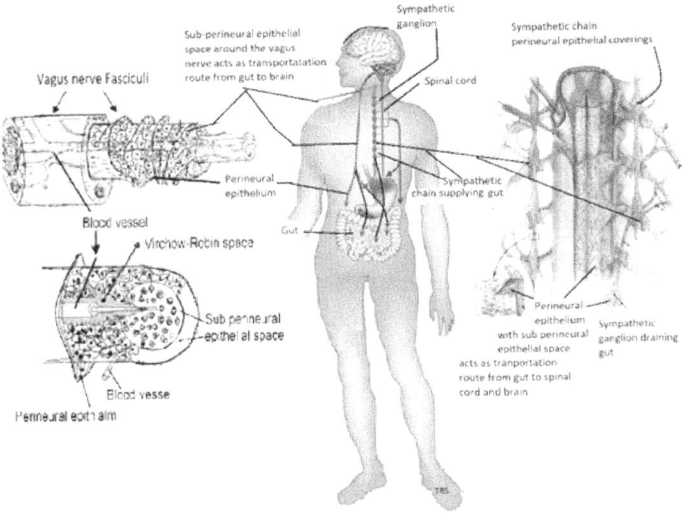

Figure 1

References

1. Baer G, Shantha TR and Bourne GH: The pathogenesis of street and fixed rabies virus in rats. Bulletin World Health Org, 38(1):119-125 (1968).33:783-794 (1965).
2. Braak H, de Vos RA, Bohl J, Del Tredici K (2006) Gastric alpha-synuclein immunoreactive inclusions in Meissner's and Auerbach's plexuses in cases staged for Parkinson's disease-related brain pathology. Neurosci Lett 396:67–72
3. Braak H, Rub U, Gai WP, Del Tredici K (2003) Idiopathic Parkinson's disease: possible routes by which vulnerable neuronal types may be subject to neuroinvasion by an unknown pathogen. J Neural Transm 110:517–536
4. Freundt EC, Maynard N, Clancy EK, et al. (2012) Neuron-to-neuron transmission of alpha-synuclein fibrils through axonal transport. Ann Neurol 72:517–524.
5. Lebouvier T, Chaumette T, Paillusson S, et al. 2009) The second brain and Parkinson's disease. Eur J Neurosci 30:735–741
6. Lee HJ, Suk JE, Lee KW, Park SH, et al. Transmission of synucleinopathies in the enteric nervous system of A53T alpha-synuclein transgenic mice. Exp Neurobiol 20:181–188.

7. Phillips RJ, Walter GC, Wilder SL, Baronowsky EA, Powley TL (2008) Alpha-synuclein-immunopositive myenteric neurons and vagal preganglionic terminals: autonomic pathway implicated in Parkinson's disease? Neuroscience 153:733
8. Rey NL, Petit GH, Bousset L, Melki R, Brundin P (2013) Transfer of human alpha-synuclein from the olfactory bulb to interconnected brain regions in mice. Acta Neuropathol 126:555–573.
9. Shantha TR and Bourne GH. Nature and origin of peri-synaptic cells of the motor end plate. Int Rev Cytol 21:353-364, 1967 (GH Bourne and JF Danielli, edited.) Academic Press, New York.
10. Shantha TR and Bourne GH. A Perineural epithelium. J Cell Biology, Rockefeller University NY, 14(2):343-346 (1962).
11. Shantha TR and Bourne GH. Arachnoid villi in the optic nerve of man and monkey. Expt Eye Res 3:31-35 (1964).
12. Shantha TR and Bourne GH. Demonstration of Perineural epithelium in vagus nerves. Acta Anatomica 52:95-100 (1963).
13. Shantha TR and Bourne GH. Demonstration of Perineural epithelium in whale and shark peripheral nerves. Journal Nature 197, 4868:702-703 (1963).
14. Shantha TR and Bourne GH. New observations on the structure of the Pacinian corpuscle and its relation to the Perineural epithelium of peripheral nerves. American Journal of Anatomy 112(1):97- 109 (1963).
15. Shantha TR and Bourne GH. Perineural epithelium: A new concept of its role in the integrity of the peripheral nervous system. Science 154:1464-1467 (1966).
16. Shantha TR and Bourne GH. The "Perineural Epithelium": A new concept. Its role in the integrity of the peripheral nervous system. Structure and Function of Nervous Tissues. Volume I. pp 379-458 (GH Bourne, Ed.) Academic Press, New York. 1969.
17. Shantha TR and Bourne GH. The `Perineural epithelium', a metabolically active, continuous, protoplasmic cell barrier surrounding peripheral nerve fasciculi. J Anatomy Lond 96(4):27-37 (1962).
18. Shantha TR and Bourne GH. The effects of transection on the nerve trunk on the Perineural epithelium with special reference to its role in nerve degeneration and regeneration. Anatomical Record (US) 150(1):35-50 (1964).

19. Shantha TR and Bourne GH. The Perineural epithelium of sympathetic nerves and ganglia and its relation to the pia-arachnoid of the central nervous system and Perineural epithelium of the peripheral nervous system. Zeitschritc fur Zellforschung 61:742-753 (1964).
20. Shantha TR and Bourne GH. The Perineural epithelium: and significance. Journal Nature 199, 4893:577-579 (1963).
21. Shantha TR and Evans JA. Arachnoid Villi in the Spinal Cord, and Their Relationship to Epidural Anesthesia. Anesthesiology 37:543-557, 1972.
22. Shantha TR, Hope J, and Bourne JH. Electron microscopic demonstration of the perineural epithelium in rat peripheral nerve. Acta Anatomica 52, 3:193-201 (1963).
23. Shantha TR, Nakajima Y. Histological and histochemical studies on the rhesus monkey (Macaca Mulatta) olfactory mucosa. Yerkes Regional Primate Research Center, Emory University, Atlanta, Georgia. Zellforsch. 1970; 103:291-319.
24. Shantha TR. Alzheimer's disease treatment with multiple therapeutic agents delivered to the olfactory region through a special delivery catheter and iontophoresis. US20120323214, US20140012182, US20150080785, WO/2015/013252A1, and WO/2009/149317A3
25. Shantha TR. CNS Delivery bypassing BBB: Drug Delivery from the Olfactory Mucosa to CNS. Drug Development and Delivery, 2017, 1, 32-37.
26. Shantha TR. Peri-vascular (Virchow-Robin) space in the peripheral nerves and its role in spread of local anesthetics. ASRA Congress at Tampa. Regional Anesthesia. March-April;1992.
27. Shantha TR. Presented at Hanoi: Rabies Cure: Nasal and Oral Route of Transmission of Rabies Virus and Possible Treatment to Cure Rabies. Rabies in Asia conference in Hanoi (RIACON) and at Ottawa. September 10, 2009. US Patent Application Publication Number: 201110020279 Al. Jan. 27, 2011.
28. Shantha TR. The Relationship of Retro Bulbar Local Anesthetic Spread to the Neural Membranes of the Eye ball, Optic Nerve and Arachnoid Villi in the Optic Nerve. Anesthesiology 73 (No 3A): A850 (1990).
29. Shantha TR. Route of transport stem cells, meningococcus, brain eating amoeba, and viruses form olfactory mucosa to the brain by

passing blood brain barrier - Drug development and delivery – under preparation 2020.

30. Staffan Holmqvist, et al. Direct evidence of Parkinson pathology spread from the gastrointestinal tract to the brain in rats. Acta Neuropathologica volume 128, pages805–820(2014).
31. Svensson E. et al. Vagotomy and subsequent risk of Parkinson's disease. Ann.Neurol. 78, 522–529 (2015).
32. Ulusoy A, Rusconi R, Perez-Revuelta BI, et al. (2013) Caudo-rostral brain spreading of alpha-synuclein through vagal connections. EMBO Mol Med 5:1051–1059.
33. Zachary D. et al. Characterizing dysbiosis of gut microbiome in PD: evidence for overabundance of opportunistic pathogens. npj Parkinson's Disease (2020) 6:11; https://doi.org/10.1038/s41531-020-0112-6.

Chapter 66

Olfactory mucosa and nerves: transportation route to the brain Transportation of various therapeutic, pharmaceutical, biochemical, and biological agents or compounds and micro-organisms including viruses and amoeba from olfactory mucosa to the brain

Route of transport of drug, stem cells, viruses, microorganisms and amoeba from olfactory mucosa of the nose to the brain

In the January 2017 issue of Drug Development and Delivery, Vol 17, No 1, pages 32-37, we published how the therapeutic agents from olfactory mucosa are transported to the brain without blood brain barrier effects through sub perineural epithelial space of 20 olfactory nerves derived from the olfactory mucosa to treat neurodegenerative diseases such as Alzheimer's and Parkinson's and other CNS diseases to replace lost and dying nerve cells and neural circuits in the damaged and/or degenerating CNS. This route of ***sub perineural epithelial space*** of Shantha and Bourne around the perineural epithelial around the olfactory nerves (and around all peripheral nerves also) from the olfactory mucosa acts as a route of transport for various therapeutic agents to treat neurodegenerative diseases, Alzheimer's and Parkinson's, as well as deliver stem cells and transport other biological agents including meningococcus, rabies virus, coronavirus, brain eating amoeba, stem cells and such.

Now, there is evolution of treating CNS diseases using stem cell replacement therapies to replenish, restore, and return the neurons, glial cell, and neuropils to curtail and cure these diseases. We want to explain, in light of various published and our own studies, the route taken by the stem cell introduced into the olfactory mucosa as therapeutic cells, including adult stem cells, immune cells (Treg, CAR-T, etc.) genetically-engineered cells, and mesenchymal cells. These can be delivered to the brain using the noninvasive intranasal olfactory mucosa delivery method along with insulin deposited on the olfactory mucosa to treat Alzheimer's and Parkinson's and other neurodegenerative diseases. The same method can be used to treat Alzheimer's and Parkinson's, and other neurodegenerative diseases. Our studies describe how exactly these stem cells and biological agents, as well as drugs, reach their destination from olfactory mucosa to be seeded in the afflicted neuronal centers to exert therapeutic effects.

Figure 1 below shows the olfactory mucosa which occupies only 3% of nasal cavity and other nerve structures on the walls of the nasal cavity. Note the olfactory mucosa (circled) is the solitary structure that is directly exposed on which the stem cells, are deposited. This is also where therapeutic agents, microorganisms, coronaviruses and amoeba are deposited and then conducted rapidly to the CNS by 20 olfactory nerves to the olfactory bulb, subarachnoid space CSF; from where they are transported to the CNS, bypassing the BBB. (Modified from Gray's Anatomy).

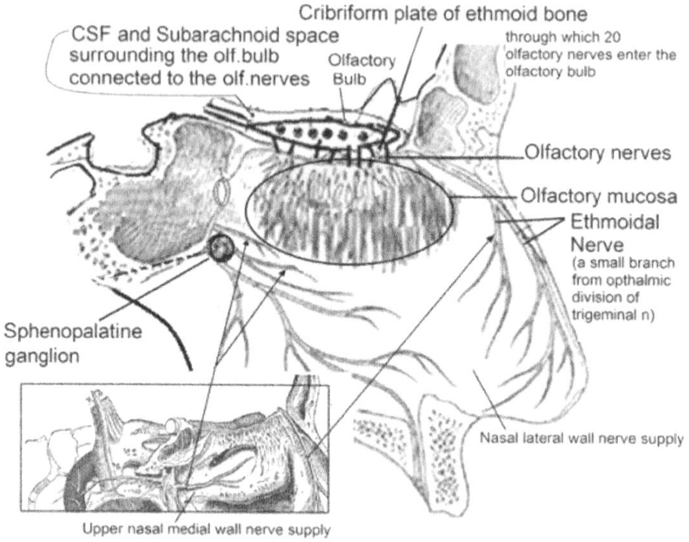

Figure 1

Figure 2 below shows the drawing of olfactory mucosa and the spaces and connections between the olfactory receptor cells that allow the stem cells as well as therapeutic agents, microorganisms, viruses, and brain-eating amoeba to be transported to the CNS through the olfactory nerves sub perineural epithelial space, to the olfactory bulb and CSF through these intercellular spaces. The dying receptor cells, 10% of them at any given time, also create tissue spaces with seeping CSF that acts as transportation route also to spread the stem cells and such. (Diagram modified from Graziadei 1971. I visited his lab, and met him in the 1960s at Tallahassee, FL, where I gave a talk on the perineural epithelial and the space it creates along with Dr. Jackson of Emory ENT dept.)

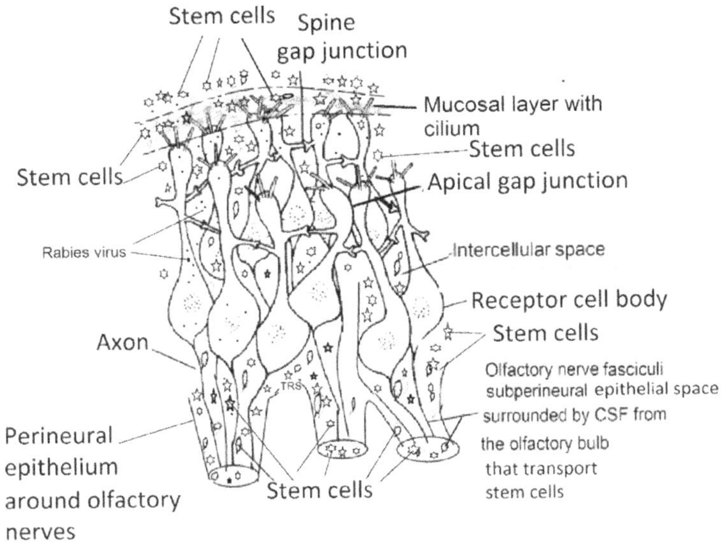

Figure 2

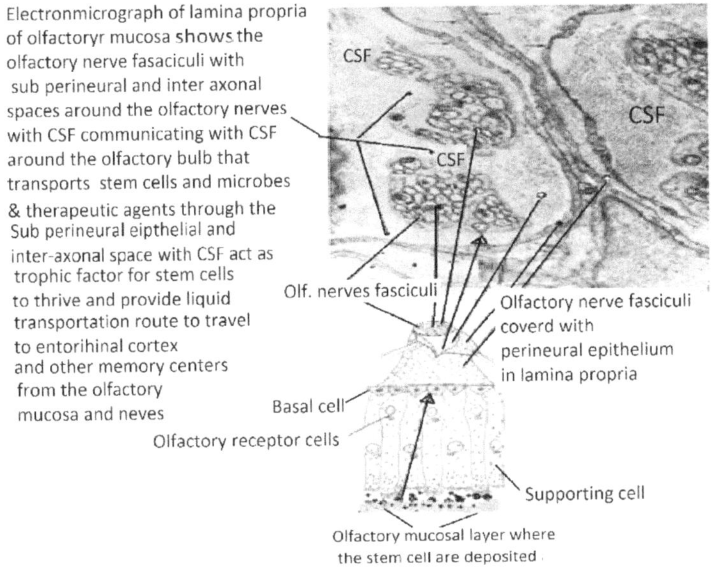

Figure 3

Figure 3 shows the olfactory mucosa and electron micrograph of olfactory nerves surrounded by multiple layers of perineural epithelium, which show distinct ***sub perineural epithelial and inter axonal space of Shantha and Bourne around the olfactory nerves*** though which the stem cells and other biological agents, including therapeutic agents, are transported and spread. These are transported and reach the olfactory bulb and the cerebro-spinal fluid in the subarachnoid space surrounding the olfactory bulb, CSF cisterns, cortical memory and dopaminergic centers, additional cortical centers, brain stem and spinal cord to treat Alzheimer's disease, Parkinson's disease (dopaminergic neurons), MS, ALS, stroke, CJD, and other neurodegenerative and CNS diseases.

The stem cells on the olfactory mucosa are primarily transported to olfactory nerves between receptor cells, then reach the sub perineural epithelial space and then on to olfactory bulb neuropil, and subarachnoid space CSF surrounding the olfactory bulb for further transportation and seeding through CSF to the rest of the neuronal centers (after Shantha and Nakajima 1970).

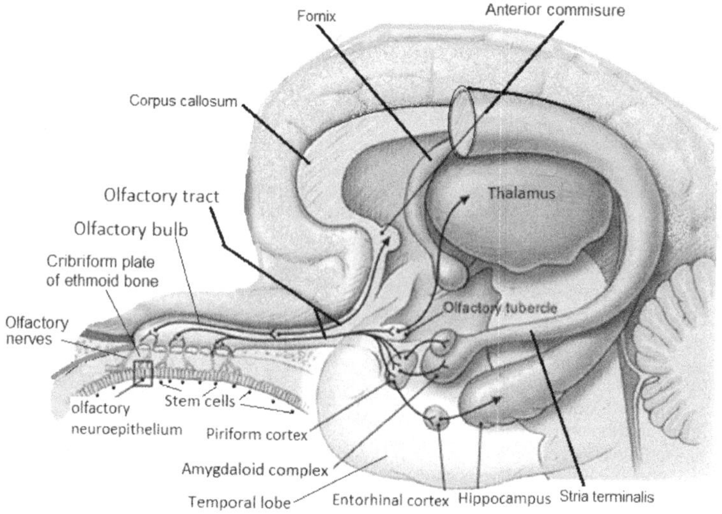

Figure 4

Figure 4 shows the route of transport of stem cells and various biological and therapeutic agents from the olfactory mucosa to various memory centers (labeled) and other areas of the brain, bypassing the BBB through ***sub-perineural epithelial and inter-axonal spaces*** to CSF content around the olfactory bulb and subarachnoid space as well as transported around the olfactory glomeruli, to olfactory tracts connected brain memory centers (labeled). These centers may receive stem cells from the subarachnoid space CSF from the suprachiasmatic and interpeduncular cisterns, Virchow-Robin space, paravascular, para axonal fibers, and glymphatic as well.

We conclude that the olfactory mucosa, olfactory epithelium, olfactory nerves, *sub-perineural epithelial and inter-axonal spaces* in the olfactory nerves, olfactory bulb, and the olfactory bulb surrounding CSF, olfactory tracts, suprachiasmatic and inter-peduncular cisterns, Virchow-Robin space, and glymphatic system (glymphatic clearance pathway) are the main necessary CSF fluid highways for the direct transport of therapeutic agents, stem cells, microorganisms, viruses including coronavirus and rabies, and amoeba as well biological agents, to the CNS, bypassing the BBB.

References

Baig AM. Pathogenesis of amoebic encephalitis: are the amoebae being credited to an "inside job" done by the host immune response? Acta Trop. 2015;148: 72-66.

Danielyan L, et al., Therapeutic efficacy of intranasally delivered mesenchymal stem cells in a rat model of Parkinson disease. Rejuvenation Res, 2011. 14(1): p. 3-16. Exp Neurol, 2014. 261: p. 53-64.

Morrison EE, Costanzo RM. Morphology of the human olfactory epithelium. *J Comp Neurol* 297(1):1-13, 1990.

Oppliger B et al. Intranasal Delivery of Umbilical Cord-Derived Mesenchymal Stem Cells Preserves Myelintation in Perinatal Bran Damage. Stem Cells and Development. 2016. 25(16):1234-1242.

Shantha TR. Alzheimer's disease treatment. US20120323214, US20140012182, US20150080785, WO/2015/013252A1, and WO/2009/149317A3.

Shantha TR. CNS Delivery bypassing BBB: Drug Delivery from the Olfactory Mucosa to CNS. Drug Development and Delivery, 17, 1, 32-37.

Shantha TR. Presented at Hanoi: Rabies Cure: US Patent Application Publication Number: 201110020279 Al. Jan. 27, 2011.

Shantha TR, Nakajima Y. Histological and histochemical studies on the rhesus monkey (Macaca Mulatta) olfactory mucosa. Yerkes Regional Primate Research Center, Emory University, Atlanta, Georgia. Z. Zellforsch. 1970; 103:291-319.

Shantha TR, Bourne GH. Perineural epithelium; structure and function of nervous tissues. Academic Press. 1969;1: 379-458.
Shantha TR, Bourne. G. The perineural epithelium: and significance. J Nature. 1963; 4893:577-579
van Velthoven, C. et al. Mesenchymal Stem Cell Transplantation Attenuates Brain Injury After Neonatal Stroke. Stroke 2013. 44:(5):1426-1432.
Wei Z, et al. Intranasal Delivery of Bone Marrow Mesenchymal Stem Cells Improved Neurovascular Regeneration and Rescued Neuropsychiatric Deficits After Neonatal Stroke in Rats. Cell Transplantation, 2015. 24: 391-402.

Chapter 67

Five life style changes and calorie restriction to prevent diseases and live longer

What is health?

The World Health Organization (WHO) defines health as **"a state of complete physical, mental and social well-being and not merely the absence of disease or infirmity."** For hundreds of years, people have been looking for the fountain of youth and to live a long life without diseases, from Ramayana epic from the Himalayas, to Ponce de Leon traveling to Florida. Ultimate findings are that there is no one drug or herb or drink that endows good health and longevity. It is life style changes along with the type of food we eat which can increase longevity/health span. These are important measures one should follow to turn the biological clock back.

Immaterial of age, adopt life style changes to have disease-free functional life and prevent early death

The latest study shows that the ability to live longer, disease-free, is in your own hand and costs nothing but discipline of following what is commonly called life style changes. These are 5 critical life styles that increase such longevity - health span, based on a study of more than thousands of people's life style methods to estimate life expectancy by levels of the lifestyle score starting from the age 50 ending at age 105 over a period of 34 years. (Li. Yanping, Pan An, Wang DD, et al. Impact of Healthy Lifestyle Factors on Life Expectancies in the US Population. Circulation. 2018; 137:345-355. Faloon W, Editorial, Life extension July 2019 pp 7-10). Adopting a healthy lifestyle could substantially reduce premature mortality and prolong life expectancy not only in the US but also all around the world. These are a must in people who are in nursing homes and extended care facilities to prevent comorbidities or reduce their effect. This would add ±10 years to your life if practiced. Li et al. defined 5 low-risk lifestyle longevity factors:

1. Never smoke (I also add never use electronic cigarettes, chew tobacco or use snuff.)

2. Maintain body mass index of 18.5 to 24.9 kg/m2 (body weight), no fat belly
3. Minimum or more than 30 min/daily moderate to vigorous physical activity – exercise
4. Moderate alcohol intake, (5-15 grams a day for woman and 5-30 grams of alcohol/day for men, that is one glass of wine, one beer, or a shot of hard liquor). I advocate green tea, curcumin milk, ginger drinks instead of alcohol.
5. A healthy diet with least animal fats, no margarine, rich in leafy vegetables (Mediterranean diet). We advocate to reduce wheat, rice and sugar—replace it with millets, add Omega 3 rich fish (mackerel, lake trout, salmon, herring, sardines, sturgeon eggs, tuna, blue fish) to the meal, replace some traditional hamburger, red meat, hotdog or pizza with meat and add vegetable or lean meat curries that include curcumin.

We add to the above list of life style changes, other longevity factors:

1. Maintain normal blood pressure (below120/80) with a stress-free life and work in dust free environment
2. Maintain blood sugar levels below 100-85 mg%.
3. Induce whole body hyperthermia >3 times a month using sauna, hot water in a tub or Jacuzzi, far infrared methods and such. Bring the body temperature to ±99.5-100.00 for at least 30 minutes
4. As you age, take micro-doses of lithium carbonate, mini-doses of metformin, and micro-doses of rapamycin as described in another chapter with one to two weeks break after 6-8 weeks dosing.

Adopting a healthy lifestyle can substantially reduce age and/or age-related diseases, premature mortality and prolong life expectancy in U.S. adults and become par with the other industrial nations. Adopting right life style changes can add ±5-10 years to your life. The researchers concluded that these 5 lifestyle changes could prolong life expectancy at age 50 years by 14.0 and 12.2 years for female and male US adults compared with individuals who adopted zero low-risk lifestyle factors. Lifestyle changes increased the life span up to 49% compared to those who did not, and those who did had 74% lower risk of dying from any illnesses (comorbidities) during life span.

Prevention of diseases should be a highest priority for national health policy, and preventive care should be an indispensable part of the US healthcare system to prevent illnesses and live healthy past 70-80-90s and at the same time save billions of dollars for the health care system. This will also save money by people being able to work longer, not spending money for hospital stay and for drugs. It will also improve a good family life. It has been widely acknowledged that unhealthy lifestyles are major risk factors for various chronic diseases and premature death. I add this may lead to hardship and poverty (Behrens G, Fischer B, Kohler S, Park Y, et al. Healthy lifestyle behaviors and decreased risk of mortality in a large prospective study of US women and men. Eur J Epidemiol. 2013; 28:361–372). The recent coronavirus pandemic has proved that unhealthy lifestyle related comorbidities lead to increased morbidity and mortality.

Beside adopting the above life style changes, I add other factors to be healthy, and live long past 80-90-100 years of age

All individuals have control in terms of what they can do to improve to live a disease-free, healthy, good functional, long life (health span) in those older years.

1. Diet: In addition to the Mediterranean diet and recommendations listed above, change to curcumin infused curried food, eat blueberries up to ¼-½ cup every day. Calorie restriction: weekly skipping meals or fasting a day or two is an important part of longevity. Avoid colas and sugary drinks; instead drink water with the meal.
2. Do not work in a dusty environment, if one has to use dust prevention protection. Avoid living next to busy expressway. Avoid stress, sleep 8 hours a day, practice yoga and meditation.
3. Take 2 intradermal BCG vaccines 4 weeks apart if you can get it. Studies show that that the BCG immunization was associated with major reductions in all-cause of mortality (nonaccidental deaths) by 20% to 60% including acute lower respiratory diseases. BCG has a remarkable record of safety over many decades of use in billions of children (Faustman, D. L. edited. The Value of BCG and TNF in Autoimmunity. 2nd edit, Academic Press, 2018, chapter 1, page 5). It is a must if one is suffering from diabetes type I, even if type II insulin dependent.
4. BCG vaccination and Rapamycin, are **contraindicated** in immune suppressed patents such as HIV-AIDS patients and those on immunosuppressants after an organ transplant, patients undergoing chemotherapy and radiation, and catechetic patients such as with anorexia nervosa.

5. Take once weekly mini dose of intermittent rapamycin (Sirolimus, Rapamune ®) - 3-4-6 mgs orally past the age 60-70-80-90s. It will prevent progression of cardio-vascular and neurodegenerative diseases, dementia, and cancers as we age. If one has coronary stent or CABG graft, taking rapamycin will keep the blood flow open for a long time. I recommend mini-doses of rapamycin to all these patients. It will have a reviving youthful effect and prolong life span and health span by many years.
6. Take bioactive curcumin 1000 or more mg a day, vitamin D3 5000 IU orally every day. Curcumin inhibits nuclear factor kappa B (NF-kB) which enhances brain and body inflammation to delay or prevent many chronic diseases.
7. Use felodipine 2.5 - 5 mg to clear the brain from debris, by autophagy (garbage disposal system within cells of the body and brain) along with telmisartan if you have elevated blood pressure.
8. Take simvastatin which has anti-neurodegenerative diseases effect besides lowering the blood cholesterol, keeping blood vessels clear.
9. Take metformin 250 mg to 500 mg after lunch and supper all your life whether you are diabetic or not. It has an anti-aging effect, raising longevity by increasing the AMPK (adenosine monophosphate kinase) activity. It reduces the insulin resistance in the cells and lowers the blood glucose especially after a meal (post prandial elevation of blood glucose).
10. Take micro doses of (300-1000 micrograms) of lithium orally. It increases brain gray matter, and endows telomeres length which is a biological clock. I get the above supplements from Life Extension Foundation.

Calorie restriction (CR) as anti-aging method—akin to taking rapamycin

Besides life style changes, taking rapamycin, metformin, and lithium, we want to add to the list fasting and calorie restriction (CR) as anti-aging mimetics. CR is the only natural intervention known to extend lifespan (Goldberg EL et al. Lifespan-extending caloric restriction or mTOR inhibition impair adaptive immunity of old mice by distinct mechanisms. Aging Cell, Volume 14, Issue 1. Nov 2014.). CR is typically defined as a 20–40% reduction in caloric intake compared to ***ad libitum feeding***, providing a limitation in calories in the absence of malnutrition (essential micronutrients are provided at 100% level).

Follow the Gandhiji way: he practiced all 5 life style changes discussed above to the end of his life. He never smoked, never drank alcohol, maintained normal body weight by calorie restriction and repeated fasting, was physically active (exercised) daily and was a strict vegetarian. The only animal fat he consumed was in goat milk and yogurt he drank and ate. He was not only a sage/mentor of nonviolence, he also practiced nonviolence against nature (mother earth) and the poor by his calorie restriction.

"Reduce" dieting can be defined relative to the subject's previous intake before intentionally restricting calories, or relative to an average person of similar body type. Mahatma Gandhiji practiced calorie restriction all his adult life and outright fasting, each lasting many days, 14 times between July 1913 to January 18th 1948 (at the age of 78) as atonement for sins of others, protest against violence, colonial British rule, self-control and discipline, and religious/ caste discriminations. His daily intake of calories was very restricted and controlled (low calorie diet). So also did civil rights activist Dick Gregory of the US. Buddha fasted 49 days. I believe many biblical leaders such as Moses, King David, Elijah, Ezra, Esther, Darius, Daniel and Jesus also fasted. I don't think any of them were obese and had a fat belly. Moses lived up to the ripe old age of 120 years. Also, many sages of Hindu religion and Buddha, the founder of Buddhism, did fast in search of answers to many riddles of life. However, the Buddha, after long fasting under the Bodhi tree, and having lived an ascetic life for years, did end up recommending that monastics not eat anything after noon. This practice could be considered a kind of intermittent fasting method of calorie restriction, which restricts eating to a specific time period.

Broad benefits of CR in aging rodents and in humans include improved cardiovascular, neurological, muscular, and skeletal health. Mahatma Gandhiji was a human example of that with no fat under skin. CR was shown to impact immune cell function by preserving thymic cellularity during aging and perhaps consequently contributing to an increased proportion of native CD8 T cells later in life (Yang K, et al. 2013). If one wants to live longer, be healthy and banish co-morbidities, one needs to adopt the calorie restriction and intermittent fasting besides lifestyle changes as described in these chapters.

References

1. Arias E. United States life tables, 2008. *Natl Vital Stat Rep.* 2012 Sep 24;61(3):1-63.

2. Blagosklonny MV, From Rapalogs To Anti-Aging Formula, Oncotarget. 2017; 8:35492-35507.
3. Blagosklonny MV. Aging and immortality: quasi-programmed senescence and its pharmacologic inhibition. Cell Cycle. 2006;5(18):2087–2102.
4. Chiang CL. World Health Organization. Life table and mortality analysis. 1979. Publisher: Geneva: World Health Organization.
5. Emerging Risk Factors Collaboration, Di Angelantonio E, Kaptoge S, et al. Association of Cardiometabolic Multimorbidity With Mortality. *JAMA.* 2015;314(1):52-60.
6. Faloon W, Editorial, Add 12-14 years to Healthy Life Expectancy. Life Extension July 2019 pp 7-10.
7. Li. Y, Pan A, Wang DD, et al. Impact of Healthy Lifestyle Factors on Life Expectancies in the US Population. *Circulation.* 2018; 137:00–00. DOI: 10.1161/CIRCULATION. AHA.117.032047.
8. Vézina C, Kudelski A, Sehgal SN (October 1975). "Rapamycin (AY-22,989), a new antifungal antibiotic. I. Taxonomy of the producing streptomycete and isolation of the active principle". The Journal of Antibiotics. 28 (10): 721–6.
9. Woloshin S, Schwartz LM, Welch HG. The risk of death by age, sex, and smoking status in the United States: putting health risks in context. *J Natl Cancer Inst* 2008;100(12):845-53.
10. Yang K, Shrestha S, Zeng H, Karmaus PW, et al. T Cell Exit from Quiescence and Differentiation into Th2 Cells Depend on Raptor-mTORC1-Mediated Metabolic Reprogramming. Immunity. 2013;39: 1043–1056

Chapter 68

Exercise

Successful healthy lifestyle with regular exercise that increases healthy longevity in people could have profound benefits for quality of life, productivity, and reduced healthcare costs. This is especially so for those in nursing homes, extended care facilities, and the aged living independently. Exercise protects against coronavirus as well as infection of the lungs and the body. Studies begin to unfold that the high morbidity and mortality due to the coronavirus infection and during flu seasons in the nursing homes and extended care facilities or at own residence may be related to lack of exercise of the residents, besides aging and age-related afflictions and depressed immune system. A review by Zhen Yan of the University of Virginia School of Medicine showed that medical research findings "strongly support" the possibility that exercise can prevent or at least reduce the severity of acute respiratory distress syndrome (ARDS-SARS), which affects between 3% and 17% of all patients with COVID-19. Based on available information, the Centers for Disease Control and Prevention estimates 20% to 42% of patients hospitalized for COVID-19 will develop ARDS-SARS. The range for patients admitted to intensive care is estimated at 67% to 85%. Research conducted prior to the coronavirus pandemic suggested that approximately 45% of patients who develop severe ARDS will have increased morbidity and mortality. One of the reasons: most of these patients are bedridden without any exercise and have functional lung involution (the shrinkage of an organ in old age or when in-active due to lack of movements and exercise).

Exercise and body defense against oxidative stress

The body produces 3 important antioxidants: superoxide dismutase (SOD), catalase (CAT) and glutathione. During exercise, SOD is secreted by the muscles and distributed all over the body, protecting cells and vital organs such as the lungs, liver, brain, heart, eyes, kidneys and such from oxidative damage due to any number of etiologies including coronavirus infection. Those who regularly exercise are protected from the oxidative damage of coronavirus due to neutralizing effects of SOD in all the tissues due to exercise. The coronavirus infection and other infections and diseases result in production of free radicals (ROS) due to massive immune system reaction and cytokine production, and consequently worsens lipid peroxidation as reflected by increased plasma malondialdehyde (MDA). Malondialdehyde is the organic highly reactive compound that occurs as the enol. It is a marker for oxidative stress increase their level, possibly through exhausting the antioxidant defense system if one does not have the SOD shield.

Studies show that the mean SOD levels were 1.74 and 2.50 U/mL and the mean CAT levels were 0.16 and 5.26 U/mL before and after exercise at the beginning of the training program ($P<0.05$ for both). After exercise at the end of the training program, while serum CAT activity decreased from 8.43 to 2.89 U/mL, malondialdehyde (MDA-lipid peroxidation) levels increased from 16.39 to 29.11 mmol/L ($P<0.05$ for both) whereas the level and the protective effects of SOD continues. That is one of the reasons one needs to exercise, to have the shield of protection by SOD during illnesses such as COVID-19, cold and flu to prevent their effect on the lungs and heart.

Every nursing home and extended care facility or the aged residing at home should have mandatory exercise protocols for all the patients with or without whole body hyperthermia and for the employees as well. Besides providing the SOD shield, everyone benefits from exercise, regardless of age, sex or physical ability such as:

1. Weight control
2. Prevent high blood pressure. No matter what your current weight is, exercise boosts high-density lipoprotein (HDL) cholesterol, the "good" cholesterol, and it decreases unhealthy triglycerides. It prevents:
    a. Stroke
    b. Metabolic syndrome
    c. Type 2 diabetes
    d. Depression, Anxiety
    e. Many types of cancer
    f. Arthritis
    g. Accidental falls
    h. It can also help improve cognitive function and helps lower the risk of death from all causes.
3. Improves mood by improving the motor and sensory input to brain and improving brain circulation
4. Improves heart, lungs and brain functions and keep them fit and functional

5. Improves muscle strength and boosts endurance to do things you like
6. Promotes 7-8 hours of sleep which is needed in nursing home residents
7. Give chance to unwind and socialize with others. Find a physical activity you enjoy, and just do it. Bored? Try something new, or do something with friends or family.
8. It may even spark physical intimacy with other residents
9. It delays the onset of dementia and may even prevent its progression and improve cognition
10. It improves lung and heart function and their health. It prevents lung afflictions due to microbial attack and environmental factors besides supplying the needed oxygen to maintain various metabolic and physiologic functions of our body and brain

Whole body hyperthermia with exercise has positive effect on the body, maintaining health and facilitating living longer, disease free.

Add to this, take micro doses of lithium, and rapamycin with metformin. One can ward off or slow down aging, age related comorbidities, and live longer.

Chapter 69

Curcumin for health and longevity

Mechanisms of curcumin action to prevent diseases such as Alzheimer's and promote longevity

Source of curcumin

Curcumin is a biologically active polyphenolic compound found in turmeric, a spice derived from the rhizomes of the plant Curcuma longa grown in India and other tropical climates. It is commonly consumed in curry food all over the Asian countries, especially in India. Turmeric and its yellow active ingredient curcumin have been used for medicinal purposes for centuries, besides being used in daily curry preparations consumed by millions and billions every day. The following are some of the biological effects of curcumin in the treatment to prolong life and to prevent, curtail, stop progression and/or cure diseases including Alzheimer's disease. Read for details on curcumin in www.indian-gold.com and forthcoming book Indian-Gold by the same author.

Curcumin to preserve health of the brain and the body

In the last two decades, curcumin has been the most researched supplement. It is available all over the world. There are hundreds of publications on the subject from Oregon state university to NIH, Web MD, Life Extension Monthly magazine, GreenMed info, WebMD, Pub MED, Wikipedia, internet postings and such.

The important aspect of curcumin is that its active ingredient in its natural form are hardly absorbed from intestines to the blood, but when it is made bioactive by oils, extract of black pepper, fenugreek, in the curry with chilies, oils, heat and such, it becomes highly absorbable into blood to have healing effects. The absorbed curcumin in the blood also crosses the blood brain barrier to have a remedial and disease curing effect on neurodegenerative diseases including Alzheimer's and Parkinson's.

University of Florida researchers have now confirmed that curcumin, like lithium, stimulates the birth of new neurons, particularly in the hippocampus, the seat of memory in the brain. All the evidence shows that curcumin enters memory centers and participates in new nerve cell formation with nerve processes and synapses, which plays a role to prevent, delay, curtail, or arrest the progression of Alzheimer's and other neurodegenerative diseases.,

Curcumin reduces inflammation caused by cytokines in the lungs due to coronavirus COVID-19 and other viral afflictions of the respiratory system. There are more than 5.8 million Alzheimer's disease patients who end up in nursing homes and extended care facilities so also many dementia patients. It is one of the comorbidities that the elderly suffers, hence we emphasize curcumin's effects to protect these patients using bioactive curcumin.

In Alzheimer's dementia-affected brains, curcumin acts as a metal chelator, and experiments show that curcumin reduces the cognitive decline in older adults. It also preserves acetylcholine neurotransmitter by inhibiting the enzyme which breaks it down at memory transport connector (acetylcholinesterase at synapses) in the brain, thus enhancing the memory and reducing the cognition decline by increasing the memory neurotransmitter acetylcholine. I believe that curcumin is effective in the treatment of neurodegenerative diseases including Alzheimer's and Parkinson's. It is anti-inflammatory, and anti-cancer, and has an anti-arteriosclerotic vascular disease effect. I take it and so do my whole family, friends and my former patients. The above studies clearly tell us that curcumin:

1. Crosses the blood brain barrier, meaning enters inside the brain to reach the neurons in the memory cells and other areas of the brain, their projections, synapses, as well as glial cells and has anti-oxidant, anti-inflammatory, anti-protein and anti-fat aggregation effects.
2. Reduces the oxidative stress by counteracting free radicals and the effect of cytokines in the turbocharged lungs due to coronavirus infection, brain and body more than any other supplement.
3. Reduces the inflammation in the brain blamed for development of Alzheimer's disease, probably by acting against microglial and astroglia cells.
4. Maintains genetic integrity and prevents telomeres shortening at the end of genes by antioxidant and anti-inflammatory effect.
5. Helps the neurons to survive and genetic material to function without aberration due to inhibition of oxidative stress and inflammation created by high nerve cell activity with spilling of glutamate by dying or dead neurons.

6. Stimulates the essential brain derived nerve growth factor (BDNF and GDNF) that promotes new connections (neuronal plasticity) between brain cells.
7. Is nontoxic, easily absorbed when combined with pepperine, oils, heat, fenugreek, and is used by millions of people every day in their food.
8. Promotes inhibition of mTOR activity, a protein associated with rapid aging, cancers, neurodegenerative diseases and such.
9. Enhances activity of the enzyme AMPK, which improves metabolism.
10. Boosts autophagy, the cellular "recycling / house cleaning" process that helps keep cells rejuvenated.
11. Protects against tissue damage caused by glycation, when sugars attach to proteins and fats.
12. Promotes inhibition of nuclear factor-kappa B (NF-kB), a protein complex associated with harmful, chronic inflammatory responses.
13. Suppresses STAT3, a protein associated with chronic inflammation and some cancers.
14. Reduces pro-inflammatory cytokines that attack the lungs that contribute to respiratory distress syndrome (ARDS-SARS).
15. Improves cardiovascular disease risk factors and components of metabolic syndrome, including support for healthy body weight, cholesterol levels, triglycerides, and blood pressure.
16. Produces protective effects on aging blood vessels and other anti-atherosclerotic activity.
17. Produces anti-glycemic effects, enhancing control of blood sugar levels in metabolic syndrome and diabetes, reduces insulin resistance.
18. Increases activity of Sirtuins, specialized proteins that regulate cellular health, associated with lifespan extension.
19. Enhances activity of the enzyme AMPK, supporting healthy metabolism.
20. Supports healthy autophagy, a natural cellular rejuvenation process.
21. Promotes protection from tissue damage caused by glycation, when sugars attach to proteins and fats in the bloodstream.
22. Has anticancer and antimetastatic activity, both helping to prevent and remove existing tumor cells.
23. Regulates apoptosis, the process of naturally "pre-programmed" cell death to maintain cell populations and prevent overgrowth.

24. Promotes protection from abnormal protein, including beta-amyloid and tau, associated with neurodegenerative diseases such as Alzheimer's disease, and has neuroprotective effects that reduce dysfunction and promote recovery from brain, spinal cord, and peripheral nerve injuries.

One animal study showed lifespan extension up to 26% with curcumin supplementation (Shen LR, et al. Curcumin-supplemented diets increase superoxide dismutase activity and mean lifespan in Drosophila. Age (Dordr). 2013 Aug;35(4):1133-42.).

When combined with vitamin D3, 5000 IU, metformin, micro-doses of rapamycin and lithium, besides whole-body hyperthermia, eating a Mediterranean diet with omega 3 rich food, replacing wheat, rice and sugar with millets in the diet can be one of the most effective ways to ward off age related co-morbidities suffered by the aged.

Bioactive curcumin

As a rule, much of ingested curcumin is rapidly conjugated in the intestines and liver, and attaches to another compound, hence the potency of most ingested curcumin is not biologically active and very little makes it into the body's tissues. Studies show that ***only free curcumin, which remains unconjugated, is bioactive.*** It is done by adding Pepperdine from black pepper, fenugreek, heating and cooking in oil with pepper and chilies in curries which converts inactive curcumin to bioactive curcumin. Bioactive curcumin is also marketed by many supplement companies such as Life Extension, who markets high grade quality supplements. We describe how we can make bioactive curcumin in the home by sautéing turmeric powder, on moderate heat, with pepper, oil, fenugreek powder and such in the forthcoming book on Indian-Gold on turmeric and curcumin.

Indian curry: Practical proof that curcumin does reduce incidents of Alzheimer's dementia

There are multiple research studies on the benefits of curcumin treating Alzheimer's disease as reported below, but still the studies do not incorporate other therapies which augment the effect of the natural herb supplement curcumin as described below.

Our discussion is also based on the daily habits of those Indians who eat curry with curcumin-based diet every day, which have fewer (4.4 times fewer) odds of getting this incurable Alzheimer's disease, a dead-end affliction of the brain with memory loss (Ng T-P, Chiam P-C, Lee T, Chua H-C, Lim L,

Kua E-H. Curry consumption and cognitive function in the elderly. Am J Epidemiol. 2006;164: 898–906. Small GW, Siddarth P, Li Z, et al. Memory and brain amyloid and tau effects of a bioavailable form of curcumin in non-demented adults: A double-blind, placebo-controlled 18-month trial. Am J Geriatr Psychiatry. 2017). This Indian diet with curcumin in the curry makes it bioactive and more absorbable into circulation, crosses the blood brain barrier, enters the brain to become an effective anti-inflammatory and have anti-dementia effects. The pepper, oils, condiments, and heat make the curcumin in the curry and other food products bioactive, to be effective against neurodegenerative and cardio-vascular diseases. This method of use in curry is nontoxic and prevents many other diseases all over the body and brain.

In India, refrigeration of food is sparse; the curcumin due to its antioxidant activity preserves the unused food to be eaten later by preventing microbial growth. A diet rich in millets with reduced intake of wheat, rice and sugar could also contribute to a low incidence of diabetes, high BP and such, and may also contribute to such low incidence of dementia and Alzheimer's among Asian Indians.

There are thousands of research papers (some estimate as many as 1,500 to 9,000) in the last 2 decades, and a dozen or so patents on curcumin, indicating research interest due to its health benefits. As a supplement, available in all food stores and over the internet, it is easy to procure and use with no adverse effects. ***All ketchup and mustard should contain this golden ambrosia—curcumin.***

Curcumin is an antioxidant and anti-inflammatory, and protects inner linings of 60,000 miles (endothelium) of the body, along with 400 miles of it in the brain.

Research shows that curcumin is a better antioxidant than alpha-tocopherol (vitamin E) (Kim DS, Park SY, Kim JK. Curcuminoids from Curcuma longa L. (Zingiberaceae) that protects PC12 rat pheochromocytoma and normal human umbilical vein endothelial cells from beta amyloid (1–42) insult. Neurosci Lett 2001;303: 57–61.). This is one of the mechanisms how curcumin is both prophylactic and therapeutic for COVID-19, Alzheimer's, cancers, cardio-vascular diseases prevention and its progression.

Curcumin reduces inflammation caused by cytokines in the in the lungs by the coronavirus, which results in increased morbidity and mortality. I believe that all residing in nursing homes and extended care facilities and

aged people should be on the bioactive curcumin along with an exercise program 5 days a week lasting a minimum of 30 minutes.

Discovering curcumin's benefits for cognition in the brain

In a study published in 2016, researchers gave older adults 1,500 mg of a specially-formulated curcumin supplement with enhanced bioavailability for 12 months. At six months of the study, the placebo group experienced a decline in cognitive function. There were no such drops in curcumin-supplemented subjects—an example of curcumin's ability to protect the brain against cognitive decline (Rainey-Smith SR, Brown BM, Sohrabi HR, et al. Curcumin and cognition: a randomized, placebo-controlled, double-blind study of community-dwelling older adults. Br J Nutr. 2016;115(12):2106-13).

Experimental studies reveal that curcumin works in four important ways to protect against neurodegenerative diseases including Alzheimer's and Parkinson's:

1. Alzheimer's patients often have abnormal insulin signaling pathways. Curcumin helps to normalize those pathways.
2. Studies in mice show that supplementing with curcumin nearly eliminated abnormal synapses, while also improving synaptic numbers and structure.
3. Curcumin not only reduces brain inflammation caused by beta-amyloid plaque but can also reverse its toxic impact on brain cells: neurons, microglia, astrocytes and oligodenddroglia.
4. Curcumin inhibits the enzyme acetylcholinesterase, which helps boost levels of the neurotransmitter acetylcholine in synapses which increases memory and cognition.

If one has gall bladder diseases, it may be a relative contraindication.

Chapter 70

Rapamycin: Revelation and revolution

Rapamycin is for aging like penicillin is for infections and its use to suppress or decrease age-related comorbidities

The age of the world population is increasing due to improved health care, vaccination, antibiotics discovery, surgical innovations, and new drug discoveries with improved nutrition and water supply over decades. This has led to an associated increase in longevity and age-related diseases with associated disabilities and dependence on care providers. This leads to the search for strategies to increase health span—*the length of time an individual lives life healthy and productively*.

Successful and effective interventions that increase healthy longevity (health span) in people using rapamycin could have profound benefits for quality of life, productivity, and reduced healthcare costs, reduced admission to nursing homes, extended care facilities, and hospitals (Goldman DP, et al. Substantial health and economic returns from delayed aging may warrant a new focus for medical research. Health Affairs. 2013;32:1698–1705). This highlights a critical question: Do lifespan-extending interventions, only extend lifespan or do they also extend **health span** (the length of healthy, high quality lifespan)? The drug rapamycin can extend lifespan as well as health span, similar to calorie restriction.

Rapamycin is a mirror image of penicillin. Penicillin, discovered in 1928 by Alexander Flaming, was made by yeast to target bacteria, while rapamycin, found in the soil of Easter Island (Rapa Nui, the local name, hence the name rapa-mycin) in 1965, was made by bacteria to target yeast (the other way around). Rapamycin was discovered by Dr. Surender Sehgal and his colleagues in 1972 from bacterium Streptomyces hygroscopicus obtained from Easter Island soil. It was originally developed as an antifungal agent marketed as Rapamune by Pfizer (Wyeth)..

How rapamycin acts in our body and genetic influence on its activity

The mammalian target of rapamycin (mTOR), [Sabers CJ, et al. 1995. "Isolation of a Protein Target of the FKBP12-Rapamycin Complex in

Mammalian Cells". *J. Biol. Chem.* **270** (2): 815–22] sometimes also referred to as the mechanistic target of rapamycin and FK506-binding protein 12-rapamycin-associated protein 1 (FRAP1), is a kinase that in humans is encoded by the MTOR gene. [Brown EJ, et al. 1994. "A mammalian protein targeted by G1-arresting rapamycin-receptor complex". Nature. 369 (6483): 756–8]. mTOR is a member of the phosphatidylinositol 3-kinase-related kinase family of protein kinases. mTOR is a complex large protein located in the cytoplasm of the nucleus. ***It operates in close communication with the nucleus of the cell and its genetic material.*** It is always there. The nucleus contains the DNA that comprises the blueprints for making each individual protein. mTOR signals the nucleus to make this protein**.** ***mTOR senses what is happening within the cell and then signals the nucleus how to respond.*** The nucleus then responds by providing the blueprints to make the specific different proteins. mTOR operates and ***functions as the central hub of the cell signaling system, akin to a computer CPU*** (central processing unit), the command and control center of the cell activities from its division, function, diseases to death.

Rapamycin binds to mTOR and inhibits its activity. Among the many things that mTOR does (when inhibited), is to regulate autophagy, causing cells to eat themselves of excess and not needed contents in the cytoplasm of all cells all over the body (house cleaning). In this process, dysfunctional cells (like cancer and senescent cells) tend to be "eaten" first.

It may have impact in reducing the cytokine storm in COVID-19 if it is administered early in the moderate to severe cases of coronavirus infection. It is administered in mini-doses-6mg loading dose and then 2-3 mg every day for one two weeks. It will save many who have the coronavirus infection.

Rapamycin intermittent pulse dosing as anti-aging and for treatment of comorbidities

Mikhail Blagosklonny, an M.D. PhD physician scientist at Roswell Park Cancer Institute in New York, has been the most vocal advocate of the use of rapamycin to extend human lifespan besides others. In addition to extension of lifespan, I would advocate administration of intermittent rapamycin to prevent, curtail, stop progression and / or cure the afflictions of the comorbidities that plague us as we age.

I separate rapamycin administration dose into 3 categories:

1. **Mega-doses rapamycin:** 15-20-40 mgs is the loading dose with reduced doses administered daily to prevent organ rejection for long periods, year after year. It is associated with an untold number of

complications such as peripheral edema, hypercholesterolemia, abdominal pain, headache, nausea, diarrhea, pain, constipation, hypertriglyceridemia, hypertension, increased creatinine, fever, urinary tract infection, anemia, arthralgia, and thrombocytopenia, but it is a life saver. The micro doses we recommend to mitigate aging and comorbidities do not result in such complications; they have very little or no such complications.

2. **Mini-doses rapamycin**: Take 6 mg loading dose and 2 mg every day for a limited time. This is given to counteract cytokine and viral storm as described above or for severe flu or in septic shock (maybe), till the symptoms and diseases start to dissipate—that is for a week or two and then stop taking after obtaining the desired results. This dose may be instituted for chronic gastro-intestinal track afflictions intermittently.
3. **Micro-doses - Pulsed (intermittent) intake of rapamycin:** take 3-6 mg depending upon the comorbidities, weight, age once every 7-10 days to prolong remaining life, to bring back some youth, prevent or inhibit or counteract or decrease expansion of comorbidities that escalate our morbidity and mortality. Take rapamycin 8-12 weeks then take one- or two-week drug holiday or stop taking if complications develop. Then restart the regimen. I take it to prevent age related diseases from neurodegenerative diseases to cancers, cardio-vascular diseases and such and recommend it as described.

I have prescribed it also for those who have cancer, cancer metastasis and those cured of cancer, neurodegenerative diseases such as Alzheimer's and Parkinson's, arteriosclerotic vascular disease, diabetes, after CABG and stent placement to the coronary blood vessels, prostate hypertrophy and prostate cancer, autoimmune diseases, gastro-intestinal track problems, obesity, and such. One of my patients with stage 4 melanoma was on it, so also an advanced prostate and scleroderma patient. I had an ASD patient on it with encouraging results. It probably acts by inhibiting brain inflammation, due to inhibition of microglial and astrocyte activity, and reducing the abnormal proteins that affect nerve cells in these patients and resetting the synaptic transmission.

While rapamycin has adverse side effects in humans who take it daily in mega doses to prevent rejection of transplanted organs such as kidneys for immuno-suppression, recent research has found that pulse dosing micro (and mini) doses—perhaps once a week or every 10 days—may confer most of the anti-aging benefits without any significant adverse side effects. Interestingly none of my patients on micro doses reported any complications, indicating the safe method of taking it to advance life and counter comorbidities of aging.

Rapamycin effects on pathways involved in cell growth

Pathways involved in the control of cell growth, multiplication, and metabolism have emerged as important players of lifespan regulation [Blagosklonny MV. Aging and immortality: quasi-programmed senescence and its pharmacologic inhibition. Cell Cycle. 2006;5(18):2087–2102.]. mTOR kinase is a key signaling node that integrates information regarding extracellular growth factor stimulation, nutrient availability and energy supplies [Ehninger D. From genes to cognition in tuberous sclerosis: implications for mTOR inhibitor-based treatment approaches. Neuropharmacology. 2013; 68:97–105.].

A number of studies in yeast, worms and flies have initially implicated mTOR in lifespan extension control - meaning living without diseases, longer life extension. Rapamycin is a *specific inhibitor of the mTOR* signaling complex, the central regulator of cell nutrient sensing and energy metabolism (Zoncu R, Efeyan A, Sabatini DM. mTOR: from growth signal integration to cancer, diabetes and ageing. Nat Rev Mol Cell Biol. 2011;12:21–35).

Rapamycin used in mega doses (15-40 mgs) is a well-known immune suppressant, used to prevent organ rejection – as in kidney transplant. However, recent ground-breaking studies highlighted the importance of nutrient sensing pathways during an immune response by showing that short-term mTOR inhibition ***using low-dose rapamycin (75µg/kg) enhanced the development of antigen-specific memory CD8 T cells during acute infection and made immune system more responsive to infections and other age-related diseases*** (Blagosklonny MV. Aging and immortality: quasi-programmed senescence and its pharmacologic inhibition. Cell Cycle. 2006;5(18):2087–2102.). Age-related decline of immune function underlies this critical vulnerability to comorbidities and exhibits reduced vaccine efficacy.

With various advances in medicine, there is still need for more advances for the treatment of sepsis, and brain and body inflammation and aging. Brain inflammation is blamed for all neurodegenerative diseases including Alzheimer's and Parkinson's. Aging is a major risk factor for a range of diseases in numerous organ systems, including cardiovascular diseases, neurodegenerative diseases, cancers, diabetes mellitus type II, osteoporosis and such. Rapamycin and rapamycin analogs provided with an antioxidant moiety prodrug are immunomodulatory agents and are useful in the treatment of restenosis prevention coronary stent, immune and autoimmune diseases. In addition to cancers, fungal growth, post-transplant tissue rejection, and immune and autoimmune disease inhibiting compositions (US 7,679,901, Zhao, Jonathon), now I advocate this for treating moderate to severe cases of COVID-19.

Rapamycin and related analogues (rapalogues) extend life and reduce morbidity and mortality from comorbidities

Ehninger D, et al. describes this in the "longevity and aging" (Dan Ehninger, Frauke Neff, and Kan Xie. Cell Mol Life Sci. 2014; 71(22): 4325–4346.) article in detail, which we have selectively incorporated here in.

- Studies identified rapamycin as the first pharmacological agent to extend maximal lifespan in a mammalian species, with effects in both males and females (Harrison DE, Strong R, Sharp ZD, et al, Rapamycin fed late in life extends lifespan in genetically heterogeneous mice. Nature. 2009;460(7253):392–395. Bitto A, et al. transient rapamycin treatment can increase lifespan and healthspan in middle aged mice. eLife, 2016; 5: e16351).
- Rapamycin decreases a subset of heart dimensional measures, such as diastolic left ventricular internal diameter (LVIDd), and also decreased overall heart mass. Echocardiography also demonstrated the expected aging-associated decrease in functional cardiac measures, such as cardiac output, ejection fraction, fractional shortening, as well as blood flow measurements and pressure gradients across the aortic, pulmonary and mitral valves.
- Immunoglobulin plasma concentrations are often robustly elevated in aged mice. Rapamycin tended to decrease plasma immunoglobulin concentrations in several cases that may reduce the body inflammation. I think it could be due to inhibition of millions of plasma cells and such on gastro-intestinal track lining and the lymphoid tissue on the wall.
- One of the well-documented age-related alterations is the strong decline in adult hippocampal neurogenesis (main memory center in the brain). Neff F. et al. determined 1-year rapamycin treatment, initiated at 4 months, *prevents the age-related decline in hippocampal neurogenesis* [Neff F. et al., Rapamycin extends murine lifespan but has limited effects on aging. J Clin Invest. 2013;123(8):3272–3291.

Bitto et al.]. This is an important effect of rapamycin to restore and maintain all levels of memory and recall.

- Additionally, mTOR controls protein synthesis (via the p70S6 kinase/ribosomal protein S6 pathway), which stimulates the translation of mRNAs (with a 5′ terminal oligopyrimidine tract—5′TOP), many of which encode for components of the translational machinery (e.g., ribosomal subunits, translation factors etc.) and stimulates translation of messenger RNA, which may play a role in memory and recall.
- Another important cellular process regulated by mTOR signaling that is controlled by rapamycin is autophagy. Autophagy (self-cleaning), a process by which the cell recycles macromolecules and organelles within their walls (cell membrane), allows for the removal of damaged cellular constituents and enables the cell to mobilize substrate under nutrient-poor conditions. mTORC1 regulates autophagy by phosphorylating and inhibiting the autophagy-initiating kinase Ulk1. In C. elegans, autophagy has been reported to be required for the lifespan extension caused by genetic mTOR inhibition. Lifespan extension, caused by rapamycin, is dependent on autophagy effects also.
- Detailed causes of death analyses in rapamycin-treated mice and controls indicated that both groups die primarily (i.e., in >80 % of cases) due to cancers, but rapamycin-treated animals do so later in life than controls. Rapamycin has well-*known* ***anti-cancer properties***, including inhibitory effects on de novo cancer formation, as well as suppression of established tumors via inhibition of cancer growth, promotion of apoptosis of neoplastic cells and/or a modification of the host response to the tumor (for example, inhibiting new blood vessel formation—angiogenesis) (Kopelovich L, Fay JR, Sigman CC, Crowell JA. The mammalian target of rapamycin pathway as a potential target for cancer chemoprevention. Cancer Epidemiol Biomark Prev. 2007;16(7):1330–1340). My advice is to take rapamycin if one had cancer as part of the treatment or prevent cancer development in mini or micro-doses. I had many patients on it.
- Studies over the past ~20 years have identified a large number of genetic manipulations that extend life in invertebrate model organisms, such as Caenorhabditis elegans and Drosophila melanogaster. Some of the pathways identified were also shown to be

involved in the regulation of mammalian lifespan that can be affected by rapamycin instructed genetic manipulation.

- Rapamycin shows extensive tissue distribution all over the body and is therefore available for mTOR inhibition in numerous organs including the brain. I believe this works, and recommend rapamycin for all those diagnosed with Alzheimer's and other neurodegenerative diseases.
- Aging is associated with a number of neurobehavioral and neurological changes, such as cognitive decline, alterations in motor coordination, balance, reduced muscle strength and such. Rapamycin has an effect on improving all these conditions with lowering of brain inflammation and reducing the neurodegeneration. We believe that injections of rapamycin in micro doses directly to cerebro-spinal fluid pool will arrest the progression of Alzheimer's and Parkinson's and bring life back to the brain of the patients (discussed in the book on Alzheimer's, and also described in another chapter in this book).
- The rapamycin protocol has therefore clear enhancing effects on learning and memory that can be employed to symptomatically improve cognition and recall in aged including motor coordination, gait, muscle strength, balance, and prevent the aging-associated decline in visual acuity which is seen in aged animals and human.
- During the course of aging, there are typical changes affecting bones and the skeletal system, including decreases in trabecular bone volume and progressive kyphotic changes affecting the spine. Rapamycin treatment significantly improved these age-related biomechanical tendon properties.
- Rapamycin treatment significantly elevated red blood cell counts in aged indicating that rapamycin effects were likely linked to aging-independent effects on erythrocyte production and/or turnover.
- One of the well-documented age-related alterations is the strong decline in adult hippocampal neurogenesis (important memory center complex in the brain). Neff et al. determined a 1-year rapamycin treatment, initiated at 4 months, prevents the age-related decline in hippocampal neurogenesis.
- There is an accumulation of polyubiquitinated and nitrotyrosylated proteins in the aging mouse brain. Wu et al. employed immunohistochemical analyses of brain sections to measure possible effects of a mTOR mutation on these aging traits. Their data showed a reduced immunoreactivity for polyubiquitinated and

nitrotyrosylated proteins in brain sections of aged mTOR mutants compared to age-matched wild type controls, (Wu JJ, Liu J, Chen EB, Wang JJ, et al., Increased mammalian lifespan and a segmental and tissue-specific slowing of aging after genetic reduction of mTOR expression. Cell Rep. 2013;4(5):913–920) suggesting that the appearance of these aging traits was suppressed in the mutants by rapamycin.

- The data showed the expected substantial age-related increase in thyroid follicle sizes. Rapamycin treatment significantly decreased thyroid follicle size in aged mice and restored its normal function.
- Shorter-term rapamycin treatment (for 6 months) yielded expression alterations of 100 genes (32 up-regulated and 68 down-regulated) in males and 1,427 genes (675 up-regulated and 752 down-regulated) in females. In summary, both studies provide initial insights of rapamycin associated gene regulation in a single organ (the liver).
- The aging traits found to be ameliorated or inhibited by rapamycin were related to:
    - immune system changes (e.g., plasma immunoglobulin concentrations, frequency of specific T cell subsets, cytokine concentrations in blood and heart, response to vaccination),
    - age-related alterations in body mass, organ size and dimensions (body weight, fat mass, lean mass, thyroid follicle size, cardiac dimension, heart weight)
    - tumors and pre-cancerous lesions, as well as neurobehavioral changes (motor activity, learning, memory and recall).

According to Bitto A, et al. 2016, rapamycin, which is already FDA approved, increases lifespan in rodents and delays age-related dysfunction in rodents and humans. Nevertheless, important questions remain regarding the optimal dose, duration, and mechanisms of action in the context of healthy aging. They showed that 3 months of rapamycin treatment is sufficient to increase life expectancy by up to 60% and improve measures of health span in middle-aged mice. This transient treatment is also associated with a remodeling of the microbiome, including dramatically increased prevalence of segmented filamentous bacteria in the small intestine. They also define a dose in female mice that does not extend lifespan, but is associated with a striking shift in cancer prevalence toward aggressive hematopoietic cancers and away from non-hematopoietic malignancies. Their data suggest that a short-term rapamycin treatment late in life has persistent effects that can effectively delay aging, reduce cancer occurrence, delay arteriosclerotic vascular disease, reduce neurodegenerative diseases such as Alzheimer's and Parkinson's, obesity and diabetes, and modulate the microbiome.

I divide administration of rapamycin into mega, mini and micro-doses depending on why it is prescribed including the millions of aging population, and those with cardio-vascular diseases, millions with coronary and other blood vessel stents for vascular diseases, cancers, neurodegenerative diseases including Alzheimer's and Parkinson's, autoimmune diseases, gastro-intestinal track afflictions and such.

Gum and periodontal disease

Experimentally it has been shown that short term treatment with rapamycin rejuvenates the aged oral cavity of elderly mice, including regeneration of periodontal bone, attenuation of gingival and periodontal bone inflammation, and revertive shift of the oral microbiome toward a more youthful composition. This provides a geroscience strategy to potentially rejuvenate oral health and reverse periodontal disease in the elderly, and at the same time restore gut microbiome. Of course, bioactive curcumin gargle is one of the most inexpensive methods of maintaining gingival and oral health.

Rapamycin effect on gastro-intestinal track diseases

The gastro-intestinal track has millions of plasma cells on the gut wall lining, along with billions of microbiome healthy bacteria and disease-causing microbes. It has a massive lymphoid structure embedded in the wall of the intestines. Mini-doses of intermittent rapamycin therapy as described here will alter the microbiome composition and improve gut health. This method should be attempted by all patients who have chronic bowel diseases such as *irritable bowel syndrome, leaky gut syndrome, Crohn's disease, ulcerative colitis, lactose intolerance, familial multiple polyposis, multiple diverticulosis* and such. Rapamycin may rejuvenate the stem cells in the gut wall crypts of Lieberkühn and thus restore the gut to normal functioning with healthy lining cells. Many of my patients on mini-doses of rapamycin reported improvement or elimination of digestive system problems. Intermittent rapamycin was the first line of treatment in my practice.

Our method of use of rapamycin

Rapamycin is one of the most important anti-aging, comorbidities controlling and other disease preventing therapeutic agents there is, besides and/or with metformin, lithium carbonate and bioactive curcumin, which also have anti-aging effects (Blagosklonny MV. An anti-aging drug today: from senescence-promoting genes to anti-aging pill. Drug Disc Today. 2007;12: 218–224.).

For anti-aging and prevention of neurodegenerative inclining Alzheimer's and Parkinson's and arteriosclerotic vascular diseases, cancer prevention and treatment, obesity and such: use 75 mcg/kg. That is about 5 mg taken orally once a week or once every 10 days.

We want to introduce micro or mini doses of rapamycin by direct delivery to cerebro-spinal fluid (CSF) as described in Chapter 46 to treat Alzheimer's and other neurodegenerative diseases including rabies and CJD, as soon as it is diagnosed. It will arrest the progression of the disease and reduce or eliminate brain inflammation by suppressing the microglia, and astrocytes activity, which is one of main factors responsible for Alzheimer's Aβ plaques, tau protein, neurofibrillary tangles, breakdown of skeletal system of the neurons such as neurotubules and their formation and inhibit other toxic substance in the brain, including alpha synuclein in and around the neurons and their processes and connection (in dendrites and synapses).

Taking rapamycin is akin to calorie restriction dieting. Mahatma Gandhiji is a classic example. His mind was sharp with no decline in his thinking and expression of his ideas to the end of his life. All overweight people should be on micro doses of rapamycin till they achieve desired weight, body mass index and waist size.

Minor complications of taking micro-doses of rapamycin

When the patient takes micro-doses (3-6 mg) of intermittent rapamycin once every 7-10 days, there are hardly any complications, as opposed to when it is administered in mega doses (15-40 mg) as anti-rejection therapeutic agents after organ transplant. With micro and mini dose, if the complications occur, they are milder such as mouth ulcers, skin infection and mild bronchitis, or mild broncho-pneumonia, which can all be treated with antibiotics and discontinuing the drug till these complications dissipate.

The following is the protocol I have experimented and evolved to slow down aging and age-related diseases. Rapamycin is part of the protocol.

1. Induce whole body hyperthermia >3 times a month. I have practiced it since 1982 for overall health, anti-aging and prevention of age-related disease. Bring the body temperature between 99.5-100 degrees F for 30 minutes or more.

2. As the time went by, I added metformin to the whole-body hyperthermia protocol. I take 250-500 mg after lunch and supper, ***though I am not diabetic***. It works by activating the AMP kinase system. It is known that diabetes typically shaves about five years off a person's life. But a large retrospective analysis found that diabetics on metformin had a *15 percent lower mortality* rate than nondiabetic patients indicating that it actually targeting aging.
3. In the last few years, I added ***rapamycin*** to slow down or inhibit mTOR command and control center of the cell to inhibit or prevent comorbidities. We started prescribing it to treat cancers, cardio-vascular diseases, patients after coronary stent, CABG, neurodegenerative diseases, Alzheimer's and Parkinson's, diabetes with complications, autoimmune diseases and such. Rapamycin enhances autophagy through activation of the ULK1 autophagy complex and activates the cells to destroy unwanted cellular metabolic debris through lysosomal activation (autophagy—self house cleaning). I take and recommend this to my former patients and friends, 3-6 mgs of rapamycin depending the comorbidities, weight, and age once a week. I have been taking rapamycin for a while and I hardly noticed canker sores (stomatitis) or impaired healing. Of course, I do take a drug holiday for a week or two and start it again.
4. I have added to the above protocol, 300-1000 micrograms of lithium, shown to prolong life, reduce incidence of dementia, increase the thickness of memory cortex of the brain, and act by inhibiting glycogen synthase kinases (GSK3). It also preserves and protects telomeres at the end of chromosomes to maintain the homeostasis of the chromosomes and their DNA content.

You do not need FDA approval for a rapamycin prescription. If a physician decides it is needed, it can be prescribed as described in the first part of the book. There are millions of cancer and heart disease patients who undergo surgery (CABG) and arterial stent placement, and they all should take micro-doses of rapamycin to prevent re-blocking, besides those who develop or are developing cognition impairment, Alzheimer's, Parkinson's, diabetes, arteriosclerotic vascular disease, and advanced cancers. ***Remember that coronary stenting and CABG do not increase longevity; they reduce pain, prevent the heart attack related morbidity and mortality. Adding rapamycin will increase longevity—think about it.***

Anecdotal anti-cancer experience with rapamycin discoverer

In any case, one imagines Surendra Nath Sehgal, the discoverer of rapamycin in 1972 with his colleagues, would be proud of what we are finding in its use for many diseases, (US Patent 4,375,464. Sehgal C. Vezina, et al., Rapamycin (AY-22,989), A New Antifungal Antibiotic II. Fermentation, Isolation and Characterization, J. Antibiotics (Tokyo), 28, 10, 727-732,1975. Vezina C, Kudelski A & Sehgal SN, J. Antibiotics (Tokyo), Rapamycin (AY-22,989), a new antifungal antibiotic. J. Taxonomy of the producing streptomycere and isolation of the active principle, 28, 721-726,1975. S.N. Sehgal, H. Baker, C. Vezina, et al., Demethoxyrapamycin (AY 24,668), A New Antifungal Antibiotic, J. Antibiot. (Tokyo), 36(4), 351-354, 1983). Thanks to Dr. Blagosklonny's efforts and his work and publications on rapamycin. It had changed my practice and spread about its use all over the world. The articles about aging and such in Time and Bloomberg Business Week on rapamycin have added to its familiarity, understanding, knowledge and use by many physicians and researchers. Streptomyces hygroscopicus is a bacterial species in the genus Streptomyces, first described by Hans Laurits Jensen in 1931 (Proceedings of the Linnean Society of New South Wales. 56: 345–370)..

Sehgal was diagnosed with cancer in 1998, and his son Ajai says that Sehgal began taking rapamycin, despite the drug not having been approved for anything yet (he made of off label use as described in the patent). He had a hunch that it might help slow the spread of his cancer and inhibit its growth, which had spread (metastasized) to his liver and other organs. His doctors gave him two years to live, but he survived for much longer, as the tumors appeared to go dormant with rapamycin therapy. The only side effect he suffered from was canker sores, a relatively small price to pay. But in 2003, after five years of taking rapamycin, Sehgal, age 70, decided to stop taking the drug. Otherwise, he told his wife, he'd never know whether it was really holding back his cancer. The tumors came back quickly with full force, and he died within months, says Ajai. On his deathbed, he said to Ajai, "The stupidest thing I've ever done is stop taking the drug (rapamycin)."

Other therapeutic applications of rapamycin and its derivatives

Since reading Sehgal's patents on cancers (US 4,885,171; 5,066,493 1991; 5,206,018 1993), and his personal experience with rapamycin, all my cancer patients, and those with comorbidities and the aged, got micro and mini doses of rapamycin (intermittent). ***It may not cure cancer, but it inhibits its growth and spread. It holds it at bay, and prolongs life, and produces almost***

pain free, good quality extended life. It is very effective in the treatment of almost 152 autoimmune diseases, including sarcoidosis (I am recommending it to a patient with excellent results), scleroderma, lupus, and sympathetic dystrophy. It may have an effect on ***phantom limb pain—a sensation that the amputated or missing limb is still attached*** after amputation. Take it before and after amputation and you will need no or fewer sessions of image therapy and psychotropic agents. I believe that it is also effective against sympathetic dystrophy and diabetic neuropathy combined with other therapeutic modalities which will augment the therapeutic effects.

We have had patients take rapamycin therapy with advanced melanoma, prostate cancer, bladder cancer, breast cancer, pancreatic cancer and such with good results. Look for details in the forthcoming book "Rapamycin: Revelation and Revolution for a Long Functional Life."

We prescribed micro doses of rapamycin to all my patients who have coronary or other arterial stents and to all those who have undergone CABG to prevent or slow down re-blocking in the lumen of the stent and CABG grafts.

Rapamycin for dogs, cats, farm animals, house pets, aquatics (fish, octopus) and such

You want to keep your pet living (any and all) long with fewer age-related diseases? Think about giving rapamycin. I have prescribed it for my ***family's middle and old age dogs*** which in a way rejuvenated and extended their lives with fewer age-related comorbidities suffering. Rapamycin use can extend the life of any and all house pets and farm animals. I believe that aging aquatic fish, turtles, octopuses and such can extend their lives in aquariums and fish tanks by using rapamycin.

Rapamycin is a 21st century ambrosia and sanjeevani:

Ambrosia is touted and mythologically labelled as a mythical food of Greek gods conferring longevity and immortality to those who consumed it. It means: To lead a long functioning life, eluding or postponing death. When rapamycin is combined with metformin, lithium, and bioactive curcumin, it could be both ambrosia and the fountain of youth that Ponce de Leon was looking for in 1513 AD in Florida. It may become the Sanjeevani herb from Himalaya procured by Hanuman, given to revive Lakshman, brother of Sri. Rama who was mortally wounded in a battle with the evil ruler of Lanka. He was revived and lived a long life with his brother Sri. Rama ruling India from Ayodhya.

mTOR activity and its relation to development of athletes and their performance

I do believe that the mTOR signaling in the genes and its activation in the cytoplasm in all the cells of the athletes is highly active, which makes these people eat and absorb more nutrients, grow bigger and stronger, and perform physically compared to average people. mTOR activity surges as one engages in physical activity. mTOR activation also facilitates athletes to perform to their fullest strength. It is activated and is the driving and energy force of the body cells to perform physically to their peak.

Chapter 71

Metformin to extend life and reduce comorbidities

Metformin does more than lower blood sugar in diabetics and non-diabetics taking it. It is anti-aging and works against neurodegenerative diseases and MS. Metformin, found in the French lilac, is the most commonly prescribed antidiabetic drug used in England since 1958, and in the United States since 1995, in doses up to 2000 mg. ***Metformin is the fourth most prescribed medication in the US,*** not only as a diabetes treatment, but also as a second-line agent for infertility in those with polycystic ovary syndrome. Now I recommend it for obesity, for aging, neurodegenerative diseases, arteriosclerotic vascular disease, MS and such.

Metformin is also one of the most promising least expensive anti-aging, life-extending drugs available besides rapamycin and lithium. The metformin antidiabetic biguanides inhibit fatty acid oxidation, gluconeogenesis in the liver, increase the availability of insulin receptors, decrease monoamine oxidase activity, increase sensitivity of hypothalamo-pituitary complex to negative feedback inhibition, and reduce excretion of glucocorticoid metabolites and dehydroepiandrosterone-sulfate. These drugs have been proposed for the prevention of the age-related increase of cancer and atherosclerosis, and for retardation of the aging process. I have prescribed it. It has been shown that administration of antidiabetic biguanides to patients with hyperlipidemia lowers the level of blood cholesterol, triglycerides, and b-lipoproteins. Biguanides also inhibits the development of atherosclerosis, reduces hyperinsulinemia in men with coronary artery disease. It increases hypothalamo-pituitary sensitivity to inhibition by dexamethasone and estrogens, causes restoration of estrous cycle in persistent-estrous old rats, improves cellular immunity in atherosclerotic and cancer patients, lowers blood IGF-1 levels in cancer and atherosclerotic patients with type IIb hyperlipoproteinemia. Recently it was shown that metformin decreases platelet superoxide anion production in diabetic patients. It is important to note that calorie restriction and metformin inhibit mTOR via AMPK, prolong life span, and delay cancer if one is cancer prone (Anisimov VN, et al. Am J

Pathol. 2010 May; 176(5): 2092–2097, AGING, November 2010, Vol 2 N 11-on rapamycin and metformin).

Metformin obstructs, hinders, blocks or diminishes many metabolic functions and factors that accelerate aging and aging associated diseases which include:

1. Protecting against DNA damage by preventing glycation.
2. Improving the poor mitochondrial function, the metabolic power house in the cell.
3. Reducing the chronic inflammation responsible for aging related diseases such as neurodegenerative diseases, cardio- vascular diseases, cancers, skeletal diseases and such.
4. Facilitating DNA **repair**, which is critical for cancer prevention, nervous system integrity, cardiovascular health and longevity.
5. Attacking the degenerative processes instigated by chronic inflammation, which can prevent the development of aging's most troubling diseases labelled as comorbidities.
6. Protecting the myelin in the brain. It is a nerve fiber laying-nerve repairing drug, thus is a must for multiple sclerosis (MS) patients to prevent its progression. It could halt the progression of MS studies (Prof. Robin Franklin and his team at the Wellcome-MRC Cambridge UK.). It may be an answer for MS sufferers.
7. Protecting against obesity, and maybe even reducing the body weight with calorie restriction. It is a very effective tool against the obesity epidemic in the US.
8. Counteracting the development of insulin resistance in diabetics, which increases blood levels of insulin that increase cholesterol, elevate blood pressure, and obesity.
9. acting as a metabolic rejuvenator, restoring the body's response to the effects of glucose and insulin to much younger physiological levels. Insulin-dependent diabetics are often able to dramatically reduce their doses of insulin, and more easily maintain stable levels of blood glucose.
10. Increasing the blood levels of DHEA and alleviates metabolic immunosuppression.

Metformin has also been shown to increase the production of known longevity-promoting signaling molecules in cells, such as mTOR (inhibited) and AMPK (activated)—all of which reduce fat and sugar storage and increase youthful functioning at the cellular level. Studies have shown that by activating AMPK, metformin specifically impacts **lifespan by increasing it**

up to 20%. And most remarkably, diabetics taking metformin were shown to **live 15%** longer than healthy individuals without diabetes. (Bannister CA, Holden SE, Jenkins-Jones S, et al. Can people with type 2 diabetes live longer than those without? A comparison of mortality in people initiated with metformin or sulphonyl urea monotherapy and matched, non-diabetic controls. Diabetes Obes Metab. 2014;16(11):1165-73).

AMPK activity declines with age, (Salminen A, Kaarniranta K. AMP-activated protein kinase (AMPK) controls the aging process-. Ageing Res Rev. 2012;11(2):230-41) making us more vulnerable to many of the diseases associated with aging, so named comorbidities. Metformin can restore this kinase back. Many recent studies show that by activating AMPK, metformin plays a major role in preventing age-related disorders including cancer, cardiovascular disease, obesity, and neurocognitive decline. One of the side effects of metformin is that it reduces B12, and testosterone levels need a supplement to restore them.

It is the most important and inexpensive anti-aging therapeutic agent there is. All the people above the age of 60s, those in nursing homes, extended care facilities, and the aged living independently, those suffering from obesity, arteriosclerotic vascular diseases, MS and other neurodegenerative diseases should take it immaterial whether a person has diabetes or not. It will prevent, curtail, and stop the progression of co-morbidities of those who are ravaged by infections including coronavirus, flu, cold and other microbial infections. Take 250-300 mgs after lunch and supper even if you are not diabetic.

Chapter 72

Lithium: anti-aging tonic for the brain and body

Johann Arfvedson, a Swedish chemist, discovered lithium in the year 1817, named from the Greek word lithos (stone). Besides curcumin, metformin, rapamycin, whole body hyperthermia, and BCG vaccination, taking lithium is one of the most important inexpensive anti-aging, dementia therapeutic agents there is. We are spending billions of dollars to find a cure for Alzheimer's disease, when we have a simple inexpensive non-prescription over the counter agent available to prevent, curtail, stop progression and / or cure Alzheimer's disease.

Lithium and GSK enzyme inhibition

Glycogen synthase is an enzyme found in all our cells of the body and brain that is responsible in glycogen synthesis. It is activated by glucose 6-phosphate (G6P), and inhibited by glycogen synthase kinases (GSK3). Lithium inhibits GSK3, and has been found to have insulin-like effects on glucose metabolism, including stimulation of glycogen synthesis in fat cells, skin, and muscles, increasing glucose uptake. Lithium is used in the treatment of bipolar disorder and was the ***first natural direct GSK-3 inhibitor discovered.***

It has been shown to have therapeutic benefit in Alzheimer's, type 2 diabetes mellitus (T2DM), some forms of cancer, and bipolar disorder. Lithium increased in vivo CD8(+) OT-I CTL immune system function and the clearance of viral infections. Lithium can possibly tame the cytokine storm by inhibiting cytokine and interleukin production.

Lithium availability for our body use

It is nature's most effective and powerful neuroprotective remedy to prevent and/or curtail Alzheimer's (AD) and other neurodegenerative diseases. Because of its superior bioavailability, lower (and safer) doses - actually micro-doses of lithium (300-1000 micrograms) are all you need to get health benefits (Nunes MA, Viel TA, Curr Alzheimer Res. 2013Jan;10(1):104-7.). That is almost 1000 to 18000 times less than lithium prescribed for bipolar patients. It is available without prescription and inexpensive.

How lithium acts in Alzheimer's disease and its neuroprotective mechanisms

Lithium salts have a well-established role in the treatment of major affective disorders (bi-polar disorder) by using bioactive therapeutic agents (extended release formulation doses are 600 mg 3 times a day (acute control) and 300 mg 3 times a day (long-term control) and takes 2-3 weeks for it to take effect. We advocate 300 to 1000 micrograms, which is far *less than* ***one thousandth of the dose*** used in bipolar patients, along with other therapeutic measures to treat aging and Alzheimer's disease.

Lithium benefits for our health:

1. Humans in areas with higher lithium in drinking water live longer.
2. Lithium helps to maintain longer telomeres, which protects the genetic material in chromosomes. Telomeres are chains of DNA at the ends of each chromosome that get shorter at each cell division in most tissues. The role of telomeres and the enzyme telomerase (telomerase controls telomere shortening) in aging and in cancer has provided one workable mechanism of aging and its prevention by using lithium.
3. Lithium has been shown to regulate genes related to healthy DNA structure; it endows longevity by lengthening telomeres at the chromosomes to maintain their functional integrity.
4. Lithium use has shown to slow the rate of brain aging as seen in psychiatry patients by reducing neuronal death.
5. People who take lithium show improved mood.
6. Lithium inhibits an enzyme GSK-3 activity. Its increased activity is blamed on many chronic diseases of older age including Alzheimer's, type II diabetes, mood disorders, cancer, and others. Lithium has been shown to inhibit overactivity of GSK-3 that results in up-regulation of insulin-like growth factor-I (IGF-I), and brain-derived neurotrophic factor (BDNF) which stimulates neural stem cells to give rise to new hippocampal neurons throughout adulthood.
7. By itself, lithium administration extended fruit fly lifespan by an average of 11%; and when combined with rapamycin and MEK inhibitor trametinib with lithium, it extended fruit fly *lifespan by an average of 48%.* You think how much life it can extend in humans!

8. Recent studies have offered evidence that lithium may also exert neuroprotective effects by inhibition of nerve cell death (apoptosis), regulation of autophagy (housekeeping-removing metabolic debris within the cell), increased mitochondrial function, and synthesis of neurotrophic factors by inhibiting glycogen synthase kinase-3 (GSK-3) that acts as effective as antioxidants.
9. By inhibiting GSK-3, lithium prevents the mechanism that causes the formation of abnormal tau proteins, and neurofibrillary tangles that destroy brain cells skeletal and transport system (neurotubules) and impair memory. Lithium disrupts the key enzyme responsible for the development of the amyloid plaques (Aβ plaques) and neurofibrillary tangles (tau proteins production) associated with Alzheimer's disease.
10. Interestingly, patients on lithium have significantly higher gray matter (nerve cell containing part) volumes in the brain, hinting new nerve cell growth activation (neurogenesis) as well, which increases the length of the dendrites with their synaptic connections needed for memory.
11. Every patient with signs and symptoms of dementia, immaterial of its cause, should be on intermittent micro doses of lithium, so also patients with repeated concussions, Traumatic Brain Injury (TBI), Chronic Traumatic Encephalopathy (CTE) and Post-traumatic stress syndrome (PTSD). All the football players, boxers, hockey, basketball and others playing contact sports should be on it.
12. Lithium treatment used in bipolar disorders in humans has been associated with
    a. increased expression of anti-nerve cell death (apoptotic) genes
    b. inhibition of cellular oxidative stress and inflammation
    c. increase of synthesis of brain-derived neurotrophic factor (BDNF and Bcl-2)
    d. brain cortical thickening with increased grey matter density of the brain surface rich in nerve cells (H. Moore et al. Lithium-induced increase in human brain grey matter. The Lancet. 356:1241 – 2 (2000)
    e. production of hippocampal (memory part of brain) neurons and its enlargement

These changes were noted due to lithium induced new nerve cells, their processes and synapses formation (neurogenesis—increased synaptic plasticity). The same changes can be expected in the aged with declining memory. Administer lithium after 250 – 500 mg intake of metformin to enhance its uptake and increase the pharmacological effects (metformin potentiation therapy) as described under metformin chapter.

Our method to treat Alzheimer's by intravenous infusion

One method to treat a diagnosed Alzheimer's disease patient is to infuse 250-500 ml of glucose / ringer's lactate / normal saline with rapid acting insulin 5-10 IU and lithium 300-1000 micrograms over a period of two hours through intravenous line as a routine health promoting anti-aging, a protocol described in the forthcoming book on Alzheimer's. We add to this regimen rapamycin, anti-oxidant Ebselen and bioactive curcumin.

Remember, micro-doses of lithium, (1000-8000 times smaller doses than used in bipolar patients) produced the same brain benefits seen in earlier mouse studies with larger doses with no toxic effects. Every one past 60s, and those in extended care facilities and nursing homes, should receive 300-1000 micrograms of lithium carbonate as supplement (Nunes MA, Viel TA, 2013).

Membranolytic synthetic polymer to deliver therapeutic agents directly into the cell through the membrane without reacting with therapeutic agents it carries and delivers

Membranolytic is the term used for any agent that carries the various therapeutic, pharmaceutical, biochemical, and biological agents or compounds through the cell membrane, without changing the therapeutic agent's effectiveness it carries though the cell membrane by direct penetration and permeation. Such a development will have wide application in the treatment of cancers, neurodegenerative diseases such as Alzheimer's and Parkinson's, microbial infection, autoimmune diseases and so on. I have used Dimethyl sulfoxide (DMSO) as a membranolytic permeation and penetration agent since 1964 with or without other therapeutic agents.

I have also use insulin, metformin and hyperthermia for enhancing the therapeutic agents' uptake through cell receptors and membrane mediation. But we need a membranolytic inert agent to allow all the drugs to enter the cell directly without the receptor's effects though the cell membrane. In our electron microscopic study of tissue, we used methacrylate to enter the cell and fix them to cut them thorough micro-microtome, and this polymer maintained the integrity of organelles and cell integrity. Such an agent is needed to carry the therapeutic agents of all kinds though the cell wall to treat hundreds of diseases including cancer. Think about it.

Chapter 73

Rapamycin for coronavirus—COVID-19

In COVID-19, patients with severe acute respiratory syndrome (SARS) and Middle East respiratory syndrome (MERS), also have rapid virus replication, a large number of inflammatory cell infiltration with production of massive amounts of cytokines, chemokines, and prostaglandins that leads to severe lung destruction aptly named cytokine storm leading to acute respiratory distress syndrome (ARDS).

Reduced counts and functional exhaustion of T lymphocytes, and massive cytokine release syndrome have been identified as adverse factors in patients affected by severe SARS-CoV-2 infection. Increased morbidity and mortality is associated with advanced chronological age, diabetes, obesity, diabetes, or cardiovascular disease and such. Severe COVID-19 can therefore mimic a state of immune senescence that can be benefited by rapamycin [J.B. Mannick, et al. TOR inhibition improves immune function in the elderly. Sci. Transl. Med., 6 (2014), p. 268ra179].

All the studies indicate that rapamycin mitigation is a mechanism for mTOR suppression-mediated longevity extension. Rapamycin also modestly alters gut microbiomes which could be linked to gut health and to immune system outcomes due to changes in plasma cells and improved stem cell from crypts of Lieberkühn and such that may play a role in longevity and health span. All these effects may play a role in crucially ill COVID-19 patients.

Rapamycin to increase T cells and prevent increased morbidity and mortality due to cytokine storm and severe progression of COVID-19

In severe COVID-19 patients, IL-2, IL-6, IL-7, IL-10, TNF-α, G-CSF, IP-10, MCP-1, and MIP-1α levels increase significantly [Zang et al.]. This cytokine storm results in immune T-cell apoptosis or necrosis, causing reduced T-cell counts. T-cells play a vital role in viral clearance, particularly through secretion of perforin, granzyme and IFN-γ and stimulating B cells to produce antibodies.

In these patients, there is potential of rapamycin to reverse T-cell senescence as reported by Mannick study [Mannick et al.] at the same time clear the virus from the respiratory system and reduce the chances of coronavirus storm. In the elderly with increased senescent PD-1+ T-cells, use rapamycin and its analog of enhanced immune function, and improved T-cell responses to produce antigenic stimulation with reduced serum senescence markers through IL-6 suppression [M. Singh, et al. Effect of low-dose rapamycin on senescence markers and physical functioning in older adults with coronary artery disease: results of a pilot study. J. Frailty Aging, 5 (2016), pp. 204-207].

In patients with ***H1N1 influenza virus, early adjuvant rapamycin therapy during a short period (2 mg/day for 14 days)*** was associated with an increased viral clearance, improvement in lung injury (i.e. less hypoxemia), and a decrease of multiple organ dysfunction. The duration of ventilation in survivors was also shortened [C.H. Wang, et al. Adjuvant treatment with a mammalian target of rapamycin inhibitor, sirolimus, and steroids improves outcomes in patients with severe H1N1 pneumonia and acute respiratory failure. Crit. Care Med., 42 (2014), pp. 313-321,]. The same results can be obtained treating COVID-19 with rapamycin. Further, a recent report shows that mTOR inhibition with the rapalogue everolimus ***improves influenza vaccine responses*** in elderly humans (Mannick et al., 2014. mTOR inhibition improves immune function in the elderly. Sci. Transl. Med. 6, 268ra179).

SARS-CoV-2 like HINI activates mTOR, and NLRP3 inflammasome pathway [X. Jia, et al. Delayed oseltamivir plus sirolimus treatment attenuates H1N1 virus-induced severe lung injury correlated with repressed NLRP3 inflammasome activation and inflammatory cell infiltration PLoS Pathog., 14 (2018), 1007428] leading to the production IL-1β, the **mediator of lung inflammation, fever and fibrosis** [P. Conti-Induction of pro-inflammatory cytokines (IL-1 and IL-6) and lung inflammation by COVID-19: anti-inflammatory strategies. J. Biol. Regul. Homeost. Agents, 34, 2020, p.1]. Rapamycin with mTOR inhibition can block the lung inflammation that leads to lung fibrosis.

Rapamycin inhibits mTOR pathway activation in the coronavirus infection, and thus IL-1β secretion. In COVID-19, the binding of SARS-CoV-2 to Toll Like Receptor (TLR), which leads to IL-1β production, could be reversed by rapamycin (Y. Zhou, Y. et al. Network-based drug repurposing for novel coronavirus 2019-nCoV/SARS-CoV-2 Cell Discov., 6, 2020). In addition, *rapamycin was recently identified in a bio-informatic drug study as a candidate for potential use in COVID-19* [Zhou Y, et al.].

When rapamycin is given at the onset of the cytokine storm phase, the coronavirus binds SARS-CoV-2 to TLR (angiotensin receptors), that activates NFκB signaling pathway and activates PI3K/AKT/mTOR signaling pathway as well as activates NLRP3 inflammasome pathway and production of Pro-IL-1β. All the above signaling pathways leading to cytokine storm are inhibited by rapamycin. Rapamycin blocks mTOR and finally inhibits cytokine storm including IL-1β and IL-6.

Also, COVID-19 disease leads to:

- CD8+ T lymphocytes senescence under cytokine storm and extensive SARS-CoV-2 replication
- Cytokine storm increases the numbers of CD8+ T cells expressing the senescent marker PD-1
- PD-1+ CD8 + T cells become unable to secrete IFN-γ, Perforin, and Granzyme, and to induce apoptosis of SARS-CoV-2 infected cells.
- Reduced T-cell counts who are more likely to progress to severe disease

Rapamycin reverses these effects in all COVID-19 patients, thus reduces or eliminates morbidity and mortality in patients with advanced age with comorbidities.

Our rapamycin protocol for COVID-19 is as follows, besides ICU resuscitative measures

1. Administer metformin 250-500 mg orally to augment and potentiate the rapamycin activity
2. After one hour administer 6 mg of rapamycin orally
3. Then take 2 mg of rapamycin daily for one to two weeks.

The dose of rapamycin can be adjusted. It is a mini-dose compared to the mega doses given as an immune suppressant in organ transplant with many complications. and the micro-doses given to mitigate aging to prolong lifespan and health span with inhibition of comorbidities.

To date, there is no one registered clinical trial of rapamycin, (Sirolimus) for the Treatment in Hospitalized Patients With COVID-19 Pneumonia. This is the time to start it instead of facing deadly consequences as a result of COVID-19 in the aged. It is "off-label, compassionate use and repurposing old drug."

Chapter 74

Magnesium

Magnesium is the second most plentiful element inside trillions of human cells and the fourth most abundant positively charged ion in the human body. Within the body's cells, it serves literally hundreds of functions related to enzymatic reactions and reactivation. The average human body contains about 25 grams of magnesium, *one of the six essential minerals that* ***must be supplied*** in the diet.

Magnesium powers our enzymes.

Magnesium is crucial to more than 300 enzyme-driven biochemical reactions occurring in trillions of cells in the body on a continuous basis. Like most vitamins, magnesium's role is primarily regulatory. It allows enzymes to function properly, which in turn enable a vast majority of the body's chemical reactions. Enzymes are the basis of the body's ability to function while supporting life. Enzymes, however, allow these reactions to occur without damaging the body's fragile tissues and organs. Yet enzymes do not function alone. Substances known as enzyme **co-factors** must regulate the functions of enzymes in order to control the rate of reactions within the body. These co-factors act as "keys" to switches within each enzyme, instructing it to start or stop activity. Magnesium is one of such elements and the most common co-factor in the body. Its presence is crucial to:

- Glucose and fat breakdown
- Production of proteins, enzymes and antioxidants such as glutathione
- Creation of DNA and RNA and to maintain their integrity and function
- Nerve cell synaptic homeostasis besides regulating cholesterol production

Thus, magnesium's function as an ***enzyme cofactor*** can be seen as analogous to the important role that our body's hormones play. The crucial difference, however, is that our body can manufacture most hormones itself using basic building blocks. *Magnesium, on the other hand, cannot be manufactured by the body,* it must be taken in food or as a supplement. In the same way that multiple bodily systems suffer in cases of thyroid malfunction or insulin resistance, such as heart beating and brain function, magnesium deficiency has far-reaching implications for the body's physiological homeostatic level of functioning.

Magnesium regulates our electrolyte balance.

Within every cell in the body, a proper balance of mineral content must be maintained. Magnesium's role in the healthy balance (homeostasis) of important minerals, such as calcium, sodium and potassium, affects the conduction of nerve impulses, muscle contraction, and heart rhythms.

These mineral exchange pumps perform one of the most vital functions of the cell membrane, regulating the electrical action potential inside and outside of the cell, and maintaining homeostasis of minerals in the body. Without constant efforts by exchange pumps, cells would be flooded with calcium and sodium moving in, and potassium and magnesium moving out as they strived to achieve an equilibrium. One such exchange pump, known as the "sodium-potassium" pump, pumps sodium out of the cell in exchange for potassium. Embedded in the cell membrane, the sodium-potassium pump is activated by magnesium inside the cell.

Magnesium deficiency

Magnesium deficiency impairs the sodium-potassium pump, allowing potassium to escape from the cell, to be lost in the urine, potentially leading to potassium deficiency (hypokalemia). Those with a known potassium deficiency, therefore, often do not respond to treatment until the magnesium deficiency is also corrected.

Similarly, magnesium's role in calcium regulation is pivotal to its role in maintaining heart and brain health. Magnesium is a known modulator of calcium, competing with calcium for entrance into heart muscle and brain nerve cells and keeping many cellular processes in balance.

The effect of magnesium on blood vessels is one of dilation, whereas calcium promotes contraction. Magnesium is also thought to antagonize calcium promotion of blood clotting.

Brain protection with magnesium L-threonate

- The mineral magnesium plays a critical role in the brain, protecting the function of synapses involved in complex cognitive processes such as learning and memory.

- Scientists at MIT developed a new form of magnesium called ***magnesium L-threonate,*** which is highly absorbable and has been shown to increase brain levels of magnesium to a much higher degree than other forms.
- Oral intake of magnesium L-threonate raises brain fluid levels of magnesium in rodents by 54%. It also increases synaptic density and increases production of NMDA receptors in brain cells (Sun Q, Weinger JG, Mao F, et al. Regulation of structural and functional synapse density by L-threonate through modulation of intraneuronal magnesium concentration. *Neuropharmacology*. 2016 Sep; 108:426-39.). Most importantly, several studies demonstrate that this boost to brain magnesium directly translates to improvement in mental function. Studies in animals and humans show that magnesium L-threonate improves and maintains cognitive function, even in older individuals with prior signs of cognitive decline.

Most importantly, several studies demonstrate that this boost to brain magnesium directly translates to improvement in mental function—cognition and memory.

We recommend magnesium L threonate 1000-2000 mgs to all nursing homes and extended care facilities residents as well as aging home residents. In the brain, magnesium is needed for the proper functioning of synapses (nerve interconnections) involved in complex tasks such as learning, cognition, and memory.

Magnesium is a critically important mineral required for the function of hundreds of enzymes in the human body, making it essential for nearly 80% of our metabolic reactions in our trillions of cells that make our body and brain. Magnesium L-threonate potentially prevents or reverses some age-related changes that contribute to cognitive decline, memory loss, dementia and other comorbidities seen in the aged.

Chapter 75

DeHydroEpiAndrosterone (DHEA)

Dehydroepiandrosterone (**DHEA**) is a hormone made from cholesterol and pregnenolone mostly in the adrenal glands, some in the testes and ovary. DHEA acts as a foundation material for production of the sex hormones testosterone and estrogen. It has some direct hormonal effects of its own throughout the body. As you age it decreases, creating comorbidity.

DHEA Functions

DHEA is a basic hormone needed to produce testosterone and estrogen. On its own, it also has many direct health-promoting effects throughout the body. Studies show that its ***low level is a predictor of earlier death*** (Barrett-Connor E, Khaw KT, Yen SS. A prospective study of dehydroepiandrosterone sulfate, mortality, and cardiovascular disease. N Engl J Med. 1986 Dec 11;315(24):1519-24. Life Extension July 2020 issue PP 46-54). Blood levels of DHEA correlate with longevity in primates (Kroll J. Dehydroepiandrosterone, molecular chaperones and the epigenetics of primate longevity. Rejuvenation Res. 2015, 20;18(4):341-6.).

Besides DHEA helping to produce sex hormones, it also has a widespread impact on health and disease. Low levels of DHEA have been tied to premature aging and shortened lifespan, along with an increased risk for:

- Cognitive impairment, and dementia
- Cardiovascular disease
- Osteoporosis, bone fractures, frailty
- Depression
- Sexual dysfunction
- Chronic inflammatory disorders

Maintaining normal levels of DHEA from young age into older age in nursing homes, extended care facilities, or at home will help fend off comorbidities and their ill effects that can be deadly (25-48%) in this population as seen in the recent coronavirus pandemic.

Aging and DHEA levels

DHEA levels drop drastically up to 80% or more by the time we reach 60-80s. An ideal DHEA-S blood level is 275 ug/dL-400 ug/dL. Most older people can achieve optimal results by taking 15-50 mg of DHEA daily besides other supplements. A daily dose is not necessary. Take it 2-3 times a week, what is called a pulsed dose.

The influence of *higher* DHEA levels studies show that the relationship between DHEA and **quality of life** is dramatic (Chen CY, Wu CC, Huang YC, et al. Gender differences in the relationships among neurosteroid serum levels, cognitive function, and quality of life. Neuropsychiatr Dis Treat. 2018;14: 2389-99.). The researchers found that in adults of various ages, *higher* DHEA levels corresponded to *better* results in three areas of besides better working memory and cognitive function:

1. **Physical health**, including levels of energy and fatigue, pain and discomfort, and sleep and rest
2. **Social relations**, personal relationships and sexual activity
3. **Environmental dimensions**, including participation in recreation and leisure activities

Chapter 76

Therapeutic agents' injection locally to treat cancers with insulin Metformin potentiation therapy

Introduction of unusual method of delivering the drugs to the site of pathology especially in the treatment of cancers

Therapeutic administration of ***"off label, compassionate use and repurposing old drug"*** discussed in the patents can be used to treat diseases to enhance the pharmacological effectiveness of the therapeutic agents, drugs or even supplements. This method we describe herein can cut the cost, reduce the dose and their adverse effects and achieve therapeutic goals for which the patient is treated. The methods and choice of therapeutic agents administered are weighed between benefits and harm in each case.

I have treated hundreds of cases of cancers of all kinds and all stages with local injection of appropriate chemo-therapeutic agents with insulin (insulin potentiation therapy—IPT) with excellent results. I have also administering lower doses parenterally, which can be easily labelled as "off label, compassionate use and repurposing old drugs" which gave the patient long, pain free, cost effective, quality life. I have instilled through the injection needle approachable to tumor site and their metastasis by injection locally with insulin the appropriate doses of chemotherapeutic and other drugs (up to 10-50% of the prescribed parenteral dose) along with administration of reduced or low dose parenteral administration therapeutic agents.

I expanded my study by adding 3-5 units of insulin into steroids injected epidurally, joints, or bursa with dramatic immediate and long-lasting pain relief and healing effects. There are thousands of local injection steroids done all over the world, and adding insulin will enhance the therapeutic effect of the steroids.

I used rapid acting insulin with Mineralocorticoids (a class of corticosteroids) instilled into external ear for the treatment of age-related deafness. I have added to the ear drops folic acid and EDTA depending upon the hearing affliction including Meiners disease and ringing in the ear. A cotton swab with insulin and an antibiotic is one of the most effective methods of rapid healing and gum infection after teeth extraction. It is a better alternative than expensive surgery and/or hearing aid.

We were the first clinic in the world dedicated to use local injection method on hundreds of cases of all types of cancer (sarcomas) effectively to prevent, curtail, stop progression of and / or cure the cancer.

Whenever I could inject a cancer or a benign tumor with a needle, they all got local injection of appropriate therapeutic agents with insulin (5-10 IU) besides systemic therapy. Insulin enhances the tumor killing effect of anticancer – anti-cell growth therapeutic agents many times. I used internal carotid artery to deliver therapeutic agents with insulin to treat brain tumors injected through internal carotid artery. I used intra peritoneal and interpleural infusion using a needle or indwelling catheter to deliver the therapeutic agents with insulin also successfully without complications and adverse effects.

The basic idea that insulin enhances effectiveness of drugs was conceived and used by Donato Perez Garcia (US Patent # 2,145,869) in 1939 to treat neurosyphilis and such. Later the idea was propagated and advanced by Donato's and Ayre (US patent 4,971, 951) and named the method "Insulin Potentiation Therapy" (IPT) which is used by many alternative cancer treatment physicians all over the world. I have used this method on most of cancer patients ***combined with local injection of therapeutic agents with insulin*** to enhance the effectiveness and to lower the dose of the chemo-therapeutic agents to mitigate the toxic effects.

The study by Dr. Alabaster and his colleagues in experiments supports the concept and the method I used, which tells us that insulin augments pharmacological effect of therapeutic agents' activity hundreds of times (Alabaster O, et al. Metabolic Modification by Insulin Enhances Methotrexate Cytotoxicity in MCF-7 Human Breast Cancer Cells. 1981, Eur J Cancer Clin Oncol, Vol. 17, No. 11, pp. 1223-1228, 1981). Every oncologist should use this method (local injection with insulin along with reduced dose parenteral administration) to treat all stages of cancers besides systemic chemotherapy. The same method can be used in treating COVID-19 in ICU and in severe cases to enhance the effectiveness of antiviral agents such as remdesivir, monoclonal antibodies, antimalarials and such. I have used it with antibiotics to treat MRSA infections successfully.

Using oral metformin as an alternative to using insulin intravenously

It is possible that use of insulin for non-diabetics is fraught with apprehension by most doctors and patients due to fear of hypoglycemia and lack of knowledge of how much to use. I developed the method where the same results as using the insulin in IPT can be obtained by patients, and physicians at home or clinic setting by use of oral metformin, a hypoglycemic antidiabetic anti-aging agent akin to insulin. I replaced parenteral administration of insulin as used by IPT physicians by oral intake of metformin hypoglycemic antidiabetic agent and named it **metformin potentiation therapy (MPT)**. This can be adopted by all practicing physicians and patients.

Hypoglycemic agents that can be used orally in our protocol instead of insulin to enhance pharmacologic effect of therapeutic agents of all kinds including anti -cancer drugs

A variety of antidiabetic hypoglycemic (to reduce sugar in the blood) compounds are known besides insulin. The best-known agents of this type include metformin, phenformin and buformin. Unlike the sulfonylureas, metformin does not induce release of insulin from the pancreas. It is thought that its effects are mediated by increasing insulin activity (reducing the insulin resistance) in peripheral tissues, reducing hepatic glucose output due to inhibition of gluconeogenesis and reducing the absorption of glucose from the intestine. Side effects associated with the use of biguanides include lactic acidosis, diarrhea, nausea, and anorexia. As we use this only as therapeutic modality with low doses, we do not expect to see these complications.

Due to massive increase in NIDDM due to morbid obesity, there is a huge market for antidiabetic therapeutic agents and many more such agents will come to the market with passage of time.

Metformin as an alternative to insulin, as oral "metformin potentiation therapy" (MPT)

We choose metformin as the ideal hypoglycemic agent to replace insulin that is used in IPT which I call **"metformin potentiation therapy" (MPT)**. As we used this therapeutic agent for a short time, we do not expect any long-term adverse effects as seen using it long term in NIDDM. It is inexpensive, easily available, its mode of actions is known and it is the fourth most prescribed drug in US. Besides, as a hypoglycemic agent, I recommend it as an anti-aging and anti-obesity therapeutic agent immaterial whether the patient is diabetic or not as described in another chapter.

Method of oral "metformin potentiation therapy"

Follow the below guide lines to use our method:

- Eat less food (calorie restriction) with fewer carbohydrates in the diet, if possible
- Take oral metformin 250-500-750 mgs depending upon body weight
- If NIDDM or IDDM, take 750-1000 mgs orally with a glass of water
- Wait 30 minutes for metformin to act in reducing the circulating post prandial glucose and exerting its hypoglycemic effect on the tissues
- Then administer oral or parenteral therapeutic agents of any and all kinds including chemotherapy agents, and antibiotics.
- Rest so that the therapeutic agents you take get into circulation.
- Any oral drug you take, take metformin orally and then ½ hour to 1 hour later, take the medication to augment the healing effects.
- One can use this method even if you take supplements including vitamins.
- If you develop tremors, sweating, increased heart rate, dizziness, or confusion, this may indicate low blood sugar levels and such. Counter it by eating a candy bar or drink a soda, eat fruit salad or ice cream to counter the hypoglycemic effects. For many years of using this method, I have never have faced these signs and symptoms.

This method can be easily adopted to treat COVID-19, any and all diseases. I advised most of my patients to take mini dose of metformin 30-60 minutes before taking most of the prescribed oral or parenteral medications especially antibiotics, hormones, supplements, anti-autoimmune diseases therapeutic agents, pain pills, arthritis meds, anti-cholesterol drugs, hormones replacement therapy, blood pressure medication, anticancer medications, rapamycin, lithium, and so on.

If one is treating coronavirus infection, bring blood sugar down to 60 mg% by using rapid acting insulin or oral metformin and then administer drugs such as remdesivir, antimalarials, vitamin C, and steroids such as hydrocortisone, or dexamethasone. You will save patients from increased morbidity and mortality.

Chapter 77

Corticosteroids: COVID-19 treatment

Indicated for severe COVID-19 cases

Headlines were all over the world from Oxford UK, about the benefit of using the corticosteroid (hydrocortisone/dexamethasone) for treating critically ill COVID-19 patients, bringing hope to those with the disease. It is not prescribed for all cases of coronavirus infection as soon as the person is diagnosed, but is only used in severe cases. We classified COVID-19 into 4 stages. The corticosteroids should be considered in late 3rd and 4th stages of COVID-19, for those who are in ICU fighting the disease, and administered only for a short time. Do not give it in the early stages of the disease to let the immune system attack the coronavirus. We have used corticosteroids to treat ARDS for decades; it is not new to us.

Our classification of "Different Stages of COVID-19" based on signs and symptoms and our treatment protocol

The coronavirus infected patients can have mild to severe symptoms as described below.

1. **First stage of coronavirus infection:** The persons are tested positive. This is the prodromal stage in the young without comorbidities. Use our mouth and nasal spray, take angiotensin receptor blockers telmisartan or losartan, take zinc oral and lozenge, use face mask, and take bed rest with nutritious diet. You do not need dexamethasone or other corticosteroids.
2. **Second stage of localized coronavirus spread:** The coronavirus spreads further from upper respiratory passages to deeper air ways, and the body's immune system inflammatory response is becoming more active, trying to combat the virus. The patient develops a sore throat, cough, aches, and fever as a manifestation of early infection. Follow the above protocol, and in addition the patient may need to take remdesivir and antimalarial drugs, and may need to start

intravenous vitamin C, thiamine hydrochloride and hydrocortisone infusion (Klenner, Marick, et al.).

3. **Third stage spreading of the coronavirus down to the air passages and some part of alveoli**: The viruses multiply and spread further down on to infect more and more cells, spreading to larynx, trachea and bronchial mucosal lining, and alveoli with fever, productive cough, burring in the chest, fatigue, and headache, due to inflammation by the immune system to rid of the virus. The patient needs to be hospitalized. Follow the above therapy. They may need the administration of dexamethasone if the symptoms such as cough become menacing with supplemental oxygen, can use vitamin C, and thiamine hydrochloride along with remdesivir zinc sulfate infusion.
4. **Fourth stage of coronavirus affecting the lungs, its surrounding microvasculature and virus battle with immune system:** The coronavirus has descended to alveoli and its vasculature which are richly endowed with angiotensin receptors II. The immune system reacts by producing cytokines which attack the virus which also attacks the lung parenchyma with pouring of inflammatory fluid to alveolar sacs and filling them, resulting in hypoxia and other systemic effects. The patient needs to be admitted to ICU. Follow the therapy as described in stage 3, intubate and provide ventilator assistance with 100% oxygen, administer corticosteroids such as dexamethasone. Administration of rapamycin may be considered to ward off the cytokines storm and out of control immune system. Think about it?

Effect of corticosteroids and COVID-19 on immune system cell

For decades we have used the corticosteroids to treat acute respiratory distress syndrome (ARDS akin to SARS). We were faced with secondary infection and many lung complications due to prolonged use of corticosteroids. Then we modified the regimen of corticosteroids administration. We gave corticosteroids such as hydrocortisone / dexamethasone until the symptoms of ARDS dissipated and then added antibiotics if needed with a very good outcome. Peter Horby, of the University of Oxford, said in a statement for treating COVID-19: "The survival benefit is clear and large in those patients who are sick enough to require oxygen treatment, so dexamethasone should now become standard of care in these patients. Dexamethasone is inexpensive, on the shelf, and can be used immediately to save lives worldwide" (June 16th 2020).

Kenner and Merrick recommended its use with high dose vitamin C, thiamine hydrochloride and hydrocortisone for septicemia and shock which is akin to severe cases of COVID-19 which responds to dexamethasone. I want these clinicians to add this triple therapy as described by Marik et al. (2016) explained in the patent with or without azithromycin and zinc. The results will be even more dramatic in saving the lives of the COVID-19 afflicted elderly with comorbidities. This method and the one used in Oxford results in suppression of immune system and inhibition of inflammatory cytokines production produced during COVID-19 affliction of the aged with comorbidities—a known effect of corticosteroids.

If one brings the blood sugar around 65 mg% by using insulin or oral metformin, and then administers dexamethasone and other therapeutic agents such as remdesivir described herein, the recovery of the COVID-19 patients is dramatic, with reduced morbidity and mortality.

The corticosteroids dexamethasone and such do inhibit the increase of immune cells (low-density inflammatory neutrophils), which become highly elevated in many of the patients whose COVID-19 condition becomes very severe resulting in Cytokine Storm. This elevation signaled a point of clinical crisis and increased likelihood of death within a few days if not treated properly according to a study by Jun Yan, M.D., Ph.D., professor of surgery and microbiology and immunology at the University of Louisville (June 2020). We believe that many of these patients recover if one adopts vitamin C, thiamine hydrochloride and hydrocortisone/dexamethasone therapy with antibiotics and antiviral such as remdesivir as the coronavirus infection advances to stage 3 and 4.

Dosing of corticosteroids

Dexamethasone 10-20 even up to 40 mg IV, and then 4 mg every 4-6-12 hours until maximum response or Hydrocortisone: 50 mg every 8-12 hours for seven days or until ICU discharge and/or symptoms of the disease improve, followed by a taper over three days. Dexamethasone dose of 40 mg have been used for multiple myeloma. Corticosteroids have a dramatic reversal of loss of vascular barrier function, i.e., capillary permeability, prevents capillary leakage of intravascular fluids into lung alveoli and stops the progression of the pathology in the lungs due to COVID-19. Corticosteroids suppress the hyperactive immune system cell, reduce inflammatory cytokines production and reduce the inflammatory damage to the lungs.

Chapter 78

Vaccination and curcumin

To enhance vaccine actions and mitigate their adverse effects

When the coronavirus vaccine comes to the market, millions will be vaccinated in addition to millions of vaccinations of children, flu vaccinations and other vaccinations. We describe how to mitigate the side effects after vaccination and to enhance their effectiveness by using bioactive curcumin and vitamin D3.

What are vaccines?

Vaccines are a biological preparation that provides active acquired immunity to a particular disease. A vaccine typically contains a disease-causing microorganism and is often made from weakened or killed forms of the microbe, its toxins, or one of its surface proteins. The agent stimulates the body's immune system to recognize the offending agent as a threat and destroy it, in future years. Vaccines can be prophylactic (to prevent or ameliorate the effects of a future infection by a natural or "wild" pathogen), or therapeutic (e.g., vaccines against diphtheria, as well as against cancer, snake bite, and Alzheimer’s which are being investigated)

The administration of vaccines is called vaccination. vaccination is the most effective method of preventing infectious diseases; widespread immunity due to vaccination is largely responsible for the worldwide eradication of smallpox and the restriction of diseases such as polio, measles, mumps, rubella and tetanus from much of the world. Vaccination and antibiotics have played a major role in expansion of world population with extended life besides improved living condition with proper nutrition and clean water along with development of antibiotics.

Terminology: vaccination and immunization

There is a technical difference between vaccination and immunization, as the National Health Service (NHS) explains: “Vaccination means having a vaccine – that is actually getting the injection, or nasal spray or oral vaccine.

Immunization means both receiving a vaccine and then becoming immune to a disease." There are many kinds of vaccines from infectious disease to cancers and neurodegenerative diseases. A detailed list is available on the internet published by CDC of Atlanta.

Vaccines are much safer as community immunity.

Natural immunity happens after you get sick with an infectious disease. But diseases can be serious — and even deadly. ***A vaccine protects you from a disease before it makes you sick.*** Did you know that some people — like infants and people with weak or failing immune systems (like people with HIV/AIDS or cancer) — may not be able to get many of the vaccines that protect us from serious diseases? The good news is that when you get vaccinated, you're also protecting the unvaccinated people around you. This is called ***"community immunity."*** You may not know that penicillin and vaccination, besides some other factors, are responsible for longevity and world population increase.

Vaccination side effects and curcumin

Most people don't have mild or serious side effects from vaccines. The most common side effects include:

- Pain, swelling, or redness where the shot was given
- Mild fever, chills
- Feeling tired, headache, muscle and joint aches, draining lymph node pain.

These side effects are a sign that your body is starting to build immunity (protection) against a disease. For children after vaccination, we recommend giving bioactive curcumin orally or mixed with milk (250-500 mg bioactive curcumin in feeding milk bottle, or mixed with semisolid or solid foods) and other food to reduce these side effects at the same time enhance the immunity development, by up regulation of immune system cells such as T regulatory cells and downregulating cytotoxic T cells effects, up regulating glucose uptake by cells and reducing the toxic production of protein, re-setting the intestinal flora – microbiomes to promote health.

Due to curcumin anti-inflammatory, anti-oxidants, anti-septic, anti-coagulant, anti-aging, anti-tumor, anti-cardio-vascular diseases, anti-neurodegenerative diseases, anti-Alzheimer's and Parkinson's effects, it helps to tame down the local inflammation at the site of the vaccination. Free radicals produced due to vaccination and immune system stimulation effects that cause flu like symptoms, fever and swelling are reduced by the bioactive curcumin with vitamin D3, thus reducing the adverse reaction of vaccination, lower the fever and local tissue reaction at the site of vaccination. One can mix the bioactive curcumin with milk and food and feed the infant before and after vaccination to ward off their ill effects.

What about serious side effects?

Serious side effects from vaccines are rare. For example, if 1 million doses of a vaccine are given, 1 to 2 people may have a severe allergic reaction. Keep in mind that getting vaccinated is much safer than getting the diseases vaccines prevent.

The difference between antitoxin and vaccine is that antitoxin is an antibody that is capable of neutralizing specific toxins that are causative agents of disease (diphtheria, tetanus, snake anti-toxin, antibodies from plasma of post COVID-19 patients etc.), while vaccine is (immunology) a substance given to stimulate the body's immune system to produce antibodies and provide immunity against a specific disease, prepared from the agent that causes the disease, or a synthetic substitute. For example, tetanus antitoxin confers immediate passive immunity lasting about 7 to 14 days and tetanus antigen active immunization lasts for 7 years or more.

What if I feel sick after getting vaccinated?

Report adverse reactions to Vaccine Adverse Event Reporting System (VAERS). Take Tylenol or Motrin, (never aspirin) with bed rest and nutritious meals with fluids. If a child develops fever, they will recover within a week. Add bioactive curcumin 250-500 mg to the feeding liquid or solid foods starting in the morning of the day of vaccination and continue for one to two weeks.

What is Reye's syndrome related aspirin intake?

Aspirin is contraindicated in children up to 8, and maybe even up to 17 years old, to prevent Reye's syndrome which starts with flu like symptoms when taken. Reye's syndrome typically affects children aged 6 to 12 and involves the swelling of the brain and damage to the liver. If left untreated, it can lead to death if you cannot provide medication designed to reduce brain swelling and help the liver rebound.

Signs and symptoms include extreme lethargy, frequent vomiting, irrational behavior, irritability, paralysis or weakness in the limbs, seizures, hallucinations, confusion, extreme lethargy, and lessened consciousness). The best treatment for Reye's syndrome is bioactive curcumin, vitamin D3, 5000

IU, vitamins C and B_1; besides other measures such as hospitalization, IV fluids (glucose, mannitol, and electrolytes), vitamin K, platelets, plasma to prevent bleeding, diuretics to reduce brain swelling, cooling blankets to reduce high fever and such. IV infusion of vitamin C, thiamine and hydrocortisone/dexamethasone may be needed as described under sepsis chapter to reduce the brain swelling and nonspecific brain inflammation besides diuretics. In unrelenting cases, mini-doses of rapamycin may be tried to reduce the brain inflammation and swelling.

Vaccination: fear of autism spectrum disorder (ASD) development due to childhood vaccines

Many studies have looked for a link between vaccination and autism. The research clearly shows that **vaccines don't cause autism**. We do not want to discuss any controversy about the preservatives in the vaccine to cause Autism spectrum disorder (ASD). Many believe that vaccines and their mercury preservative can cause ASD. So far there is no substantiated evidence to that effect. It is important to note that there may be a brain inflammation type of reaction due to non-specific protein content of the vaccine that may trigger some form of ASD like symptoms. The benefits that mass vaccinations provide far outweighs any presumed or real ill effects it can bring about.

BCG vaccination and curcumin and its effects on other vaccinations

Scientists from India have found that BCG – the world's only vaccine against tuberculosis (two billions have received it since its development) can be made more effective if nano-particles of curcumin, the main component of the popular kitchen spice turmeric, are used in tandem with the shot according to team leader Gobardhan Das and Anand Ranganathan from JN University's Special Centre for Molecular Medicine in Delhi. "We have shown that a booster dose of BCG together with nano-curcumin gives superior vaccine efficacy. This opens up a new area of vaccine research and has huge implications for other vaccinations as well. We are ready for clinical trials."

I think that the nanocurcumin not only creates an environment conducive for priming and activating the cells that fight against the enemy (immune system T cells) but also enhances the production of two key immune cells that battle with the bacteria. We advocate giving the children bioactive curcumin 250-400 mg in milk or food after vaccination. It has anti-oxidants, anti-inflammatory, anti-cancer, anti-cardio-vascular diseases, anti-neurodegenerative diseases, anti-scarring, anti-septic, anti-coagulant and anti-fat-protein clump accumulation effects besides it making vaccination more effective. If one fears autism after vaccination, they should give bioactive curcumin with vitamin D3, which may prevent or inhibit the brain inflammation and its related effects of vaccination.

Bladder cancer with BCG vaccine, rapamycin and curcumin

According to Kamat et al. (Cancer Res. 2009 Dec 1;69(23):8958-66) curcumin potentiates the anti-tumor effects of BCG vaccination against bladder cancer. Curcumin acts by down regulating NF-kB and upregulation of tumor necrosis factor-related apoptosis-inducing ligand (TRIAL) regulators that has anticancer effect. It is interesting to note that the curcumin alone reduced the bladder tumor volume, but a significantly greater reduction was observed when BCG and curcumin were used in combination showing that the curcumin potentiates the anti-tumor effect of BCG vaccine. Further, a peptide, uniquely produced by subsets of dual T and B lymphocytes, was found to amplify auto-reactive cells T cell attacks on various tissues and may prevent and/or lower these tissue toxic proteins. I add that using intravenous fluid at ±100.00 degrees Fahrenheit and filling the bladder with BCG vaccine with appropriate doses of chemo-therapeutic agents, insulin, (now I will add rapamycin) and curcumin will be very effective in curing, curtailing, or curbing bladder cancer. Now I have included mini-dose rapamycin once a week besides using with bladder filling fluid to suppress the tumor growth by inhibiting the nutrition uptake and utilization by cancer cells in a patient with bladder cancers with metastasis.

Vaccination and sudden infant death syndrome (SID)

The death of a child due to sudden infant death syndrome, choking, by parents (suffocating) who are inebriated and falling on the sleeping child, lung infection, childhood infections, traumas, autism, and such have been blamed on vaccination. The autopsy of the dead baby of one of the leaders of the antivaccine movement showed that the baby accidentally suffocated while co-sleeping with her mother, who blamed vaccination for the death, and raised money to pay for their campaign, becoming a fake star in the anti-vaccine world. *I attribute the population explosion and longevity of life all over the world to development of vaccination, antibiotics, and improved food and water resources along with other advancement in medicine and advanced communication (internet and cell phones).*

BCG vaccination to reduce body and brain inflammation to improve overall health

Billions of people are BCG vaccinated all over the world. It is one of the simplest methods to prevent, curtail, or tame down bodily inflammation that triggers and/or contributes to many of the comorbidities in the young and old. It may cure or curtail type I as well as type II diabetes (Faustman DL edit. The value of BCG and TNF in Autoimmunity. 2018. Academic Press). It is known that that IL-2 expands T regulatory (Treg) immune cells populations and these cells are neuroprotective [Baek H, Ye M, Kang GH, Lee C, et al. Neuroprotective effects of CD4+CD25+Foxp3+ regulatory T cells in a 3xTg-AD Alzheimer's disease model. Oncotarget 2016;7: 693 47–57]. In the mouse model, it activates and cause amplification of Treg cells by peripheral IL-2 treatment that decreased plaque formation in the brain and restored cognitive function [Dansokho C, Aucouturier P, Dorothé EG. Beneficial effect of interleukin-2-based immunomodulation in Alzheimer-like pathology. Brain 2016; 139:1237–51. Gofrit ON. et al. Medical Hypotheses 123 (2019) 95–97]. These Tregs cells kill or inhibit the autoreactive T (microglial cells in the brain) cells and thus prevent the body and brain inflammation that contributes to cardio-vascular diseases, cancers, neurodegenerative diseases such as Alzheimer's and Parkinson's and other afflictions of the skeletal system and organs.

BCG vaccination in elderly people may prevent the development of AD and other comorbidities and/or delay their onset by a few years.

The above simple protocols ought to be adopted in nursing homes, extended care facilities, and the aged living independently to maintain health and curtail the progression of many other diseases including dementia (Alzheimer's), arteriosclerotic vascular disease, cancers and such.

BCG vaccine action:

a) BCG is shown to increase TNF (tumor necrosis factor) which is beneficial in autoimmune diseases both by eliminating the autoreactive T cells that attack an individual's healthy tissues—in the case of type 1 diabetes, pancreatic islets—and by inducing production of regulatory T cells (Tregs) that could prevent an autoimmune reaction by destroying autoreactive T cells. It acts against the inflammation producing autoreactive microglial cells to reduce the brain inflammation that is one of the most important contributors to neurodegenerative diseases including Alzheimer's and Parkinson's.

b) BCG vaccine produces shifting of the process of glucose metabolism in the neurons from oxidative phosphorylation, the most common pathway by which cells convert glucose into energy, to aerobic glycolysis, a process that involves significantly greater glucose consumption by all cells including nerve cells, thus providing energy to neurons due to insulin resistance as you age.

c) There is likelihood that BCG vaccination produces TNF by immune system cells which acts against cytokine producing to reduce body and brain inflammation, thus acting as a vaccine against many comorbidities aging population suffer including Alzheimer's.

"The clinical effects and the proposed mechanism demonstrated are exciting and add to the emerging consensus that the BCG vaccine can have a lasting and valuable impact on the immune system," commented Mihai Netea, Ph.D., professor in the department of internal medicine at Radboud University Medical Center in The Netherlands.

BCG vaccination is ***contraindicated*** in the immune compromised such as HIV AIDS, those undergoing chemotherapy and radiation, or immune system suppression due to any number of reasons.

How to grow virus, microbes and unicellular organisms fast to test and produce vaccines and therapeutic agents

We have a simple and efficient method to enhance the growth of viruses, microbes or unicellular organism in the egg yolk, on a Petri dish, test tubes or in vats. This simple method may help us to augment the rapid development and production of active and passive vaccines such as coronavirus. Pharmaceutical companies and research laboratories, please contact me if you want to use our method.

Part IV

Abridged biography of the author

Chapter 79

Abridged biography of the author

Dr. T. R. Shantha has been a member of the faculty of Emory University School of Medicine, Medical College of Georgia, Grady Memorial Hospital, Georgia Baptist Hospital, Columbus medical center, and is presently a visiting and Emeritus professor at JJM and BMC Medical Colleges. He has published more than 100 research articles since 1962, in peer-reviewed reputable journals, including Nature (7 papers), Science, NEJM, J Urology, Anesthesia, Anatomy, Exp. Eye Research, American J of Physiology, etc. He discovered Terbutaline as a treatment for Priapism, which is now used all over the world as the first line of treatment in the emergency rooms and by urologists. He is the discoverer of local injection of chemo-therapeutic agents with low dose insulin for all size tumors. He also discovered administering metformin before administering any therapeutic agents and even supplements with enhance the absorption and effectiveness.

He developed a simple airway that can hold the jaw and tongue protruded without airway obstruction during and after anesthesia (used 351 million times/year). He also has developed devices for the treatment of obstructive sleep apnea and snoring and received 6 patents on these topics. He has multiple patents issued and published. He has won numerous awards for his academic contributions, including AMA, GAPI and VSNA distinguished physician-scientist awards. He was one of the nominees for the Nobel Prize in Physiology and Medicine in 2007 for his and Dr. Bourne's research work on the membranes of the nervous system and its associated sub perineural epithelial space, a pathway to the brain and spinal cord from peripheral nerves discovered at Emory University. At present he is working on the treatment of Alzheimer's, Parkinson's, sleep apnea, snoring, prostate hypertrophy and cancer; migraine, autism, myopia, anti-aging methods and such.

Dr. T. R. Shantha's Bio Data (CV)

Degrees, fellowships and diplomate: year conferred

LMP: University of Mysore, 1955
M.D.: University of Calcutta, R.G. Kar Medical College, 1958
Ph.D.: Emory University School of Medicine, 1962
F.Z.S: Fellow: Zoological Society of London 1964
F.R.M.S: Fellow: The Royal Microscope Society, England 1965
FACA: Fellow: The American College of Anesthesia. 1973, No. 5780
D.A.B.A: Diplomate: American Board of Anesthesiology, 1973, No. 6106
D.A.P.M: DABA sub specialty pain management 1992, No. 41208.
Diplomat of the American Academy of Pain Management
Diplomat, Clinical Electro Medical Research Academy, 1996,
Certified in Hyperbaric Medicine, May 1999 at PRMH-South Carolina.
In the past he has been a member of several professional and research organizations.

Dr. T. R. Shantha with Prof. G.H. Bourne: the discovery of perineural epithelium (PE) and sub perineural epithelial space — its role in health and disease at Emory University

1. Perineural epithelium (PE): J Cell Biology, Rockefeller University NY, 14(2):343-346 (1962). This first research paper was reviewed by Dr. George Palade who won the Nobel Prize (on endoplasmic reticulum 1974)
2. Facial and trigeminal PE. Acta Anatomica 51:112-124 (1962).
3. Perineural epithelium. J Anat Lond 96(4):27-37 (1962).
4. Perineural epithelium. Acta Anatomica 52:95-100 (1963).
5. Pacinina corpuscle PE. Ame J.Anatomy 112(1):97- 109 (1963).
6. PE in vagus nerves. Acta Anatomica 52:95-100 (1963).
7. PE significance. J Nature 199, 4893:577-579 (1963).
8. Vagus N-PE. Journal Nature 199, 4893:577-579 (1963).
9. Autonomic ganglion PE. J **Nature** 198, 4868:607-608 (1963).
10. Whale and Shark PE. Nature 197, 4868:702-703 (1963).
11. Ele. Mic. PE. Acta Anatomica 52, 3:193-201 (1963).
12. Sympathetic N-PE: Z. fur Zellforschung 61:742-753 (1964).
13. PE & Wallerian degen. Anat. Record (US) 150(1):35-50 (1964).
14. PE & Pacinian corpuscle. Acta Anatomica 60,199-206, (1965)
15. PE & Perineural epithelium. Science 154:1464-1467 (1966).
16. PE & Pacinian corp. Am. J. of Anatomy, 118: 461-470 (1966).

17. PE & Pacinian Cor-Olfactory bulb. Nature, 209:1260 (1966).
18. PE & Motor end plate-PSC. Int Rev Cytol 21:353-364, 1967
19. PE & Rabies passage. Bulletin of the WHO-33:783-794 (1965).
20. PE—Rabies spread. Bull WHO, 38(1):119-125 (1968).
21. PE on Muscle spindle. Histochemie 16: 1-8, (1968).
22. "Perineural epithelium" 98 pages review, in Structure and Function of Nervous Tissue, Academic Press, G. H. Bourne. Edi. 1970.
23. Epidural anesthesia. PE-Villi.Anesthesiology.37:543-557, 1972
24. PE & Subdural blood patch. NEJM. 325(17):1252-1253 (1991).
25. PE and Virchow-Robin space in Peripheral nerves. ASRA Congress at Tampa, Regional Anesthesia 1992
26. Olfactory mucosa transport to brain by passing BBB through sub-perineural epithelial space. Drug develop. and delivery J-F 2017,
27. PE & Coronavirus route of transfer to brain though sub perineural epithelial space: "coronavirus-COVID-19 prophylaxis and treatment" Patent pending (2020).

Other publications

1. LSD-25 and BOL-148. Acta Neuropath 3:29-39 (1963).
2. Autonomic neuron in PNS, Nature 198, 4868:607-608 (1963).
3. LSD-25 and BOL-148. Acta Histochem Bd 17:259-267 (1964).
4. Olfactory histochemistry. J. Nat. Can. Inst 35(1):153-165 (1965)
5. Macro photography. Stain Tech 40(5):309 (1965).
6. Olfactory glomeruli. Histochemie 5:125-129 (1965).
7. Golgi neurons. cortex. La Cellule 65(2):201-209 (1965).
8. Golgi in neurons. Acta Histochemica 22:155-178 (1965).
9. Golgi-IV Cerebellum. Z. Zellforschung 68:699-710 (1965).
10. Peritoneal cells. Histochemie 5: 331-338 (1965).
11. Golgi. III. olfactory bulb. Expt Cell Res 40:292-300 (1965).
12. Double central canal SC. Nature; 209:729 (1966).
13. Golgi in spinal cord neur, Ann Histochemie 11:337-351 (1966).
14. Golgi sympathetic GC. Cytologia 31:132-143 (1966).
15. Hypothalamus, Z Zellforsch 79:76-91 (1967).
16. On spinal cord. Expt Brain Res 3:25-39 (1967).
17. Area postrema. Acta Histochemica 27:42-54 (1967).
18. Cerebellum. Acta Histochemica 27:129-162 (1967).
19. Olfactory bulb. Histochemie 10:224-229 (1967).
20. Monoamine oxidase. Brain Res 6:570-586 (1967).
21. Coconut—coronary. South Medical Journal 70:766, 1967.

22. Tongue—taste buds. Acta Anatomica 69:183-200 (1968).
23. Nucleus basalis. Acta Histochemica Bd 30:96-108 (1968).
24. Subfornical organ. Histochemie 13:331-345 (1968).
25. Choroid plexus. Histochemie 14:149-160 (1968).
26. Cerebral cortex. Acta Histochemie 30:218-233 (1968).
27. Synthesis of vitamin C. Histochemie 18:293-302 (1969).
28. Hypoglossal nucleus. Acta Histochemie 32:18-36 (1969).
29. Sleep left side down and keep the doctors away, Inspirator International, Vol 11/1997, Edition 4/5, page 24-28.
30. Risks of Inhalation Insulin in Life Extension, Sept. 2007 issue in USA 78-83. This research resulted in withdrawal of the drug by Pfizer saving thousands of patients from future lung cancer.
31. Inhalation Insulin and Oral and Nasal Insulin Sprays for Diabetics: Panacea or Evolving Future Health Disaster: Part 1; Townsend Letter December 2008; Pages 94-98
32. Inhalation Insulin and Nasal Insulin Sprays for Diabetics: Panacea or Evolving Future Health Disaster: Part 2 Townsend Letter January 2009; Pages 136-140
33. Health Risks of Inhaled Insulin, Life Extension, February 2015, Vol. 21, Issue 2, pp 62-68
34. Pulmonary delivery of Insulin to treat diabetes—a debacle. Drug Development & Delivery, May 2016 pages 52-57
35. Transport of therapeutic agents from olfactory mucosa to the brain by passing BBB, Drug development and delivery Jan-Feb Issue of 2017, vol 17, # 1, 32-37.
36. Stem Cell and Microorganism from the Olfactory Mucosa of the Nose to the CNS Bypassing the Blood Brain Barrier (BBB) - Part II 2021

Publications on the structure and function of the eyes:

1. Radial bands in optic N. Nature 4860:1215-1217 (1962).
2. Radial bands—optic N. Exp Cell Res 32:196-199 (1963)
3. Corneal endothelium. Acta Ophthal 41: 683-688 (1963).
4. Acid phosphatase Acta Histochemie Bd 18:317-327 (1964).
5. Monoamine in eyes. J Histo and Cyto12:281-287,1964).
6. On Sclera. Acta Anatomica 58:396-305 (1964).
7. Arachnoid villi—optic nerve. Expt Eye Res 3:31-35 (1964). Before this discovery was made, we did not know how the optic nerve CSF pressure is released during coughing, bearing, lifting weight, straining, etc., can create visual disturbances.

8. B-glucuronidase. Histochemie 3:413-421 (1964).
9. oxidative enzymes. Acta Anat 57:193-219 (1964).
10. Descemet's Mem eye. Nature 202, 4939:1347-1348 (1964).
11. Golgi apparatus EYE. J Anatomy 1965, 99(1): 103-117.
12. choroid—pia-Ara. Acta Anat 61:379-398 (1965).
13. Oxidative & dephos. Enzymes. Am J Ophthal 60(1):49-55 (1965).
14. Cyclic 3', 5'—nucleotide, Histochemie 7:177-190 (1966).
15. Phosphorylase. Histochemie 7:80-95 (1966).
16. Metabolic effects. Amer J. Physiology 210:1249-1256 (1966).
17. Corneal endothelium, Acta Ophthal. 41: 683-688 (1963).
18. Histochemistry. Amer J Ophthalmology 60(1):49-55 (1965).
19. Local Anes. spread. Anesthesiology 73 (No 3A): A850 (1990).
20. Glaucoma—metformin. Life Extension J 2017. P 86-95
21. Eye—***Axenfeld loops***. Function: under publication 2021

Publications in anesthesiology and oral airways since 1972:

1. Epidural Anesthesia. Anesthesiology 37:543-557, 1972.
2. Hiccup. Anesthesia and Analgesia 52:822-824, 1973.
3. Anesthesia and Analgesia 52:279-280, 1973.
4. Priapism. Anesthesiology 70: 707-709 (1989).
5. Priapism. Journal Urology 141:1427-1429 (1989).
6. Priapism. Human Sexuality pp 63-67 (April 1990).
7. Priapism. Anesthesiology 73(No 3A): A850 (1990).
8. Horner's. Anesthesiology 62(No. 3A): A192 (Sept 1990).
9. Anesthesis: Surgi Lapar & Endoscopy 1(3):173-178 (1991).
10. Causalgia. Anesthesia and Analgesia 73:502-510 (1991).
11. Epidural cath. Anesthesia and Analgesia 73:502-510 (1991).
12. Subdural blood patch. NEJM. 325(17):1252-1253 (1991).
13. Subdural space: Brussels, Reg. Anesthesia 17(3S):85 (1992).
14. Virchow-Robin space in peripheral nerves. ASRA Congress at Tampa, Regional Anesthesia 1992, World Congress of Anesthesiologists, June 1992.
15. Unilateral bronchospasm. World Congress of Anesthesiology, held in Hague, June 1992.
16. Unilateral bronchospasm. Anes. Analgesia 74:291-293, 1992.
17. Sub-cricoid intubation. B. J. of Anesthesia 68:109-112 (1992).
18. Epidural space, Anesthesiology 77:216-218 (1992).
19. Inter-pleural Analgesia, Pain Digest 1992, Vol. II, 18-27.
20. Inter-pleural Analgesia, Pain Digest, 1992, Vol. II, 3-17.

21. Tetracycline Pleural. Pain Digest, 1992, Vol. II, 28-29.
22. Stellate block. Int. Symp. reg. Anesa, 1 1996, Auckland, NZ.
23. Cervical Plex B. Int. Symp. reg. Anesa, 1 1996, Auckland, NZ.
24. Occipital Blo. ASRA 1997. Reg. Ane. Vol.22, # 2S, 109, 1997
25. Developed new oral airway used during 351 million times a year under anesthesia around the world, New Oral Airway presented at Annual ASA meeting at SF.2018, **has 4 patents issued**.
26. Devices for snoring and obstructive sleep apnea (6 patents)

Review Articles:

Methods Achievements Exp Path 1:52-95 (E Gajusz and G Jasmin, Eds.) S. Karger, Basel/New York, 1966.

Perineural epithelium in Structure and Function of Nervous Tissues. Volume I. pp 379-458 (GH Bourne, Ed.) Academic Press, NY. 1969. For this work, he was nominated to receive Nobel Prize in Physiology and Medicine for the year 2007.

The Chimpanzee Vol. I (GH Bourne, ed.) 3 chapters, pp 188-237. 238-305. 306-368 published by Karger, Basel, New York, 1969.

Structure and Function of Nervous Tissues. Vol. II (GH Bourne, Ed.). Academic Press, New York. 1970.

Books published:

Scientific/Medical Books: From Emory University and Yerkes Regional Primate Center of Atlanta.

1. A stereotaxic atlas of the brain of the java monkey (Macaca Irus). S. Karger, Basel, Switzerland, 1967.
2. A stereotaxic atlas of the brain of the cebus monkey (Cebus Apella). Oxford University Press, London, 1967.
3. A stereotaxic atlas of the brain of the tree shrew (Tupaia Glis). The Williams and Wilkins Co., Baltimore, 1968.
4. Enzyme histochemistry of the rhesus monkey brain (Macaca Mulatta). The Academic Press, New York, London, 1969.
5. Your breast and its care. Fredric Fell Publishing NY. 1974.
6. AIDS-HIV: A prescription for survival. Int. Pub. House.1991.
7. Coronavirus – COVID-19: Prophylaxis and Treatment. Patent Pending Methods. Amazon books, Shan Publishing. House 2020

Books on Jesus:

42 years of Jesus study resulting in the following books:

1. Resurrection Journey: Jesus' last 40-Days. Amazon books, Shan Publishing House. 2012
2. Jesus' out of body experience: Amazon books, Shan Publishing House 2014.
3. Jesus Christ and 3. Amazon. Amazon books, Shan Publishing House 2019.
4. How Jesus died and rose from the dead: The Truth. Amazon books, Shan Publishing House, 2019.
5. Last 40 days of Jesus' life, From Sepulcher to Mount of Olives: Amazon books, Shan Publishing House 2021.

Other books under preparation

1. Sleeping left side down (SLD): for health and longevity 2021
2. Freedom and Equality for all: Civil rights movement, people who made it happen in USA. Needs a publisher, 2021
3. How to prevent, delay, or curtail Alzheimer's disease
4. Indian-Gold: Book on turmeric and curcumin
5. Rapamycin—To prevent aging, comorbidities, and live longer
6. M. K. Gandhi: Fasting and its effect on his body and brain
7. Judas the Faithful: Under preparation
8. The world's most positive thinker: Under preparation
9. Rani from Bangla: Gripping story girl trapped in Bangladesh. Movie rights available on this and any of the Jesus publications.

Research Presentations:

Presented dozens of research papers from Emory University in research meetings since 1962—too many to name. Latest are:

1. Ft. Lauderdale, FL: HBO-HBA for CP and the Brain-Injured, Hydrogen Peroxide and Ozone for Cancers etc. 2004
2. ICIM: on Cancers, Chemo and Radiation. October 9, 2004
3. Las Vegas: HBO For Cancers and Aging. 2004
    a. Cancun Insulin Potentiation Therapy (IPT) conference 2004:
    b. Discovery of Insulin and IPT Amazing History
    c. High Dose Methotrexate Therapy Using IPT
    d. Local Injections of Tumors
4. Two and Three Cycle IP: June 28 - July 1, 2004
5. HBO LV -ACAM and anti-Aging Seminar. 2005/2006
6. Hanoi: ROACON: Rabies in Asia conference in Hanoi on September 10th 2009. Rabies treatment to cure rabies. 2009

7. Ottawa: RITA XX: Rabies Cure, Canada. October 19th 2009.
8. San Francisco: New Oral Airway. Presented at ASA annual meeting, Nov. 2018. Received 4 patents on the device.

Dr. T. R. Shantha's lists of patents issued and published

1. US20120156202: eye—macular degeneration (AMD)
2. US20120101033: retinitis pigmentosa treatment
3. US20110294730: method of treating glaucoma
4. US20160074480: New methods of treating tuberculosis
5. US20120323214: Alzheimer's disease treatment
6. US20120157377: methods to enhance night vision
7. US20120179090: transmucosal delivery of drugs
8. US20120128683: autism treatment (expanded now)
9. US20120003296: treating dry eye syndrome
10. US20110020279: rabies cure (next patent in progress)
11. US20110021974: retinitis pigmentosa treatment
12. US20110217260: enhancing eyelash and eye brow
13. US20150080785: Alzheimer's disease treatment
14. US20090304776: transmucosal delivery—drugs—insulin
15. US20110052678: method for treating AMD
16. US20100172865: methods of enhancing hair growth
17. US20150073330: Alzheimer's disease
18. US20140012182: Alzheimer's disease treatment
19. US20140035460: high efficiency light bulb
20. US20120203141: suppressing pain in digits
21. US20120066910: rotating utility knife scraper
22. U9254219: snoring and obstructive sleep apneas
23. 5195965: treatment of human vaginal warts
24. 9072613 snoring and obstructive sleep apnea treatment
25. 8226228: cooling and heating eyeglasses
26. US20110299027: cooling and heating eyeglasses
27. 8206336: transdermal local anesthetic patches
28. US20120234331: snoring and obstructive sleep apnea
29. US20120234332: snoring and obstructive sleep apnea
30. US20140184111: electric motor drive for energy savings
31. 8853978: electric motor drive capture for energy savings. It will save between 40-50% of electricity. Device is available.
32. US20120255561: snoring and obstructive sleep treat.
33. D635392: wok—Advanced, practical, easy to cook

34. 7883488: transdermal local anesthetic patches
35. US20090312706: transdermal local anesthetic patch
36. US20110081333: treatment bags under eyes
37. 7954900: leg supporting device for vehicular travel
38. US20110020252: method of long human skin tanning
39. US20110089725: leg supporting device—vehicular travel
40. US20110080739: light with replaceable light source
41. D885558: oral airway device
42. US20110166498: transdermal patch—electromotive force
43. US20110155143: anti-snoring and sleep apnea
44. D849233: oral airway device
45. D849234: oral airway device
46. US20090311311: transdermal anesthetic patch
47. US20090045715: modular lighting apparatus
48. 7905626: modular lighting apparatus
49. 7883487: transdermal anesthetic patches
50. 7804233: light bulb and method of use
51. 5735817: sphenoidal stimulation—pituitary and brain
52. 5792100: sphenoidal stimulation—pituitary and brain
53. US Patent Filed on 7-14-20: Coronavirus prophylaxis and treatment

Patents applied for and/or under preparation:

Patents on CNS diseases: Funds for Parkinson's treatment research are actively sought. If successful we can also treat PTSD, TBE, stroke, Alzheimer's (applied), senile brain atrophy, addiction, MS, ALS, depression and neurodegenerative—CNS diseases.

1. Coronavirus killer device to neutralize the viruses as they are floating in the droplets and aerosolize besides wearing a mask.
2. Prevention and treatment: prostatic hypertrophy and cancer
3. Bladder cancer, interstitial cystitis
4. Prevent breast cancer and breast pain
5. Migraine: New easy methods of treatment
6. Myopia treatment device
7. Energy-box—Power Cube—to save energy
8. Modified socks and underwear for the aged
9. Tinnitus, Meniere's syndrome, deafness
10. Obesity, hair loss, dandruff
11. Prepuce condom—circumcision and STID
12. Wine saver after opening the bottle to drink later

Charitable works by Dr. T. R. Shantha and his family:

1. Donated $35,000 of lab equipment and supplies to the Morehouse School of Medicine, 1983.
2. Donated $20,000.00 for the addition of the Tower Building at Georgia Baptist Medical Center in Atlanta, Georgia.
3. Contributed for India-America Cultural society of Atlanta.
4. Built Totada Ramaiah Gowramma Memorial Girl's High School (1968-2010), Hosadurga, Chitradurga district costing thousands of dollars. Thousands of girls have graduated from the school since its opening in 1988.
5. Major contributor for the remodel of Rameswara Temple and Sangameswara Temple in Janthikolalu (birth place of Dr. Shantha).
6. Volunteer Physician at Hands of Hope in McDonough Georgia for years treating the need and needy.
7. Volunteer Professor, 1970 -1975, Emory and Grady Hospital of Atlanta.
8. Volunteered as Clinical Professor for Medical College of Georgia for 5 years at GBH.
9. Provided jobs to hundreds of farmers and educated in Tamil Nadu through building sugar industry and two hotels through Appu organization.
10. Guided two of his physician daughters (Jessica and Erica) to provide charitable medical service in central America, Haiti, India, Sierra Leone.
11. Planning to set up "King Center for Vision" to serve eye problems of the needy and propagate the nonviolence message of Basavanna, Gandhiji and M.L. King, Jr.
12. Donated 2.2-acre river front land in Camden, in the memory of Hon. John Lewis to Brown Chapel AME Church, Selma, AL, where voting march started, led by John Lewis, donated by Dr. T. R. Shantha and Dev, Usha, Andy, Jessica, Erica and Lauren.

Distinctions and Awards:

1. Kapoor Chand Gold Medal by Maharaja of Mysore in 1951, for Center (highest) rank in SSLC examination.
2. Recognized by the Georgia Society of Anesthesiologists in 1969 and 1970 as an outstanding resident.
3. Named Who's Who in American Men of Science and in Outstanding Atlantans (Heritage publication, 1978-79)
4. Received A.M.A. Physician Recognition Award, 1970-1996, 1997-2000
5. Declared on 23rd of July, 1993 Dr. Shantha Day by DeKalb County, for contribution in research and philanthropic work.

6. Honorary citizen awards by the City of Atlanta, New Orleans, Knoxville, Topeka Kansas, Kansas City of Missouri, and appointed **as *Kentucky colonel*** by the Kentucky Governor.
7. Selected as the distinguished physician and scientist by Amer. Ass. of Indian Physicians, 2005 (TX) and VSNA Org.
8. Nominated for Nobel Prize in Physiology and Medicine twice.
9. Proclaimed Dr. T. R. Shantha, day in DeKalb County, Georgia on 27Th June 1993, for his dedication and services in the county by CEO of the county Liane Levetan.

Selected research contributions by Dr. Shantha from Emory University School of Medicine, Medical College of Georgia, Grady Memorial Hospital, Georgia Baptist Medical Center and Columbus Medical Center

1. Dr. Shantha described pia-arachnoid mater of CNS and optic nerve extending as supra-choroid lamellae of the eye.
2. Dr. Shantha described the pia-arachnoid membrane of the central nervous system and named it **perineural epithelium.**
3. Below the sub perineural epithelial space is the sub-perineural epithelial and inter-axonal space that acts as a highway for transport centrifugally and centripetally of various micro molecules from periphery to sub arachnoid space of the CNS and back.
4. He showed that the onion ring-like coverings of the Pacinian corpuscle are the extension of pia-arachnoid mater of the central nervous system.
5. He along with Dr. G.H. Bourne demonstrated the origin of cells on the motor end plate called peri-synaptic cells of the motor end plate are extension of perineural epithelium.
6. Dr. Shantha showed the presence of Pacinian corpuscles on the olfactory bulb which sub serve pressure sensation within the cranium.
7. Dr. Shantha showed that multiple layers of this perineural epithelium surrounded the sympathetic chain of sperm whales akin to humans.
8. Dr. Shantha demonstrated the presence of Arachnoid Villi in the optic nerve of man and other primates.
9. Dr. Shantha showed the presence of intercellular pores in the corneal endothelium, which transmit nutrients to the avascular Descemet's membrane and cornea.
10. He demonstrated a double central canal in the spinal cord of normally functioning rats.
11. He showed the presence of nerve ganglion cells far away from the spinal cord in the peroneal nerve.

12. He published 12 papers on the Golgi complex of the entire CNS and eye.
13. He developed a subarachnoid blood patch (not epidural blood patch) to stop CSF leakage headache and prevented seven spinal surgeries, published in NEJM.
14. He found Pacinian corpuscle in the dorsal and ventral root of the spinal cord that carry pressure sensation from the epidural space.
15. He showed the presence of Arachnoid Villi in the dorsal and ventral spinal roots and their role in epidural anesthesia.
16. He developed a new method to block the occipital nerve for neuropathic pain and occipital headaches.
17. He showed the presence of the Virchow–Robin space in peripheral nerves as seen in the CNS which plays an important role in the spread of local anesthetics in peripheral nerves.
18. He showed the mechanisms of the action of LSD & BOL138 in the central nervous system.
19. He wrote monographs on neurons (almost 300 pages), and chimpanzee brains, published by Karger of Switzerland.
20. He found a new simple drug for the treatment of **Priapism -Terbutaline** (painful persistent penile erection), which was published in the Journal of Urology and quoted in ER handbooks and urology. He also used it to prevent premature contractions and delivery on his own wife with triplets and carried the pregnancy from 22 weeks to 36 weeks.
21. He developed and perfected a sub-cricoid retrograde method of intubations for difficult intubations which was published in the British Journal of Anesthesia.
22. He was the first person to induce peritoneal hyperthermia in Georgia Baptist Hospital (now Atlanta Medical Center) for the treatment of massive terminal candida infection of the peritoneal cavity
23. Four patients, who were previously bed-ridden, have returned to work due to his method of treatment of Scleroderma. He presented these cases in 2005 research meeting in Florida
24. His treatment using glucose facilitated therapy and others resulted in complete remission of lupus and reduction in symptoms of other autoimmune diseases.
25. Many chronic Lyme disease patients who were disabled have gone back to work due to Dr. Shantha's method of treatment.
26. He showed that daily large consumption of coconut in the food greatly increases the risk of death by heart attack.

27. He developed and treated hundreds of cases with local injection of therapeutic agents with insulin.
28. He developed metformin to facilitate increased activity of all therapeutic agents including supplements to prevent, curtail, stop the progression of and / or cure diseases.
29. He used insulin with steroid to inject the spine (epidural, facet joints, ligaments, back pain), any and all joints and bursae, sore muscles and tendons pain, which was effective in relieving pain rapidly (unpublished).
30. He developed intravenous use of therapeutic agents with insulin for the treatment of sympathetic dystrophy which was effective (unpublished).
31. He discovered new oral airway that can be used in 350 million anesthesia patients every year. He developed devices for prevention of snoring and obstructive sleep apnea and obtained 6 patents.
32. He described the structure and function of Axenfeld loop in the eye. It was described a century ago, but no one knew its real structure and function.

Khabar News 2005

CommunitySpotlight

Dr. Shantha recognized as the 'most distinguished physician and scientist' by AAPI

Describing himself as "one of the longest residing Indians in Atlanta," Dr. T.R. Shantha adds that there were only seven Indians here when he arrived in 1960 to attend Emory University, where he completed his doctoral and postdoctoral studies in record time. Triply board certified and highly acclaimed, he has been unanimously selected by AAPI (American Association of Physicians from India) this year as the Most Distinguished Physician and Scientist. The author of 125 original research articles and 6 books, Dr. Shantha is apparently the only scientist on record to publish 31 research papers in a 3-year stretch at Emory University School of Medicine. A discoverer of many new histological, anatomical and histochemical features in the body, his research has been quoted in various medical textbooks. He owns 3 patents and has others (including 3 on SIDS) in developmental stages. Dr. Shantha has discovered a simple drug (Terbutalene) for priapism that's used in emergency rooms and by urologists everywhere. Based on one of his patents, devices have been developed to treat uterine bleeding in the elderly. Among numerous other honors, he has received the American Medical Association's Physician Recognition Award more than once. Dr. Shantha volunteers every month at Hands of Hope, a charity clinic that provides free medical care to hundreds of sick people who have no insurance. He has funded the building of a high school for girls in Karnataka, his home state in India. For his research, philanthropy and medical work, a "Dr. Shantha Day" was declared by DeKalb County.

Nobel Prize Nominating letter to Karolinska Intravenous. Stockholm, Sweden. Dr. Peter Medawar, a Nobel Laureate, was also nominated to receive the Nobel Prize for our discovery of membranes of the peripheral nervous system and their role in health and disease.
Published in AAPI (Association of Physicians from India in US >50,000 members) monthly magazine about Shantha's Nobel Prize nomination and other awards dated July 27th 2006 Journal

Members in The News---------------------
Dr. T. R. Shantha

Ministry of Education
Republic of Belarus

Institution of Education
"Yanka Kupala
State University
of Grodno"

LABORATORY OF BIOCHEMISTRY
OF BIOLOGICALLY ACTIVE
SUBSTANCES

22, Ozeshko Str. , Grodno, 230023 Belarus
tel./fax: +375 (152) 48 68 10
tel.: +375 (152) 48 50 68
www. nil. grsu.by
E-mail: l.nefyodov@grsu.by

№ 22-N

Karolinska Institutet
Nobel Committee Physiology and Medicine
Box 270
SE-177 Stockholm, Sweden

Ladies and Gentlemen,

Thank you for your trust to nominate the candidate for the Nobel Prize.

I want to use this opportunity this year and nominate for the Nobel Prize in Physiology and Medicine Prof. Dr. Totada R. Shantha, a prominent scientist. He made an important contribution to the development of science, which you can see from the documents which are attached.

Please accept my application, list of Dr. Shantha's publications, list of his scientific achievements and copies of his most important publications.

Sincerely,

Head, Department of Biochemistry
and Laboratory of Biochemistry
of Biologically Active Substances
Professor, PhD, M.D., D.h.c.

Leonid I. Nefyodov

Laboratory of Biochemistry of Biologically Active Substances
22, Ozeshko Str., Grodno, 230023, Belarus
tel/fax:+375 (152) 48 68 10; www.nil grsu.by; E-mail: l.nefyodov@grsu.by

27th July 2006

Index